Revising the Revisions: James Hutton's Reputation among Geologists in the Late Eighteenth and Nineteenth Centuries

by

A. M. Celâl Şengör
İTÜ Avrasya Yerbilimleri Enstitüsü
ve
Maden Fakültesi Jeoloji Bölümü
Ayazağa 34469 İstanbul
Turkey

THE GEOLOGICAL SOCIETY OF AMERICA®

Memoir 216

3300 Penrose Place, P.O. Box 9140 ▪ Boulder, Colorado 80301-9140, USA

2020

Copyright © 2020, The Geological Society of America (GSA), Inc. All rights reserved. Copyright is not claimed on content prepared wholly by U.S. government employees within the scope of their employment. Individual scientists are hereby granted permission, without fees or further requests to GSA, to use a single figure, a single table, and/or a brief paragraph of text in other subsequent works and to make unlimited photocopies of items in this volume for noncommercial use in classrooms to further education and science. Permission is also granted to authors to post the abstracts only of their articles on their own or their organization's website providing that the posting cites the GSA publication in which the material appears and the citation includes the address line: "Geological Society of America, P.O. Box 9140, Boulder, CO 80301-9140 USA (https://www.geosociety.org)," and also providing that the abstract as posted is identical to that which appears in the GSA publication. In addition, an author has the right to use his or her article or a portion of the article in a thesis or dissertation without requesting permission from GSA, provided that the bibliographic citation and the GSA copyright credit line are given on the appropriate pages. For any other form of capture, reproduction, and/or distribution of any item in this volume by any means, contact Permissions, GSA, 3300 Penrose Place, P.O. Box 9140, Boulder, Colorado 80301-9140, USA; fax +1-303-357-1070; editing@geosociety.org. GSA provides this and other forums for the presentation of diverse opinions and positions by scientists worldwide, regardless of their race, citizenship, gender, religion, sexual orientation, or political viewpoint. Opinions presented in this publication do not reflect official positions of the Society.

Published by The Geological Society of America, Inc.
3300 Penrose Place, P.O. Box 9140, Boulder, Colorado 80301-9140, USA
www.geosociety.org

Printed in U.S.A.

GSA Books Science Editors: Joan Florsheim, Christian Koeberl, and Nancy Riggs

Library of Congress Cataloging-in-Publication Data

Names: Şengör, A. M. Celâl, author.
Title: Revising the revisions : James Hutton's reputation among geologists in the late eighteenth and nineteenth centuries / by A.M. Celâl Şengör, İTÜ Avrasya Yerbilimleri Enstitüsü ve Maden Fakültesi Jeoloji Bölümü, Ayazağa 34469, Istanbul, Turkey.
Description: Boulder, Colorado : The Geological Society of America, Inc., 2020. | Series: Memoir ; 216 | Includes bibliographical references and index. |
Identifiers: LCCN 2020043127 (print) | LCCN 2020043128 (ebook) | ISBN 9780813712161 (hardcover) | ISBN 9780813782164 (ebk)
Subjects: LCSH: Hutton, James, 1726-1797--Influence. | Geology--History--18th century. | Geology--History--19th century.
Classification: LCC QE11 .S46 2020 (print) | LCC QE11 (ebook) | DDC 551.092--dc23
LC record available at https://lccn.loc.gov/2020043127
LC ebook record available at https://lccn.loc.gov/2020043128

Cover: James Hutton. Etching by J. Kay, 1787. John Kay's famous caricature showing James Hutton in front of an outcrop at the Salisbury Crags near Edinburgh that had taken the shape of the heads of his opponents. The second from the top, directly facing an amazed Hutton, is most likely Jean-André Deluc, with his pointed nose and prominent chin. The one at the bottom is possibly Francis Lord Jeffrey (1773–1850), jurist and one of the founders of the *Edinburgh Review* and its editor between 1803 to 1829 (Dean, 1992, p. 126), with his bulbous forehead, bushy eyebrows and a suggestion of a double-chin. In between is probably Robert Jameson with the suggestion of a slight ski-jump nose with a rounded end and pursed lips. Detail of National Portrait Gallery image NPGD18643, © National Portrait Gallery, London, npg.org.uk.

Contents

Dedication v

Preface ix

Acknowledgments xi

Abstract 1

Introduction 2

Science and Historiography of Science: The Case of Geology 2

Revisionist Claims Concerning Hutton's Place in Geology 4

The British Empire 9
    Before Lyell 9
    After Lyell and before Geikie 38
    Appendix to Hutton in the British Empire: Hutton in the *Memoirs of the Wernerian Natural*
        History Society 45

France 49
    Before Lyell 49
    After Lyell and before Geikie 61

The Russian Empire 70
    Before Lyell 70
    After Lyell and before Geikie 72

Germany 74
    Before Lyell 74
    After Lyell and before Geikie 77

Austrian Empire (Austro-Hungary After 1867) 88
    Before Lyell 88
    After Lyell and Before Geikie 89

Switzerland 92
    After Lyell and Before Geikie 92

Italy 94
    Before Lyell 94
    After Lyell and Before Geikie 97

The United States of America 101
    Before Lyell 101
    After Lyell and Before Geikie 105

Discussion 111
    Hutton Regarded as the Founder of Modern Geology (A chronology from 1829 to 1897) 111
    Hutton's Theory Regarded as Something New (A chronology from 1787 to 1878) 113
    Hutton Regarded as the Founder of Metamorphic Geology 114
    Evaluation of the Historical Record. 114

Conclusions 121

References Cited. 122

Notes 137

Indices 151
    Geographical Localities 151
    People. 155
    Subjects 161
    Titles of Publications 169

*Dedicated, with deep gratitude, to three great scientists,
who are also great historians and philosophers of science:
Edward Harrison, Steven Weinberg and David Deutsch*

We look to Scotland for all our ideas of civilization.

—Voltaire

Diligence and accuracy are the only merits which an historical writer may ascribe to himself; if any merit indeed can be assumed from the performance of an indispensable duty. I may therefore be allowed to say that I have carefully examined all the original materials that could illustrate the subject which I had undertaken to treat.

—Edward Gibbon

The impression produced upon the public mind by the new doctrines introduced and established, from time to time, by physical science ... are always liable to be underrated by historians.

—Sir Charles Lyell

What a man believes upon grossly insufficient evidence is an index into his desires—desires of which he himself is often unconscious. If a man is offered a fact which goes against his instincts, he will scrutinize it closely, and unless the evidence is overwhelming, he will refuse to believe it. If, on the other hand, he is offered something which affords a reason for acting in accordance to his instincts, he will accept it even on the slightest evidence. The origin of myths is explained in this way.

—Lord Bertrand Russell

Preface

This book resulted from my revolt against what I think to be a recent aberration in the historiography of geology, namely considering James Hutton's place in the history of geology exaggerated and his contributions limited. I wrote it with the greatest reluctance, because polemical writings are seldom useful in science, as Lyell told Darwin once[1]. My decision to write it anyway was precipitated finally by a number of favorable reviews by geologists I greatly respect of some of the revisionist publications that I thought were simply bad history of science; bad and thus misleading. While thinking about this issue, I remembered what Thomas Paine (1737–1809) wrote in 1776, in the introduction to his best-selling pamphlet *Common Sense*: "A long habit of not thinking a thing *wrong*, gives it a superficial appearance of being *right*,"[2] which sent a shudder down my spine, because, as my countryman Publilius Syrus (*floruit* 85–43 BCE; he was from Antioch, present-day Hatay in southern Turkey, and not Syria, in the present political geography) affirmed in his *Sententiae*, "judex damnatur cum nocens absolvitur" (judges are condemned if the guilty go free). Young geologists reading those revisionist publications would get what I think a completely distorted view, not only of the history of geology, but also of what geology is about. That reputable geologists would not rise against flagrant misrepresentations of the history of their subject is something I find disconcerting and hard to understand. Could it be a result of lack of interest in the subject among geologists, as the late Victor Eyles wrote already in 1966? Small samples of what I mean by bad and misleading historiography I already gave in my rebuttal of Alexander Meier Ospovat's (1923–2010) unfounded claim that Sir Charles Lyell had distorted Werner's image in his *Principles of Geology* (Şengör, 2002) and in my essay review of Martin John Spencer Rudwick's (1932–) *Bursting the Limits of Time* (Şengör, 2009). While giving examples of what I criticize in the present book, use of the names of authors I chose as illustration (naturally among a larger number) of what I think bad historiography was inevitable, although naturally no offense is intended. My strong words are directed against the statements and what I perceive to be their causes, not against personalities. I repeat here what I wrote at the end of Şengör (2001), quoting the British idealist philosopher Francis Herbert Bradley (1846–1924) as perfectly reflecting my state of mind also in writing this book: "I am afraid that I always write too confidently—perhaps because otherwise I might not write at all. Still I don't see that in doing so one can do much harm, or run the risk of imposing on anyone whose judgement is of any value" (Şengör, 2001, p. 39).

I tried to make this book as factual as I could make it by allowing the authors of the past to speak for themselves, to express their judgements in their own words. From that viewpoint, this book is not so much a polemical, but a documentary publication. It does not confine the documentation to those works displaying a first-hand acquaintance with Hutton's writings, because my purpose is to document the overall ambience concerning Hutton's ideas in geology at any given time between the temporal limits I set to my essay. The book does not in any way evaluate Hutton's theory scientifically; it simply documents its history of reception by geologists. If I tried also to write a geological assessment of Hutton's theory, this book would have been more of a geological publication than one on the history of geology. Such an appraisal should perhaps

---

[1] "I rejoice that I have avoided controversies, and this I owe to Lyell, who many years ago, in reference to my geological works, strongly advised me never to get entangled in a controversy, as it rarely did any good and caused a miserable loss of time and temper" (Darwin, 1887, p. 89).

[2] I am quoting from the 200th Anniversary Edition published in 2003 by Signet Classics, New York, under the title *Common Sense, Rights of Man, and Other Essential Writings of Thomas Paine*, p. 3. The italics are Paine's.

Şengör, A.M.C., 2020, Preface, *in* Şengör, A.M.C., Revising the Revisions: James Hutton's Reputation among Geologists in the Late Eighteenth and Nineteenth Centuries: Geological Society of America Memoir 216, 2 p., https://doi.org/10.1130/2020.0216(00).

be done indeed and one day I might do it myself. However, even the present book was written in such haste, amidst so much other work on current geological problems—largely because the admonitions of Paine and Publilius have been weighing so heavily on my conscience for some time—that it would have been quite impossible for me to undertake an evaluation of Hutton's theory, which would need to cover almost the entire spectrum of the science of geology during the past two centuries, with the only exceptions of paleontology and biostratigraphy. Even paleontology might not justifiably be excluded given Hutton's interest in finding fossils in metamorphic rocks and his prescient views on natural selection and the evolution of species, but that aspect of his speculations has received hardly any attention. Moreover, an attempt to evaluate Hutton's theory scientifically would have destroyed the homogeneity of this book and blurred the focus of its purpose. I am keen that this book maintains a sharp focus to fulfill its limited purpose, namely documenting what geologists thought about Hutton's geological ideas in the century following his death. For the same reason, I avoid getting too deeply into the philosophy of geology as it relates to Hutton's work. What little I do discuss is only to clarify certain points as they pertain to how Hutton was understood by the posterity.

Having said that, I must mention an important point made by Professor Gian Battista Vai of Bologna in his very detailed review of an earlier version of the typescript of this book: Vai annotated profusely not only my text, but also the excerpts I quote from the receivers of Hutton, pointing out that many of them were not terribly well-versed in the history of geology in Italy and that many ideas ascribed to Hutton had already been expressed by Italian scientists before him, going as far back as Leonardo da Vinci (for a convenient summary of these, see Vai, 2009). While this is certainly true, the difference of Hutton's theory from the ones Vai reminded me of is that Hutton's theory was comprehensive, embodying many of the ideas earlier expressed and synthesizing them in one all-encompassing theory of the earth. That is why its impact has been so profound. Since this book is about the reception of Hutton's ideas and not a critique of the receivers, I did not get into many of the informative points Vai made.

My contention that social considerations are not very helpful in our understanding of the development of science are so incompatible with the current trend in the historiography of science that in the Conclusions section, I had to digress somewhat from my central theme to justify my claim. This book may be considered as a kind of appendix to my earlier publication, *Is the Present the Key to the Past or the Past the Key to the Present? James Hutton and Adam Smith versus Abraham Gottlob Werner and Karl Marx in Interpreting History*, also published by the Geological Society of America as its Special Paper 355 in 2001.

All translations from non-English languages in it are mine, except those that I was able to find published in the literature (although I did not actively search for them). But even those I always quote after a check with the original and, where needed, with indication of my corrections. In translating from Italian and Russian, I gratefully acknowledge the help, respectively, of my friends Professors Gian Battista Vai in Bologna and Boris Alexeich Natal'in in İstanbul. For Latin, Mr. Kutsi Aybars Çetinalp rendered decisive assistance. I used many *Wikipedia* articles for dates of people and, for some, their portraits and an overall view of their lives, but I quote them here only after checking from an independent source. Most of the portraits I reproduce come from my own collection that accumulated over the years.

A.M. Celâl Şengör
İstanbul, 4 February 2020

REFERENCES CITED

Şengör, A.M.C., 2001, *Is the Present the Key to the Past or the Past the Key to the Present? James Hutton and Adam Smith versus Abraham Gottlob Werner and Karl Marx in Interpreting History*: Geological Society of America Special Paper 355, x + 51 p., https://doi.org/10.1130/SPE355.

Şengör, A.M.C., 2002, On Sir Charles Lyell's alleged distortion of Abraham Gottlob Werner in *Principles of Geology* and its implications for the nature of the scientific enterprise: The Journal of Geology, v. 110, p. 355–368, https://doi.org/10.1086/339537.

Şengör, A.M.C., 2009, Essay Review: A Rankean view of historical geology and its development: Earth Sciences History, v. 28, p. 108–134.

Vai, G.B., 2009, Storia breve della geologia in Italia, *in* Cavalli Sforza, L.L., ed., *La Cultura Italiana*, v. 8 *Scienze e Tecnologie*, Pievani, T., ed., Unione Tipografico-Editrice Torinese, p. 305–345.

Acknowledgments

This book was written mainly between March 2019 and July 2019 during the brief intervals that my other work allowed me. During that limited time, I was unable to consult many whom I would have consulted had I had more time, but I did ask the help of Kennard Bork of Granville in Ohio, Feza Günergun of İstanbul, Ludmilla J. Jordanova of Durham, Irena Malakhova of Moscow, Boris A. Natal'in of İstanbul, Alastair H.F. Robertson of Edinburgh, Kenneth L. Taylor of Oklahoma, Hugh Torrens of Keele, and Gian Battista Vai of Bologna, which they most generously rendered. Boris A. Natal'in kindly researched the Russian geological literature of the nineteenth century for me and found interesting references to Hutton. Without him, my review of the Russian literature would have been confined to two items only! The İTÜ history of science post-graduate student Mr. Kutsi Aybars Çetinalp not only translated a number of Latin texts for me, but helped to locate and obtain Becher's and Stahl's publications. Without his help, my knowledge of Becher and Stahl would have been much more lacunar. Over the years my numerous conversations with my late friend Donald B. McIntyre in Scotland, Dennis R. Dean in Chicago (and in İstanbul), and the late David R. Oldroyd in Sydney (and both in İstanbul and in Kocaköy near Assos) not only enlarged my knowledge of the literature about Hutton and related topics, but allowed me to test my thoughts by submitting them to their criticism. I also discussed Hutton's ideas and their relations to the geology and the geologists of Great Britain with my teachers John F. Dewey and the late Kevin C.A. Burke when I was their student in Albany. Raised in the glorious British tradition of geology and being two of the giants of geology in our times, it was gratifying to see that their views of Hutton coincided entirely with mine. More than two decades ago, Donald B. McIntyre took me on a long field trip to show me all the Hutton localities in Scotland and even gave me copies of certain relevant Hutton documents from the archives of the University of Edinburgh, which he had earlier prepared for himself. I must, however, absolve all of the people named above from any responsibility for what is published in this book. Last, but not least, I thank my colleagues and students at the İTÜ Eurasia Institute of Earth Sciences and the Department of Geology for tolerating my long absences from the office and my wife Oya for allowing me to spend a large amount of money in buying old books that I did not have, but needed, to write this book.

My typescript was reviewed by two outstanding geologists who are also great intellectuals, Professors Brian Wernicke and Victor R. Baker. I could not have wished for better reviewers. Their extremely detailed, learned, and long reviews each could easily develop into their own papers. I am deeply grateful to the book editor of the Geological Society of America, my scholarly friend Professor Christian Koeberl, for his very competent handling of my typescript and the Geological Society of America book assistant Jon Raessler for his multifarious help.

// *Revising the Revisions: James Hutton's Reputation among Geologists in the Late Eighteenth and Nineteenth Centuries*

A. M. Celâl Şengör*

İTÜ Avrasya Yerbilimleri Enstitüsü ve Maden Fakültesi, Jeoloji Bölümü, Ayazağa 34469 İstanbul, Turkey

ABSTRACT

A recent fad in the historiography of geology is to consider the Scottish polymath James Hutton's *Theory of the Earth* the last of the "theories of the earth" genre of publications that had begun developing in the seventeenth century and to regard it as something behind the times already in the late eighteenth century and which was subsequently remembered only because some later geologists, particularly Hutton's countryman Sir Archibald Geikie, found it convenient to represent it as a precursor of the prevailing opinions of the day. By contrast, the available documentation, published and unpublished, shows that Hutton's theory was considered as something completely new by his contemporaries, very different from anything that preceded it, whether they agreed with him or not, and that it was widely discussed both in his own country and abroad—from St. Petersburg through Europe to New York. By the end of the third decade in the nineteenth century, many very respectable geologists began seeing in him "the father of modern geology" even before Sir Archibald was born (in 1835). Before long, even popular books on geology and general encyclopedias began spreading the same conviction. A review of the geological literature of the late eighteenth and the nineteenth centuries shows that Hutton was not only remembered, but his ideas were in fact considered part of the current science and discussed accordingly. The strange new fashion in the historiography of geology has been promulgated mostly by professional historians rather than geologists and seems based on two main reasons: (1) a misinterpretation of what geology consists of by considering methods rather than theories as the essence of the science, and (2) insufficient attention to the scientific literature of geology through the ages. In only one case, the religious commitment of a historian seems a reason for his attempt to belittle Hutton's contribution and to exalt those of his Christian adversaries, hitherto considered insignificant. To write a history of geology it is imperative that extra-scientific considerations such as religion or political ideology or even the mental state of the scientist(s) examined must not be mixed, overtly or covertly, into the assessment and the writer should have a good knowledge of, and experience in doing, geology. Social considerations may tell us why science is done or not done in a society, but they cannot tell us anything on the origin and evolution of its *content*. In understanding the intellectual development of geology, in fact science in general, sociological analysis seems not very helpful.

---

*sengor@itu.edu.tr

Şengör, A.M.C., 2020, Revising the Revisions: James Hutton's Reputation among Geologists in the Late Eighteenth and Nineteenth Centuries: Geological Society of America Memoir 216, 150 p., https://doi.org/10.1130/2020.0216(01).
© 2020 The Geological Society of America. All rights reserved. For permission to copy, contact editing@geosociety.org.

INTRODUCTION

This book has a very restricted purpose, which consists in documenting the Scottish polymath James Hutton's (1726–1797; Fig. 1) reputation among geologists during the late eighteenth and the nineteenth centuries before the publication of the first edition of Sir Archibald Geikie's *Founders of Geology* in 1897 with a view to putting on record that he was generally considered by the international geological community as the founder of modern geology, i.e., the science of the structure and history of the planet earth, even before Sir Archibald was born in 1835. It is thus a "history of reception" of Hutton's ideas by the geological community between the date of his first geological publication in 1785 and essentially the end of the nineteenth century. It documents that the knowledge and influence of Hutton's geological writings were widespread both in his own country and elsewhere, even years before Sir Charles Lyell (1797–1875) became his champion in his *Principles of Geology* (1830–1833), and has remained so ever since, contrary to some recent revisionist historians' claims. In what follows I have chosen four popular revisionist accounts among a multitude, as it is impossible to review them all here. But the four I have chosen I consider typical and have been extensively quoted by the others.

SCIENCE AND HISTORIOGRAPHY OF SCIENCE: THE CASE OF GEOLOGY

A part of the revisionists' claims is based on what I consider a misunderstanding of what science is. The science of geology includes the *theory of the earth*[1] and the *methods* by which that theory is tested. Establishing a true theory of the earth is the goal of geology and the methods are the means by which such a theory is continuously improved[2]. The establishment of accurate knowledge and correct understanding of the structure, behavior and history of the earth is an unchanging, yet ultimately unattainable goal, but the methods used to strive toward it continuously improve as new ideas and technologies become available. Methods do not constitute the *goal* of the science, but its *means* to attain its goal. Similarly, an architectural monument does not include the tools used for its construction, neither the scaffolding that once embraced it, yet without them, or without the people who developed them, it could not have been built. By contrast, teaching architecture does include the tools used in constructing buildings. Similarly, nobody will credit the developers of submarine sounding devices during World War II, nor indeed their first widespread civilian employers such as William Maurice "Doc" Ewing (1906–1974) since then, nor the creators of the World Wide Standardized Seismograph Network (WWSSN), for having invented plate tectonics, although they were critical for its invention by J. Tuzo Wilson (1908–1993) in 1965 and testing by his followers in the subsequent years. However, teaching plate tectonics ought to mention the tools and their inventors and users without which there could have been no plate tectonics.

Geological theory, the essence of the science, has a multi-tiered structure: at the uppermost, most general, level it states by

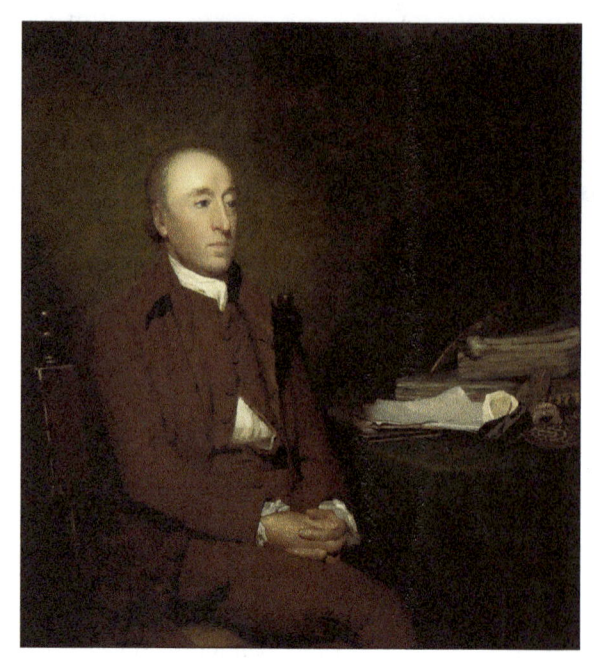

Figure 1. James Hutton by Sir Henry Raeburn (1790; for the dating, see McIntyre, 1997, p. 143). Sir Henry Raeburn, *James Hutton, 1726–1797, Geologist*. National Galleries of Scotland. Purchased with the aid of the Art Fund and the National Heritage Memorial Fund 1986.

which fundamental cause(s) the planet evolves and generates its structure (e.g., according to our present understanding, it is the heat loss of the planet that runs the earth engine). At a lower level, the details of the mechanism(s) brought about by the fundamental cause are specified (e.g., plate tectonics, mantle convection, atmospheric and hydrological circulation, the entire geostrophic cycle etc.). Below that, the consequences of those detailed mechanisms in creating the structures of the rocky rind of the planet are defined, which range from the formation of the largest features of the earth's lithosphere (earth's crust, continents, ocean basins, cratons, mountain ranges, rift valleys, alluvial plains, etc.) to the origin of minerals and their atomic structure. The more restricted theories in this lowermost level are not necessarily dependent for their success on the correctness of the overarching theory of the earth. The theory of turbidites, for example, requires the presence of underwater slopes and abundant supply of clastic material, irrespective of how the water bodies, the slopes, and the clastic sources are created, as such creation can be accomplished equally well in the framework of the now-abandoned magmatic uplift or the thermal contraction theories or indeed in that of the currently accepted plate tectonics. Theory aims at *explaining*, i.e., *understanding* the planet (or parts of it) and its (their) history. Methods are *work tools for collecting information* with a view to achieving that understanding. It would help us in appreciating the difference between methods and theories and their relations to science in general if we reflect on Albert Einstein's insightful comment about them made during his inaugural address before the Royal Prussian Academy of Sciences on 2 July 1914 in Berlin:

The method of the theoretician requires that as a foundation he needs general conditions, so-called principles, from which he can deduce inferences. His activity is therefore divided into two parts. First, he has to find those principles and, secondly, to develop the implications derived from the principles. For the fulfillment of the second task, he receives at school excellent tools. So, if the first of his tasks is already accomplished in a field or a complex of contexts, he will not lack success if he has sufficient diligence and understanding. The first of these tasks, namely the setting up of the principles that are to serve as a basis for deduction, is of a completely different kind. There is no learnable, systematically applicable method that leads to the goal. Rather, the researcher must, so to speak, somehow discern from Nature those principles by considering larger complexes of empirical facts with a view to finding general features that can be formulated sharply.

Once this formulation is achieved, a development of inferences commences, which often yields previously unsuspected connexions, reaching well beyond the facts, on the basis of which the principle had been developed. As long as a principle to function as a basis of deduction is not found, the individual observations would be of no use to the theoretician. He cannot even make use of the empirically established general laws. He must stand helpless in front of the individual results of empirical research until principles are found that can serve as a basis of deductive developments. (Einstein, 1914, p. 740)

Lord Bertrand Russell, one of the great philosophers of the twentieth century, agreed:

As a rule, the framing of hypotheses is the most difficult part of scientific work, and the part where great ability is indispensable. So far, no method has been found which would make it possible to invent hypotheses by rule. (Russell, 1946, p. 566–567)

What Einstein and Russell meant was echoed a century later by another great physicist and a superb historian of science, Steven Weinberg (1933–):

Descartes and Bacon are only two of the philosophers who over the centuries have tried to prescribe rules for scientific research. It never works. We learn how to do science, not by making rules about how to do science, but from the experience of doing science, driven by desire for the pleasure we get when our methods succeed in explaining something. (Weinberg, 2015, p. 214)[3]

Einstein's one-time colleague in Berlin, the great physicist Max Planck (1858–1947), also emphasized the importance of *understanding* during the learning process:

It is less relevant what is taught in schools, than how it is taught. A single mathematical expression, truly understood by a pupil, has more value for him than ten formulae which he learnt by rote and which he also knows how to use following instructions, but without understanding what they mean. The school should not teach the routine in a discipline, but consequent, methodical thinking. ... a theory can never be replaced by simple routine, which, in unusual circumstances, would fail helplessly. That is why the first requirement for attaining sound achievements is a thorough elementary education, whereby the fullness of the material taught is less important than how the material taught is handled. (Planck, 1949, p. 275)

From the viewpoint of geological theory, James Hutton's overall interpretation of geological phenomena was a novelty and, even when viewed from the early twenty-first century geology, it looks modern, including its errors, because Hutton made the sorts of errors we can today make and do indeed make. As the Lithuanian-born Russian geologist and historian of geology, Sergei Ivanovich Tomkeieff (1892–1968) said (1948, p. 275), "To-day the main ideas of Hutton are more alive than ever and present-day geologists, often without knowing it, are still following the pathway which his genius enabled them to find." Thus, Hutton had accomplished what Einstein defined as the first task of a theoretician and evidently derived from it the pleasure Weinberg mentions[4]. I think Hutton's service to geological understanding rather than to its methods and data storage was first emphasized by Tomkeieff's superb paper on Hutton and the philosophy of geology (1948). After having pointed out that Abraham Gottlob Werner and William Smith (1769–1839) had given geologists valuable methods, he continued "But science does not consist only of methods. It needs comprehension (theory) and this comprehension was provided by Hutton and Hutton alone[5]. Without him geology would certainly not have developed when and as it did" (Tomkeieff, 1948, p. 256). Two-and-a-half decades later, another geologist-historian, François Ellenberger (1915–2000) restated, with his typical Gallic enthusiasm, Tomkeieff's judgement:

One is tempted to ask oneself, by what miracle this modest ex-physician, amateur chemist and Scottish agronomist could have conceived a theory of the earth so original and so remarkable, ahead of his century, and even the succeeding one. (Ellenberger, 1972, p. 268)

By framing his "theory of the earth" that still forms the theoretical basis of geological science, Hutton did indeed prove his great ability as a scientist.

The intellectual ancestor of Hutton's theory of the earth was Steno's theory developed mostly in Tuscany[6] (Stenonis, 1667, 1669)[7]. Every development in geological research between Steno and Hutton consisted of elaboration of Steno's methods and explanatory theory with three important exceptions: Immanuel Kant's (1724–1804) theory of the origin of the Solar System (Kant, 1755; expanded and improved by Pierre Simon le Comte de Laplace {1749–1827}, *An IV*, i.e., just a year before Hutton's death independently of Kant, whose theory had attracted essentially no interest and was forgotten until Alexander von Humboldt resurrected it in his *Kosmos*, 1845, p. 59, 1850, p. 32, 356, 549, 575), Abbot Anton Lazzaro Moro's (1687–1764) igneous theory (Moro, 1740) and Abbot René Just Haüy's (1743–1822) theory of crystallography[8] (Haüy, 1784). Georges-Louis Leclerc Comte de Buffon's (1707–1788) claim that the earth was much older than can be computed by using the mythology of the *Old Testament* (de Buffon, 1778; see especially also Roger, 1962, particularly p. LXIII–LXVII and the comparative table on p. LXV) was really not so much a novelty, because the ancient Mesopotamians and Egyptians (Hosmann, 1729, p. 156; see Verbrugghe and Wickersham, 1996), the Hindus, Chinese (Jackson, 2006, p. 6–8), and the Greeks also entertained similar, in fact more extreme, views (e.g., Ussher, 1645) which had driven Theophilos of Antioch (115–180 CE) to erect the first Biblical chronology to combat them (Hosmann, 1729).

REVISIONIST CLAIMS CONCERNING HUTTON'S PLACE IN GEOLOGY

What I say in the preceding paragraphs about Hutton's accomplishment in geology, is what every geology student learns as an undergraduate in the whole world. However, what geology students learn about Hutton's position in the development of geology as a science has recently been questioned by some historians, who, to use the American geologist, and historian and philosopher of geology Victor Baker's (1945–) title of his review of Rudwick's *Worlds Before Adam*, have generated a history of the science of geology "turned upside down" (Baker, 2008, p. 406[9]).

We read in that revisionist literature, for example, such astonishing statements as "Hutton appears as a revolutionary only against his immediate background, a stagnation in British geomorphic thought in the late eighteenth century" (Greene[10], 1982, p. 30); or, "Such information serves to clear the ground only of Hutton's exaggerated originality and is not a judgement on the adequacy or power of the theory in his own version or in Playfair's refined and theologically muted form. It merely restores Hutton to the context—fluvial denudation theory and natural theology—in which he worked and indicates that the elements from which he fashioned the theory were not new and that his reputation has been enlarged by attributing to him the definitive solution to questions that remained unsettled" (Greene, 1982, p. 31). The first statement would make one think that Hutton created only a geomorphological theory, whereas the second would be similar to denying Einstein the startling originality of his theory of special relativity, because it is based on Lorentz's transformations and Poincaré's law of relativistic velocity addition (and, shall I add, the Ephesian Heraclitus' discovery of the importance of reference frames in the late sixth century BCE?). Similarly, the general theory of relativity could not have been developed without Minkowski's space. The claim that Hutton's theory did not settle many questions (such as granite's endogenic nature?! The geostrophic cycle?!) would make any geologist chuckle, as Hutton had won the debates on erosion/denudation/transportation/deposition, the igneous origin of granites, basalts, and related rocks and the immensely long history of the earth hands down on an international scale by the end of the second decade of the nineteenth century[11]. A quotation by Greene, about the granite debate of the mid-twentieth century in the same place, from H.H. Read (1957, p. 169[12]), with a view to supporting his statement that the granite problem had not yet been resolved even in the twentieth century, misses the whole point, because the controversy Read was talking about (and its predecessor in the late nineteenth century: Prestwich, 1886, p. 428–435 with still older harbingers such as Cotta, 1848, p. 102, and Vézian, 1862; see below) was one that took place entirely within what one might call the Huttonian camp[13]! For example, the problems implied by Fyfe (1988, p. 339) when he wrote, "It was interesting that at the Edinburgh conference [i.e., the first Hutton Conference on the Origin of Granite, convened in Edinburgh in 1987 celebrating the bicentennial of the first printing of Hutton's epoch-making 1788 paper], questions were discussed concerning granites which have been discussed for at least 100 years…. After forty years of chemical analyses, isotopic analyses and laboratory experiments, it is indeed strange that the book is not closed. But it is also possible that the tools needed for the job have not been used or have not been available or that the scale of observations has been too confined, that we have studied only pieces of the puzzle" have been all within the Huttonian framework. The question that Read was trying to answer was whether the source region of granitic magmas was continental crustal rocks or basaltic melts and whether granite could be generated by metamorphism alone (see Read, 1943, 1944, 1951, for superb presentations of his views; all of them reprinted in Read, 1957). After Hutton, the only alternative for the idea that granite is emplaced as a magmatic rock, whatever its source material, was that it could be essentially metamorphic, replacing the country rock in which it is now found without going through a complete magmatic stage (even the most enthusiastic granitisers could not do without some melting). These two alternatives are summarized very competently in the first and the last papers in Gilluly (1948) and by Cäsar Eugen Wegmann (1896–1982) in the preface he wrote (Wegmann, 1957) to the second edition of Eugène Raguin's (1900–2001) *Géologie du Granite* published in the same year as Herbert Harold Read's (1889–1970) *The Granite Controversy*. However, the lifespan of the latter "metamorphic" interpretation of granite was short (about two decades) and it really never formed a serious challenge to Hutton's magmatic interpretation[14]; in fact, geologists such as Wegmann considered Hutton to be the father also of the "metamorphic" granitisation idea, since he ascribed to ancient sedimentary rocks the rôle of the source of the granites! Wallace Spencer Pitcher (1919–2004), a former student and younger colleague of H.H. Read, provides a superb summary of the history of granite research in his *The Nature and Origin of Granite* (Pitcher, 1997, p. 1–18) emphasizing the complete Huttonian triumph.

This example nicely shows that a modern science historian, not trained as a scientist himself, had little appreciation of the context of something in the history of geology he was quoting from the geological literature with the hope that it might support his revisionist ideas on Hutton. Because he clearly did not understand what the mid-twentieth century granite controversy was about, he made a bad choice that really supports the opposite of what he wished to illustrate[15]. In fact, Greene might have cited T. Sterry Hunt's neptunism to a much better effect for his cause, because Hunt, an extremely distinguished geochemist, claimed (as a loner, to be sure), in the framework of his crenitic hypothesis, that Werner's theory of geology was not dead in the late nineteenth century! For Hunt's ideas, see below.

Much more baffling is Rachel Laudan's (1944–) statement that "Hutton made a limited but significant contribution to causal geology which his successors recognized" (Laudan, 1987, p. 137), which led one of her geologist reviewers to complain that she was being unjust (Sarjeant, 1988, p. 307). I agree, although I think that the late Bill Sarjeant was understating the affront. Having read Laudan's book, I could not convince myself that she understood what geology was about and how it proceeded in its business ever since Anaximander's first geological speculations in the early sixth century BCE, despite the fact that she had read geology for an undergraduate degree in Bristol, before she did postgraduate work in history and philosophy of science in University College London. I cannot see how Hutton's theory of the earth, which was developed as an answer to the simple question of "how continuous denudation fails to eliminate the land completely?" can possibly be considered in the "cosmogonic tradition[16]." Hutton had been brought up to think that land is immobile (I do not think as a chemistry enthusiast and a trained physician he would have been aware of Theophrastus'[17] or Jean Buridan's[18] dynamic earth models; if he were, we have no evidence to show it). What he saw in his 13 years of farming in Slighhouses in Berwickshire led him to see a contradiction in the rate of erosion of the soil and the idea of an immobile land (see especially Jones, 1985, p. 592–593; Withers, 1994, p. 46). Like any good scientist since Anaximander, he observed, beginning with 1753 or perhaps even earlier (Playfair, 1805, p. 44; Jones et al., 1994), with a view to answering his questions and speculated building models with a view to testing them.

A concrete event in that line is cited by the notable Scottish geologist and Hutton scholar, Donald B. McIntyre (1923–2009), about Hutton's discovery, in 1765, "a bed of oyster shells above Kinneel [Kinneil] some feet deep appearing in the side of the bank, ~20 or 30 feet above the level of the sea, which corresponds with old sea banks.... There are many other marks of a sea beach upon a higher level than the present, but I mention only those which I can give with certainty (Hutton, 1795, v. 2, p. 166)" (McIntyre, 1997, p. 111–112). McIntyre mentions, quoting Playfair, another, earlier field excursion to the north of Scotland Hutton took with Sir George Clerk-Maxwell of Penicuik (1715–1784), the great-grandfather of James Clerk Maxwell, in 1764 (McIntyre, 1997, p. 121), inspired by some new mineral findings by the Welsh mining engineer John Williams (Torrens, 2004a; see below).

Laudan's "wave of scholarship" placing Hutton firmly in the "cosmogonic tradition" (p. 128) was fighting against straw men it itself erected. Although Hutton certainly had a unified view of nature, his geology was very little indebted to his physics or chemistry or contemporary astronomy (notwithstanding the many statements to the contrary in the recent literature), because it was based dominantly on geomorphology and geometric relations of rock bodies (soil erosion, unconformities, xenoliths in intrusive bodies, contact metamorphism, etc.) and their temporal implications (especially cross-cutting relationships). Yes, heat was an engine for running geology, but this had been suspected ever since Empedocles (ca. 494–434 BCE) and Theophrastus; Steno mentioned it as a geological engine and Lazzaro Moro founded an entire theory of earth behavior on the assumption of a central fire in the earth. Hutton's predecessors in this issue, however, could not do what he did, simply because they did not think about the geometric relationships of rock bodies and their temporal implications. Steno, Hutton's most important predecessor, for example, certainly did think, following Descartes, of geometric relationships of rock volumes to one another and made very remarkable temporal deductions from them[19], but his attention was confined to sedimentary rock layers and did not involve the cross-cutting relationships of igneous rocks. Hutton's predecessors' means of investigating the internal architecture of the earth's rocky vest and to establish its phases of construction were thus inadequate. Steno had already given geologists the means to be able to solve the problem of the interpretation of the geometry of rock masses[20] (Şengör, 2009a, 2016) and they were among the most fundamental means that Hutton employed to understand the history of the surface of our planet. Laudan's statement that "Hutton would have found the tracing of sequences of unique events advocated by historical geologists a pointless exercise" flies in the face of what Hutton himself wrote about what he calls "annals of a continent":

> The low flat island of Pladda had formed part of the high island of Arran; no proposition in natural history, concerning what is past, is more certain. The island of Arran had been extended beyond Pladda. But how far are we to go? Might we suppose that the rock of Elza (Ailsa), that stupendous mass, to have been also part of Arran? This rock is a mass of granite-porphyry, and there is also whinstone mixed in that mass: this I have only from information and from specimens, having never been upon that singular island. If we shall now suppose the softer strata of marl and sandstone of Arran, which had extended far into the sea and among which the porphyry mass of Elza had been injected, in like manner as the Holy Island is at Lamlass, to be worn and washed away in the natural operations of the atmosphere and sea, the solid mass of Elza would remain a monument of what had been in Arran; while the little island of Pladda is the intermediate step by which we may remount to this view of high antiquity.

> By thus ascertaining the first step in our cosmological speculation, we advance with some degree of certainty into the annals of a continent which does not now appear; and in tracing those operations which are past, we foresee distant events in the course of things. We see the destruction of a high island in the formation of a low one; and from those portions

of the high land or continent which remain as yet upon the coast and in the sea, we may perceive the future destruction, not of the little island only, which has been saved from the wreck of so much land, but also of the continent itself, which is in time to disappear. Thus Pladda is to the island of Arran what Arran is to the island of Britain, and what the island of Britain is to the continent of Europe. (Dean, 1997; Hutton, 1899[1997], p. 260–262)[21]

Instead of relying on a "wave of scholarship" of the sixties and the seventies of the last century, had Laudan read Hutton himself, perhaps she would not have come to such extraordinary conclusions; or at least that reading might have led her to include in the "wave of scholarship" she bathed in, Wolfgang Blei's publications (1977, 1981, p. 221 and 290), in which she would have found documented the historical character of Hutton's theory of the earth, a point made earlier by Dott (1969), which she cites. Her lack of understanding of how geology proceeds with its task is apparent in her very first chapter where she separates a historical and a causal aim of geology[22]. Anybody who has done any geology would realize that historical geology cannot be done without taking into account the causes of the processes that create that history. So, the geologist aims at learning and understanding *both at once*. The Werner student Alexander von Humboldt thought that this was important enough to emphasize and wrote that "in understanding Nature, being is not to be divorced absolutely from becoming" (von Humboldt, 1845, p. 63) and a page later: "the geognost cannot understand the present without grasping the past" and further, on the same page: "Being can only be recognized in its compass and inner being as *something that has come into existence*" (italics von Humboldt's). I know this to be true from my own experience of doing both local and global geology for the past four decades, but we can read the same from the pen of one of the greatest geologists who ever lived, namely Eduard Suess (1831–1914):

In Europe the great local variations in the characters of the deposits which furnish the remains of terrestrial animals far more marked in this continent than in the United States obliges the investigator to trust almost exclusively, certainly to a far greater extent than in the case of marine deposits, to organic remains for his chronological and stratigraphical conclusions. But it is scarcely necessary to remark that, while the characters of the terrestrial fauna afford a most valuable passive criterion, yet *it is the physical causes of faunal transformations which will, when once they are recognized, form the only true basis for a delimitation of chronological periods*. (Suess, 1883, p. 15–17; italics mine)

Laudan writes that Werner "drew on both the Becher-Stahl cosmogonic tradition[23] and the work in mineral taxonomy to construct a geological ('geognostic') theory that opened the way to a separation between causal and historical geology" (Laudan, 1987, p. 17). Had Werner really drawn on the work of Becher and Stahl, he would have worried about causes, since these people were doing what is now considered chemical research—although I very much doubt whether Werner drew on the work of Becher and Stahl, as I document in the last footnote. Nevertheless, Werner's history was full of causes governing both deposition of sediment and retreat of the sea. So, there really could not have been a separate "causal aim" and a "historical aim" in geology. Nobody who cannot understand the causes, can understand, or even simply reconstruct, a history.

What Laudan should have separated would have been pure description from interpretation. But, as the history of what we today would consider an object of pure observation, the *concept* of "stratum" or "bed," shows, even that is hardly possible (Şengör, 2009a, 2016). Indeed, Werner described the minerals and the succession of what he thought were strata, but many things he thought were sedimentary strata were not and a number of his contemporaries, even former pupils (e.g., Voigt, 1789, comparing the lava flows alternating with sediments in Italy, Iceland and Germany), told him that, but to no avail. He attempted to interpret the latter historically in a causal framework, following many predecessors from Anaximander, through some of the early Christian church fathers, to De Maillet, von Linné and de Buffon, who tried to explain the entire geology by the regression of a once universal ocean—and failed[24]. Neither undertaking was original with him, except the extent to which he put to use the method of identifying minerals using their external characteristics and the degree to which he perfected the method. (The invention of that method was not original with him; Werner learnt it mainly from the physician and mineralogist, his teacher in the University of Leipzig, Johann Carl Gehler's {1732–1796} *De Characteribus Fossilium Externis* {On the external characters of fossils[25] published in 1757 in Leipzig}, but even before him Agricola {1494–1555} and even the Lutheran preacher in St. Joachimsthal {present-day Jáchimov in the Czech Republic}, Agricola's and Martin Luther's friend, Johannes Mathesius {1504–1565} in his sermons published in his widely read posthumous *Sarepta* {1578}[26], published guides to mineral identification using their external appearances.) Werner's stratigraphy had been introduced before by Lehmann and Füchsel (see Fitton, 1839, p. 37 {441}[27] for one acknowledgment of this in the English geological literature) on the basis of Steno's principles (in fact, Werner even went backward from them and from Steno, by denying Steno's rule of original horizontality, possibly through a misinterpretation of what de Buffon had written about inclined beds: e.g., de Buffon, 1749, p. 246–247; the only place where de Buffon speaks of deposition, as an exception because by currents, on an inclined surface is in 1778, p. 107; otherwise, he always insisted that deposition in an aqueous medium consistently generated horizontal beds parallel with one another: "Thus these laminae of coal took their form by two combined causes: the first is by deposition, always horizontal, from water...": de Buffon, 1778, p. 109[28]). That is why Werner could not understand mountain-building.

Laudan's practice is unfortunately characteristic of many of the recent revisionist writings in historiography of geology that seek to belittle Hutton's contribution (for a similar criticism, see Torrens, 1998). The most unexpected statement about Hutton's position in the history of geology was published by Martin Rudwick (2005, p. 159, note 43), who used to be a geology faculty member in Cambridge, one of the early supporters of continental drift among English geologists and with a reputation as a fine teacher (John F. Dewey, written communication, 2019). He maintains that the "tradition" of regarding Hutton as "founder" was initiated by Sir Archibald Geikie (1897) and perpetuated by Sir Edward B. Bailey (1967) and McIntyre and McKirdy (1997). Nothing can be further from the truth[29]. Hutton had long been regarded as the father of modern geology, even before Sir Archibald Geikie was born as I document below. I find it shocking that a tsunami of protest against this dislocation of historical fact did not arise from amongst geologists. Rudwick says for his account of Hutton he is indebted to modern scholars, such as Davies (1967, 1969), Dean (1992), Porter (1977), Laudan (1987) and Gould (1987). I do not think these historians, with the exception of Laudan and Gould, would agree with his statement about the reputation of Hutton in the nineteenth century. Let us see what Davies himself says in his publications: In his excellent book *The Earth in Decay* (Davies, 1969), he wrote that Hutton's place *only in the history of thought concerning fluvial erosion* had to be revised. "Hutton's theory," Davies then added, "represents an enormous advance upon the fanciful, bibliolatrous theories of most of his contemporaries. Any reassessment of his geological work must leave him as one of the outstanding figures in the history of the science" (Davies, 1969, p. 194). Some historians of geology, clearly Rudwick included, somehow managed to misread this clear statement as implying that Davies thought that Hutton's great fame in geology had been undeserved. Davies, annoyed by the misunderstanding piled upon his work, repeated his appreciation of Hutton as an earth scientist much more forcefully in 1985 and agreed, just like Ellenberger had done in his 1972 paper, with Sergei Tomkeieff's verdict that "within the pantheon of science, Hutton should be accorded a place but little removed from that held by Sir Isaac Newton" (Davies, 1985, p. 388). It is clear that Davies' great book may not be enlisted into the service of any attempt to belittle Hutton's position in the history of geology.

Rudwick's statement shows the great danger of relying on second-hand information (why did he not quote much of the original geological literature citing Hutton in the late eighteenth and the nineteenth centuries, with which both the departmental and the Main Library in Cambridge are so well-stocked?). That is why I have read this "modern scholarship" always with an eye on the original documents. The long quotations below are given to confront the reader with the original statements made by geologists in the past and not my filtering of them. As an example of the modern scholarship distorting what Hutton said we can take the one-time postgraduate student of Martin Rudwick, Ralph Grant's[30] statement that he treats Hutton's theory as part of the tradition of cosmogony "as Hutton explicitly acknowledged in chapter three of that work" (Grant, 1979, p. 23). Now, in that chapter, Hutton acknowledges nothing of the sort. He in fact says quite the opposite:

> Now, if I am to compare that which I have given as a theory of the earth, with the theories given by others under that denomination, I find so little similarity, in the things to be compared, that no other judgement could hence be formed, perhaps that they had little or no resemblance. I see certain treatises named Theories of the Earth; but, I find not any thing that entitles them to be considered as such... (Hutton, 1795, p. 270).

I wish Grant had read Hutton himself *in the context of the problem Hutton was trying to solve*. Grant refers to the passage I quoted, but thinks that the difference Hutton saw between his theory and older ones was not scientific but philosophical or teleological. The rest of Hutton's chapter belies this, because Hutton there outlines how to do proper geology and accuses his predecessors of not having done that. He calls the English theologian Thomas Burnet's (1635?–1715) theory of the earth, for example, a physical romance that cannot be considered in a serious view (Hutton, 1795, p. 271). Grant (1979, p. 24) says that Roy Porter identified Hutton as one whose speculative interests make him typical of the cosmogonical tradition. But even that is not exactly correct, because Porter (1977, p. 150) says that "Scotland at the turn of the century witnessed the last stand of the old traditions of cosmogonical theory and of the natural history of the earth." Grant somehow took this sentence describing the situation in Scotland in general and confined it to Hutton and eliminated the extremely important component "natural history of the Earth." By contrast, Porter, in his book, strongly emphasizes the natural history aspect of the geological work in Scotland and underlines that although in that tradition, Hutton was nevertheless an outsider with respect to the Scottish geological milieu of his time, because his ideas were so original (e.g., Porter, 1977, p. 156). Much more critically, Porter observes that Hutton's "was a cosmogonical vision which denied traditional cosmogony" (p. 196). All in all, Porter's assessment of Hutton is quite excellent (see also his fine assessment of the state of geology in Great Britain during the Enlightenment: Porter, 1980). Had he not been taken in by Hutton's repeated, yet what I think to be disingenuous references to the wisdom and order in nature[31], his analysis of him would have been what I should have said about Hutton myself. Had Porter been trained as a geologist and had he not tried to understand Hutton's geology from his social context but instead approached him from the angle of the geological problem he had set himself to solve, he might have placed a different emphasis on his narrative of Hutton's geology; as a consequence, he missed many, even important, aspects of Hutton's activity as a geologist: for example, he says that Hutton was not famed as a collector (p. 156). That cannot be true, not only because Faujas de Saint-Fond reported on his rock and mineral collection saying he was the only one in Edinburgh to own such a collection and Princess Dashkova asked him, as an expert, to organize her collection (see below) but also his friend and the elegant expositor of his theory of the earth, John Playfair (1748–1819), in his biography of Hutton, mentions his collection (Playfair, 1805, p. 89–90). People in fact used to send him unsolicited rock and mineral samples from all over Great Britain clearly owing to his renown as a collector (on Hutton's geological collection, see also Jones, 1984).

Porter says that Hutton's theory was deductive, which is entirely true. He also underlines that Hutton placed great emphasis on observation, on testing his deductions—which is how science has always been done. Porter clearly understood what Hutton was doing. This brings us to one of the worst offenders in the historiography of Hutton's geological work, namely the late Stephen Jay Gould (1941–2002). In his *Time's Arrow Time's Cycle* (Gould, 1987) and its predecessor short article entitled *Hutton's purpose* (Gould, 1983 [1990]), Gould shows that he neither understood Hutton's way of doing science (which has since been called critical rationalism by Popper: see Şengör, 2001, cf. chapter 3 in Hutton, 1795) nor did he realize that Hutton did in fact do geological history as I mentioned above. His knowledge of Hutton is very superficial (he does not ask what problem Hutton was trying to solve and with which observations he had started and when; neither does he realize that Hutton's world history was not cyclical, but had a spiral shape: Dott, 1969; Blei, 1977, 1981; Şengör, 2001[32]):

> Most revealing are Hutton's methodological statements about those quintessential data of history—sequences of events in time. He does not view them, in any sense, as components of narrative interesting in themselves, but only as data to use in establishing general theories of timeless systems. (Gould, 1987, p. 81)

What is really revealing in Gould's statement quoted above is his complete misunderstanding of what Hutton had said. He also clearly knew nothing of Hutton's influence in Great Britain and in the rest of Europe:

> Hutton was never a considered a major figure by continental geologists. I don't even think he even had much influence upon the great flowering and professionalization of British geology following the founding of the Geological Society of London in 1807. (p. 66)

The present book is one long documentation of the colossal fallaciousness of Gould's statement. Gould's knowledge of the history of science in general is also faulty (he does not seem to realize that all science begins with conjectures that usually take the form of deductions from existing hypotheses or from those one has generated oneself). Gould's criticism of Hutton's way of doing science (see, e.g., his statement: "Hutton presents his theory as the a priori solution to a problem of final causation, not as an induction from field evidence" Gould's p. 76) would apply equally well to Einstein's way of doing science, as the quotation from Einstein's inaugural speech in front of the Royal Prussian Academy of Sciences on 2 July 1914 in Berlin, quoted above, makes clear. I do not think I need to tell my readers whether that way of doing science was rewarding or not!

Many more examples similar to the ones I selected from recent revisionist publications concerning the historiography of Hutton's geology can be cited, but overdoing it becomes tedious and would fatigue the reader; I suppose the ones I chose above do make my point. However, not all recent publications about Hutton are as defective: Dean (1992, 1997), McIntyre (1997), and McIntyre and McKirdy (1997), for example, are excellent, which I think faithfully reflect Hutton's place in the history of geology. My own book (Şengör, 2001) I think throws a new light on Hutton's accomplishment as pointed out by the distinguished historian of geology, the late David Oldroyd, in his essay review of it (Oldroyd, 2003). I show below that the general opinion of the geological community about Hutton during the late eighteenth and especially in the nineteenth centuries was precisely what every geology student has learnt in the twentieth century and still learns in the twenty-first as an undergraduate in the whole world and that the attempted revisions by historians like the ones I cite above are not in the least supported by the data at hand; having read the attempted revisions with some care, I am at a loss to understand what they are based on; as Gibbon once expressed in the marginal notes he entered into his own copy of the second edition of his *Decline and Fall*:

> To what cause, error, malevolence, or flattery shall I ascribe the unworthy alternative? (Bury, 1910, p. xxxv)

I shall not attempt in this book to answer anything like Gibbon's question revealing his deeply felt irritation[33], but argue that turning the history of the science of geology upside down is not warranted and that it needs to be turned right-side up again, lest the revisions turn the history of our science into a play of Chinese whisper! I have come to think, and hope to document below, that the main flaw of the revisionists is their deficient use of the geological literature that was written during the time of their study of the history of geology and, in case of some, even a misapprehension of how geology as a science operates *and has always operated*. That is why in what follows I cite the literature of the last quarter century on the history of geology very sparingly and only where it is absolutely necessary[34].

In the paragraphs below I review examples from the geological literature, country by country in Europe, where geology was flourishing, in chronological order, from the late eighteenth century to just before the publication of Sir Archibald Geikie's book in 1897, although Sir Archibald had expressed as early as 1858, and also frequently thereafter, views akin to those published in his more famous work (see Geikie, 1858, p. 245; 1871a[35], 1871b, 1874[36], 1875a, p. 96–103; 1875b, especially p. 198; 1882a,

p. 286–311[37], 1888[38], 1892[39], 1893a, 1893b). In the first edition of his textbook, Geikie (1882b) mentions Hutton very briefly, in fact surprisingly so, essentially in a few fleeting sentences, as the one who, with his friends, helped to create scientific geology (p. 6) and argued for soil erosion by means of rain and the drainage of waters on the surface of the earth (p. 341) and for the fluvial origin of valleys (p. 371).

The reason I chose the 1897 date as my younger limit is simply because the 1897 book has been much more widely read and quoted internationally, also by non-geologists, than the earlier texts by Geikie I just cited. In the review of the evidence that follows I try to quote, as much as possible, the most widely used textbooks, a few papers and larger monographs, and some better-known popular accounts at any given time, as they best reflect the average view prevailing in the science at the time of their publication, not forgetting, however, Read's (1943, p. 64) very apposite point that "many text-books are out-of-date as soon as they are published, but text-books, though suspect to the specialist, are nevertheless the source of basic authoritative information for the general enquirer." Following Gibbon's exemplary practice "I have always endeavored to draw from the fountain-head; that my curiosity, as well as a sense of duty, has always urged me to study the originals, and that, if they have sometimes eluded my search, I have carefully marked the secondary evidence, on whose faith a passage or a fact were reduced to depend" (Gibbon, 1776[1910], p. xii). In the quotations below, a number of authors clearly cite Hutton at second hand. But that does not matter for my purpose, because what I am interested in is the fact that *they felt the necessity of citing him*. The numerous quotations below show clearly that Hutton's theory has always been greeted as a novelty, in some cases even quite independently of Playfair and Lyell, and, as early as the thirties of the nineteenth century, he was hailed as the father of modern geology both in Great Britain and on the continent, even some years before Sir Archibald Geikie was born (on 28 December 1835 in Edinburgh). I always quote verbatim the original source (if it is not in English, I translate it) rather than paraphrasing it, even in the case of second-hand quotations where their source is cited literally. This procedure makes cumbersome reading indeed, but it documents what was actually written, and in which words, thus removing me from between the original source and my reader allowing him or her to develop an independent judgement. Many of the sources I cite are difficult to obtain, even in our days of the ever-increasing resources of the internet; hence the long quotations.

I separate the statements about Hutton that I cite below into two groups: those published before the publication in 1830 of the first volume of Sir Charles Lyell's *Principles of Geology* and those published after that date, but before the appearance of Sir Archibald Geikies's *Founders of Geology* to show that even before Lyell so famously began ventilating Hutton's views, they already had been recognized internationally as novel and important for a satisfactory understanding of geology. Lyell certainly helped enormously to establish Hutton's views, as Tomkeieff (1948) rightly insists, and despite Fitton's (1839) friendly protest (see below), but he did not cause its initial widespread impact on the continent, despite the Napoleonic blockade.

THE BRITISH EMPIRE

The British Empire was Hutton's home country in the language of which he published his research and thus obtained probably the widest readership. He lived at the apogee of the Empire; only the 13 Atlantic coast colonies in North America split off to form the United States of America during the two final decades of Hutton's life, but the language of the new nation remained English and many of its traditions British (except the presence of a formal aristocracy).

Before Lyell

The immediate reaction to Hutton's geological ideas came from within the Empire; initially, some were positive, others negative, but even far away from Great Britain, within the Empire, his ideas found favorable echoes very early on. It was finally a countryman of his, another Scot, Sir Charles Lyell (1797–1875), who secured the widest acceptance of Hutton's ideas in geology internationally through the immense influence his publications and other activities had in the science, despite the fact that his agreement with Hutton's geological views was not complete. However, even if Sir Charles had never existed, the trend in the dissemination and approval of Hutton's ideas before him forecasted their eventual triumph. In retrospect from the development of the science of geology to our own day, that was inevitable (see Tomkeieff's superb discussion on this point in his 1948 paper).

In Edinburgh, James Hutton was much respected, among other things, as a geologist and was invited by Princess Yekaterina Romanovna Vorontsova-Dashkova (1743–1810; Fig. 2)[40], the future director of the Imperial Academy of Arts and Sciences in Russia (the present Academy of Sciences of the Russian Federation) and, at the time, resident in Edinburgh for the education of her son, Pavel Michailovich, Prince Dashkov, ForMemRS, (1763–1807), to catalogue her Derbyshire minerals[41]; afterwards, the Empress Catherine the Great (Yekaterina Alexeevna after her conversion to Orthodoxy; originally, when she was still a German Lutheran princess, Princess Sophie Friederike Auguste von Anhalt-Zerbst-Dornburg: 1729–1796; ruled 1762–1796) extended to Hutton an invitation to move to Russia (Appleby, 1985). Invitations were simultaneously extended to the physician and chemist Joseph Black (1728–1799; Fig. 3) to "establish a medical and chymical seminary at St. Petersburg" and to the engineer, chemist, and inventor James Watt (1736–1819) to act as Master Founder of Iron Ordnance to the Empress (Anderson and Jones, 2012, p. 39). Appleby states that "Had the trio been tempted over to Russia, it would have dealt a serious blow to the scientific achievements of the late Scottish Enlightenment" (1985, p. 403).

The earliest written reaction to Hutton's theory of the earth within the British Empire that I am aware of is the letter

Figure 2. Princess Dashkova by Dmitry Levitzky (1784) in the Hillwood Estate, Museum & Gardens, Washington, D.C. Photographed by Edward Owen.

Figure 3. Joseph Black (1728–1799). Engraving by J. Rogers after Sir Henry Raeburn. Blackie & Son (17--); National Library of Medicine no. 101410573. Public Domain Mark 1.0.

Dr. Joseph Black wrote to Princess Dashkova on 29 August 1787 (Fig. 4) in response to a query by her (I suppose this letter is also the first document about Hutton's theory that reached the Russian Empire). After having pointed out that Hutton had read two papers to the "Philosophical Society" in Edinburgh, Black adds:

> which are soon to be published in the first volume of their Transactions already printed. The first paper is on the cause of rain, the other is a Theory of the terrestrial globe or an explains [sic] the formation & arrangement of Fossils. His reasonings and opinions on this subject have great merit & are much admired. He desires me to express his gratefull [sic] acknowledgements for your kind invitation. It is very tempting on many accounts, but he is not easily set in motion. His attachment to the friends he lives with was always strong by affection & time has made it stronger by habit.
>
> Your Excellency may perhaps have a Curiosity to know something of his Theory of the Earth. I shall here give you the general Principles of it. There are two grand operations going on perpetually in Nature. One of these is the graduall [sic] & slow demolition of the elevated parts of the Earth's surface by the action of air, water & frost. By the repeated impressions of these agents the hardest Rocks & Mountains are slowly moldered down into Rubbish & dust. The materials are carryed [sic] down into the Valleys by Torrents, Brooks & Rivers and are first deposited by the rivers during their course towards the Sea & thus form extensive & fertile plains, and afterwards carryed [sic] into the ocean by the same rivers which are constantly changing their beds & washing away a part of the soil especially when they are swelled with floods. When this matter is brought down to the ocean or other large collections of water, wherever there is not sufficient current to keep it in motion it is deposited at the bottom & forms new stratified matter. This operation is going on perpetually & has produced so much effect that every elevated part of the surface of the Earth which we now see was once covered with other matter to a great depth. Were it to go on without being counteracted the inequalities of the earth's surface would be levelled in the course of time to a perfect plain [, and mostly covered up by the sea or by collections of stagnating water]. But there is Remedy in nature to prevent such a great change in the constitution of the Globe. There is also another cause acting as constantly to prepare new rocks & mountains or as Dr. Hutton expresses it "new Land or new Worlds" in place of those which we see at present wearing down by time & in place of many others which have existed & have been demolished before them. This other cause is subterranean fire the agency of which, tho [sic] not so obvious, can be demonstrated to produce its effects [constantly] during very long periods of time under a great many parts of the surface of the Globe. This is evident from the hot springs & Volcanic mountains & Islands, some of these burning, some extinguished which occur in such a multitude of different places besides a number of stones & Rocks found everywhere & which bear so exact a resemblance to the productions of subterranean fire found near Vulcanos that natural Historians have ventured to declare they must have been produced in a similar manner. The extensive agency of subterranean fires appears also from the history of Earthquakes which proves that those fires exist at a great depth & have very extensive communications. The Earthquakes which within these very few years desolated Calabria shook a circular extent of the

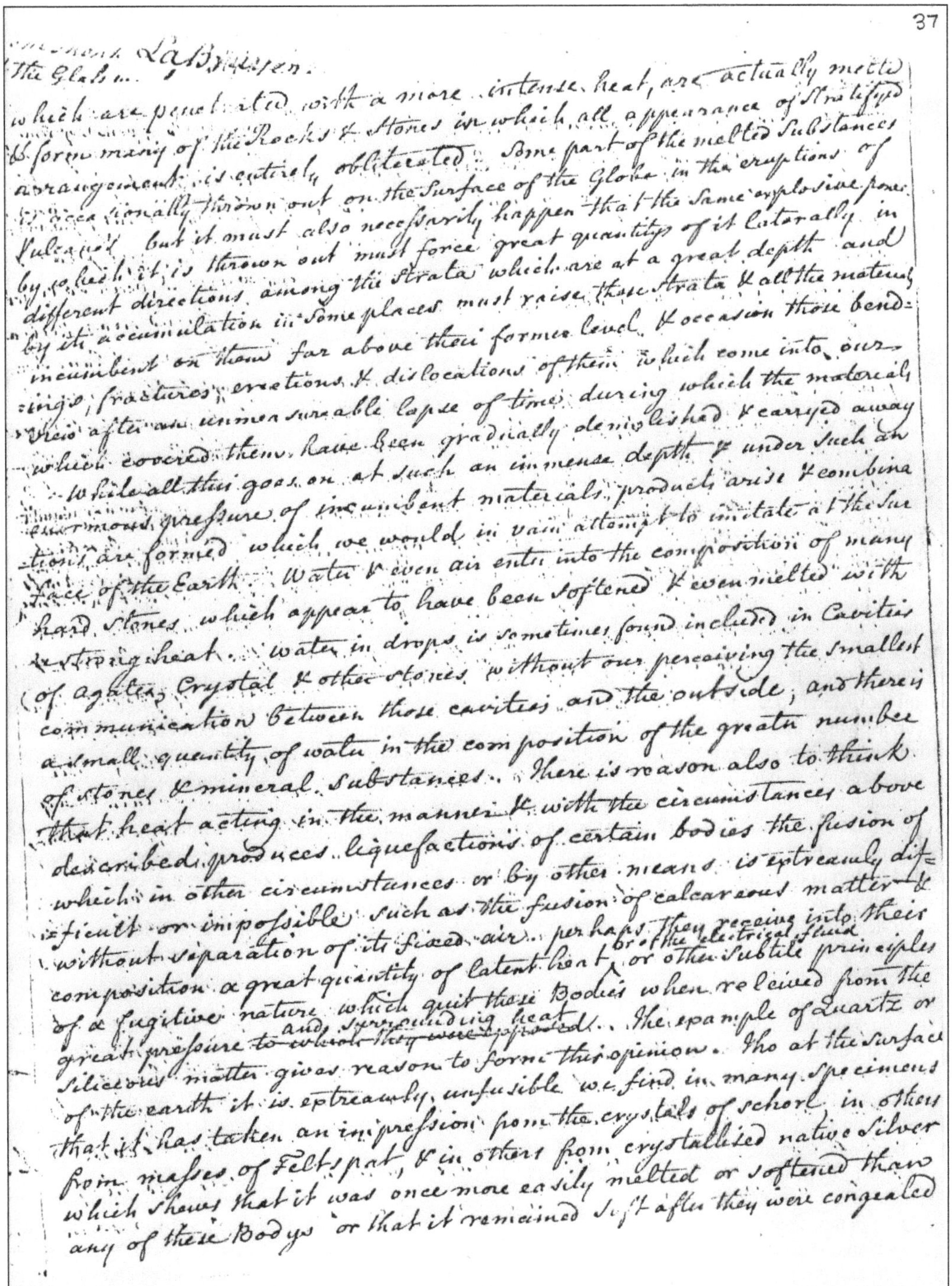

Figure 4. A page from the rough copy in Black's hand of the letter to Princess Dashkova explaining Hutton's theory of the earth (Centre for Research Collections, University of Edinburgh, ref: Gen. 873/III/36–39).

earth's surface 144 Italian miles in diameter; the place from which the elastic matter exerted its force could not therefore be much less than 70 miles below the surface; but supposing it to be only 60 miles, the pressure to which it was exposed is enormous and must have exceeded the pressure of the atmosphere at the surface of the earth more than a 100,000 times, vizt:—five miles depths of air equally dense with the air at the surface of the earth would have a pressure equal to that of the atmosphere; five miles of water would press 850 times as much, equals 4250 atmospheres; five miles of earth and stone two-and-a-half times as much as the water, equals 10,625 atmospheres; fifty miles of d^o 160,250 atmospheres. (The Italian mile is to the English as 10–12). The earthquake which destroyed Lisbon affected the whole Island of Britain & the islands of the West Indies; at what an enormous depth below the sea fires & communications have been situated by which such an effect could be produced. At such depths these fires act in silence [and darkness] during successions of centuries & under the pressure of the immense load of incumbent Earth & water their most general effect is to harden & consolidate into stones of different kinds the materials of former land which have long since been carryed [sic] into the ocean & deposited at the bottom of it in the form of regular but loose stratified matter; the strata of sand are hardened into sandstone. Those which contain clay mixed with other materials are formed into schistus's of different kinds. Collections of shells & corals which are formed at the bottom of the Sea & afterwards covered with stratified matter are consolidated into Limestones & Marbles, while the collections of vegetable inflammable matter of which great quantitys [sic] are [also] carried into the Sea by many rivers are compacted & indurated into strata of Pitcoal [sic]. It is also evident that these fires must act with all the different degrees of force; a great part of the materials [upon] which they act must be situated beyond the reach of their greatest power & must therefore suffer but little change; others receive so much heat as to be softened by it & made to approach to a state of fusion & afterwards when the heat leaves them become dense and hard & are otherways changed in many respects tho [sic] they still retain some marks of their stratified texture; others again which are penetrated with more intense heat, are actually melted & form many of the Rocks & Stones in which all appearance of stratified arrangement is entirely obliterated. Some part of the melted substances is occasionally thrown out on the surface of the Globe in the eruptions of Vulcanos but it must also necessarily happen that the same explosive power by which it is thrown out must force great quantity of it laterally in different directions among the strata which are at a great depth and by its accumulation in some places must raise these strata & all the material incumbent on them far above their former level & occasion those bendings, fractures, erections & dislocations of them which come into our view after an immesurable [sic] lapse of time during which the materials which covered them have been gradually demolished & carryed away to sea, so all this goes on at such an immense depth & under such an enormous pressure of incumbent material products arise & combinations are formed which we would in vain attempt to imitate at the Surface of the Earth. Water and even air enter into the composition of many hard stones which appear to have been softened & even melted with strong heat. Water in drops is sometimes found included in cavities of agates, crystals, & other stones without our perceiving the smallest communication between those cavities and the outside, and there is a small quantity of water in the composition of greater number of stones & mineral substances. There is reason also to think that heat acting in the manner & with the circumstances above described produced liquefactions of certain bodies the fusion of which in other circumstances or by other means is extremely difficult or impossible such as the fusion of calcareous matter & without separation of its fixed air[42] perhaps they receive [also] into their composition a great quantity of latent heat or of the electrical fluid or other subtle principles of a fugitive nature which quit these bodies when relieved from the great pressure and surrounding heat. The example of Quartz or siliceous matter gives reason to form this opinion. Tho [sic] at the surface of the earth it is extremely unfusible we find in many specimens that it has taken the impression of the crystals of schorl, in others from masses of Feltspat [sic], & in others from crystallized native silver which shows that it was once more easily melted or softened than any of these bodys [sic] or that it remained soft after they were congealed. We find also that it has often penetrated wood or Chalk or Limestone or shells in a manner which shows that it must have had a great degree of fluidity & tenuity & have been in a state of totally different from that in which we find it in those parts of the globe which are within our reach. Dr Hutton is of the opinion that Granite is one of those stones which have been melted by subterranean fire at a great depth and that the grains of different matter of which it is composed have been formed in most cases by a spontaneous segregation or crystallization in consequence of fusion & of an exceedingly slow refrigeration. The mountains of granite therefore have been accumulated in the places where they are found by explosive protrusions of the melted matter in a lateral direction at a great depth & in such quantity as to have raised the stratifyed [sic] matter which was superincumbent and in many cases to have burst thro [sic] a number of these strata. This idea accords very well with the phenomena observed by M. Saussure in his instructive Voyages dans les Alpes & it agrees with what Dr. Hutton has seen in examining the Granite mountains of this country. According to this idea of the Granite mountains it is improper to call them primary mountains & the elevated strata which appear inclined toward them or resting on their sides the secondary mountains. It is certain by the plainest proofs that the stratified matter which is called secondary or which is supposed to have been brought together at a later period than the accumulations of granite was in reality preexisted to those in the greater number of cases.

In this System of Dr. Hutton there is a grandeur & sublimity by which it far surpasses any that has been offered. The boundless preexistence of time and the operations of Nature which he brings into our View, the depth & extent to which his imagination has explored the action of fire in the internal parts of the Earth strike us with astonishment. And when we consider the view he gives us of a great river such as that of the Amasons [sic] descending into thousand streams from the high country of the Andes & forming those immense & level plains thru [sic] which it flows in the greater part of its course the mind is expanded in contemplating so great an Idea and the length of time which the change thus imagined (I may say demonstrated) must have required; the short lived bustle of Man's, remotest reach of History or Tradition, [or of the inquisitive Antiquarian] appear as nothing when compared with an object so great. Nor has this system this merit alone. It is founded on the efficacy of Powers which we see actually existing and dayly [sic] producing similar effects and it is supported by innumerable facts observed in consequence of an accurate & judicious study of Fossils[43] for a great number of years. Dr. Hutton had formed this system or the principal Parts of it more than 20 years ago & he has found reason to be more & more confirmed in it by his Study of Fossils ever since that time. Other authors have observed a great part of the phenomena on which this system is built, and have perceived some of the truths deducible from them; but none has perceived them in such a comprehensive manner, or employed them so ably to explain the whole of these subjects. The Paper he has published in our Transactions is but a specimen; he is pping [i.e., preparing] a larger work. (Edinburgh University ref: Gen. 873/III/36–39)[44]

The abstract Black sent to Princess Dashkova embodied in his letter is the most succinct and competent statement of his friend Hutton's theory; it is certainly easier to follow than Hutton's own abstract (Hutton, 1785) and by virtue of the fact that Black himself was a great scientist and that he was in daily contact with Hutton in Edinburgh because of their close friendship, it enjoys a very high degree of authority[45]. The Nobel laureate Scottish chemist Sir William Ramsay (1852–1916) writes in his posthumously published biography of Black that Hutton and Black, "almost inseparable cronies" (Ramsay, 1918, p. 117), "were most intimate friends; every afternoon was passed together, and many evenings" (Fig. 5). The great Scottish historian and sociologist Adam Ferguson of Raith (1723–1816), Black's first cousin and his first biographer (see Ferguson, 1805), wrote that, after having listed such friends of Black, in two separate lists, as William Cullen (1710–1790), Alexander Monro secundus (1733–1817), James Watt (1736–1819), David Hume (1711–1776), Adam Smith (1723–1790), John Home (1722–1808), Alexander Carlyle (1722–1805), and John Clerk of Elden (see 1728–1812), "At the head of either list, however, in respect to Black's habits of intimacy, ought, perhaps, to have been placed James Hutton, who made up in physical speculation all that was wanting in any of the others" (Ferguson, 1805, p. 113). It was mainly with Black that Hutton discussed his theories of geology (p. 114). In fact, Hutton's biographer Playfair wrote that "Several years before the time I am now speaking of [i.e., 1783], he had completed the great outline of his system, but had communicated it to very few; I believe none but his friends Dr. Black and Mr. Clerk of Elden." (see Playfair, 1805, p. 51; see also Anderson and Jones, v. 2, p. 906, footnote 6). "Black," say Anderson and Jones "had a deep understanding of Hutton's revolutionary theories of the formation of the earth and was the first person to give them public support by introducing them into his lecture course. He summarized them in a masterly fashion in an uncharacteristically long and detailed letter to Princess Dashkova..." (2012, v. 1, p. 41). One should bear in mind this abstract by Black when reading the reactions to Hutton's ideas below. Black's statements above[46] are such that one can have them read by a first-year geology undergraduate today with a view to making him or her understand how basically the earth operates.

The physician, chemist and science writer Thomas Beddoes (1760–1808) was an early follower of Hutton and when he returned to Oxford as Reader in chemistry in the spring of 1788,

Figure 5. James Hutton (left) in conversation with his friend Joseph Black by John Kay from his *Kay's Originals*, v. I (1880), p. 57 (National Portrait Gallery image NPGD18646, © National Portrait Gallery, London, npg.org.uk). The caption "philosophers" would translate into our present terminology as "scientists."

he wrote to his former teacher Joseph Black in Edinburgh to ask for the conveyance of his gratitude to Hutton and to request a granite sample from him (dated 23 February 1788):

I shall take the opportunity of bringing his admirable paper on the theory of the earth before the class and am very desirous of being able to illustrate it as much as possible. Now could he without infirming his own collection give me a specimen of the granite which he has figured? If he could spare me a morsel of anything else I should be infinitely obliged to him… (quoted from Torrens, 2017, p. 82).

Torrens (2017), in his excellent paper, reviews at some length Beddoes' "Huttonian" activities in geology including those he incorporated into his "natural history of the earth and atmosphere" lectures at Oxford. The case of Beddoes very nicely shows Hutton's almost immediate influence on science education in the UK. Dean (1992) also deals briefly with Beddoes' contribution to the spread of Hutton's ideas.

The reaction to Hutton's 1788 paper within the British Empire was dealt with satisfactorily and at length in Dennis Dean's excellent book (Dean, 1992), so I do not need to cite individual authors that he cites and discusses. There were both supporters and opposers and in what follows I quote only those writers Dean either did not cite or did not emphasize their comment that Hutton was the founder of modern geology. Before I do so, however, let me emphasize that both the yea-sayers and the nay-sayers were united in thinking that what Hutton was saying was something new, very different from the numerous "theories of the earth" that preceded it[47]. That some of the recent revisionist publications lump into one genre Hutton's theory with the older "theories of the earth" is an old mistake, as von Zittel already pointed out in 1899 (p. 100):

Hutton's work, rich in content, aroused little interest, partly because of its title, identical with those of the worthless publications of the time and partly because of its abstract way of presenting things that is not always easy to understand.

Von Zittel himself had no qualms about writing that Hutton's ideas "formed the basis of our modern views" (von Zittel, 1899, p. 103).

Dean dealt with the next detailed account of Hutton's theory, the first in print, namely with that by the autodidact Welsh mining engineer and archaeologist John Williams[48] in some depth, as it appeared in the first edition of Williams' *The Natural History of the Mineral Kingdom* (Williams, 1789). Williams opposed Hutton's theory, fearing that its acceptance would lead to atheism, because he believed that Hutton was defending an eternal world. Williams finished his preface with the pathetic entreaty "Let us turn our eyes from the horrid abyss and stretch our hands, and cry, Save, Lord, or we perish!" (Williams, 1789, p. lvii–lxii). Some of the arguments Williams advanced against Hutton's theory in this first edition was to be taken over by the Swiss naturalist and Christian apologist Jean-André Deluc (but without acknowledgment; see below).

However, when a second edition of the same work was prepared for the press in 1810 by the Scottish physician and botanist and an early editor of the *Encyclopaedia Britannica*, James Millar (1762–1827), Williams' anxious attack on Hutton's theory had all but disappeared and the discussion of the theory, in terms only of the considerations in support of it, was shifted to the end of the second volume and reviewed together with the theory of Abraham Gottlob Werner (1749–1817; Fig. 6), the famous professor of mineralogy and mining in the Mining Academy in Freiberg in Saxony. Millar wrote, at the end of the brief summary of Williams' life in the preface to the second edition: "…and what is not unusual in such characters, the warmth of his enthusiasm, on the peculiar subjects of his contemplation, seems at times to have led him too far into visionary speculations" (p. viii). No wonder, Millar abandoned Williams' enthusiastic objections to Hutton's theory in God's name![49]

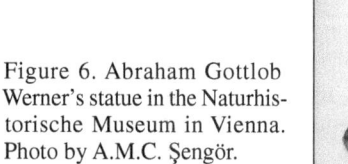

Millar lists the following observations in support of Hutton's theory (I summarize).

1. Much of the exterior crust of our globe is made up of materials of an ancient date. The strata we see are composed of the debris of the strata of the past.
2. The stratified rocks we see have formed at the bottom of the sea and became consolidated there.
3. Strata are seen almost everywhere in various attitudes away from the horizontality which they most likely had during their deposition. This must be due to later deformation caused by uplift.
4. Veins of unstratified material were filled by the injection of formerly fused materials.

For the evaluation of Hutton's ideas, Millar simply quotes Playfair's assessment in his biography of Hutton.

Millar next reviews Werner's theory. He says that Werner's "theory may be considered as the reverse of the preceding; for according to it, all mineral substances have been in a state of solution or suspension in water. According to this theory, all rocks are divided into three great classes, primary, transition, and secondary; and it is supposed that the characters of these different classes of rocks warrant the conclusion, that they have been formed at different periods" (p. 565)

Millar lists the observations usually cited in support of Werner's interpretations as follows (again, I summarize).

1. The primitive class of rocks occupies the highest eminences; they are crystalline and devoid of remains of organic beings. These observations are cited in support of the view that these rocks had formed first. All classes of rocks are disposed in layers called formations. Their sequence is fixed and is the same in every quarter of the globe: hence the designation *universal formations*.
2. The transition rocks, both of mechanical and crystalline nature, were laid down at lower elevations indicating a drop in the level of the universal ocean. They contain no land animal debris or remains of plants, but few indications of former marine animals.
3. In the so-called *secondary rocks*, there is much more clastic material and a marked diminution of chemical deposition. This is held to show that the sea level had dropped further and more land had become exposed. The secondary rocks abound in fossils and coal was a product of this episode. As sea level had dropped very considerably, sea basins had become separated from one another and the sedimentary rocks deposited in them formed independent formations as opposed to universal formations. The present topography is accounted for by the differences in resistance of various rocks to waste

Figure 6. Abraham Gottlob Werner's statue in the Naturhistorische Museum in Vienna. Photo by A.M.C. Şengör.

and decay and by stating that the present surface in part represented the irregularities of the original substratum.

After having pointed out that in all theories of the earth the origin of mineral veins plays a prominent role, Millar recommends, for further reading to learn more about the two opposing theories Hutton's and Playfair's books, the fourth volume of Thomson's chemistry[50], also the fourth volume of Murray's chemistry[51] and his book comparing the Huttonian and the Wernerian views and finally the vol. 55 of the *Journal de Physique*[52]. He passes no judgement on the two theories he just summarized.

I cite Williams' books' two editions, because Dean dealt with its first edition, but not the second one in any detail and Forster (1821), whose book I discuss below, quotes the second edition verbatim and concludes, somewhat surprisingly, with a verdict on the two theories similar to that given in the first edition (minus the religious fervor).

A very surprising defect in Dean's book is the scant attention he pays to Jean-André Deluc's (1727–1817; Fig. 7; the Genevan "reader" to Queen Sophia Charlotte of Mecklenburg-Strelitz {1744–1818}, the wife of King George III of the British Empire) important book *An Elementary Treatise on Geology* published in 1809 both in French and in English. As Breislak (1811a, p. XIII) rightly observed, Deluc's book contains "rather a critical commentary on Hutton's system than an elementary treatise on geology, as the title promises." That book's contribution to geology has been minimal, because, although Deluc touched upon important problems and significant weaknesses in Hutton's theory, the alternative solutions he offered were impossible given the knowledge of the day. His reputation of being an avowed advocate of the veracity of the *Genesis* narratives and the ensuing myth in the *Bible* certainly did not contribute to his credit. However, the fact that he devoted an entire book to refute Hutton and Playfair concerning significant issues in geology, in addition to his earlier statements which Dean discusses, underlines the importance of Hutton's ideas. In my view, the great significance of Deluc's book is his emphasis on the difference between the Baconian view of how to do science (which he defends, but does not abide by) and what we might anachronistically call the critical rationalist view (on this issue see especially Şengör, 2001).

In the first chapter of his book, entitled "Preliminary Discourse on Geology" Deluc rightly starts by saying that of all the sciences, that which is called geology is the "most extensive and the most complex" (p. 1).

This appellation was thus prematurely bestowed on the science, in consequence of the inconsiderate haste against which the immortal Bacon repeatedly cautioned those who devoted themselves to the study of nature. The mind is all times so eager to draw inferences, that it will not stop to collect all the data necessary for deducing legitimate conclusions, respecting the objects on which it is employed, but is unfrequently rash in proportion to the importance of the subject, while this very circumstance ought to preclude all precipitate decisions. Hence arises a considerable obstruction to the real advancement of science, the progress of which is much less retarded by ignorance than by error. (p. 1)

This admonition looks commendable until one realizes that Deluc is defending an impossible research program, because, it had become clear by his time that no amount of data could be adequate to formulate what he calls "legitimate conclusions." Both Hume in his *Treatise on Human Nature* (1739[1978]) and Kant in his *Kritik der Reinen Vernunft* (1781)[53] had pointed out that one cannot make an adequate number of observations to erect a true picture of anything, because "there can be no *demonstrative* arguments to prove, *that those instances, of which we have had no experience resemble those of which we have had experience*" (Hume, 1739[1978], p. 89, italics Hume's) and that any complete description would require an infinite number of observations. Hutton, as he laid out in his three-volume *Principles of Knowledge* (1794a, 1794b, 1794c), agreed with that point of view, as I showed in Şengör (2001). Einstein was of the same opinion, as the quote above from his inaugural lecture in the Royal Prussian Academy of Sciences shows and Popper formalized that point of view in his immortal *Logik der Forschung* (Popper, 1935 and many translations and editions thereafter). Hutton was in complete agreement with what Popper was to write a century and a half later:

Figure 7. Jean-André Deluc (or De Luc). Oil painting. Credit: Wellcome Collection. Attribution 4.0 International (CC BY 4.0).

We only understand the limits of a thing, in knowing what it is not. (Hutton, 1788, p. 279)

A view he reiterated six years later:

We shall so far understand the nature of external things, in knowing what they are not. (Hutton, 1794a, p. xxxii)

Hutton's most "Popperian" statement that I know is the following:

Matter of fact is that upon which science proceeds, by generalization, to form theory, for the purposes of philosophy, or the knowledge of all natural causes; and it is by comparison of these matters of fact with any theory, that such a theory will be tried. (Hutton, 1795, p. 301)

Did the geologist Eduard Suess (1831–1914) not also emphasize how our geological theories had to change because of repeated tests which showed that what we used to think the earth's geology is by realizing that it was not what we had thought it was?

In what a wonderful manner Nature refutes our assumptions! ... After we have given up a geometric system and accepted the one-sidedness of the movement, we find a dominant uniform northward striving in many mountain ranges, considered old or young, from the Cordillera to the Caucasus. We would like to formulate a law of flow of the upper part of the earth towards the pole. But this is also wrong. Farther to the east follow some dislocations along the meridians, then the moving force turns south in the mighty high mountain ranges of inner Asia. We thus obtain a picture of the face of the earth which does not at all correspond to our expectations of regular beauty, but so much more to the truth. (Suess, 1875, p. 145–146)

The famous German geologist Hans Wilhelm Stille (1786–1966) always emphasized that his orogenic phase theory was valid "until the opposite is proven" (not that he paid much attention to the opposite proofs).

However, already in his fourth numbered paragraph (p. 2–3), Deluc betrays what he defended in his two introductory paragraphs, by accepting the Biblical myth outlined in *Genesis* as undisputed fact (and thus being afflicted by Bacon's "idols of the cave," so going against a philosophy he allegedly defends)! That faith in the veracity of a primitive myth mars Deluc's entire account of geology, as he repeatedly selects interpretations of geological phenomena agreeable to the Biblical account, to the point of dividing the geological history into two unequal periods: an earlier period during which the geological events did not resemble those operating now and a later one (after the Deluge) during which processes were operative that were identical to those now active (see his paragraph 42 on p. 36–37 in the English edition of *An Elementary Treatise on Geology*; in the French edition p. 28–29)[54]. That he does repeatedly by refusing to acknowledge well-grounded observations, contrary to his own recommendation in the opening paragraph of his book. Why does he do that? He reveals the reason in his paragraphs 5 and 6 (p. 3–4):

5. No essential information, therefore upon this subject ["history of the earth and of mankind" as Deluc specifies in his paragraph 4, p. 3], was wanting to men; nor indeed can it be conceived that such knowledge should have been withheld from them, if they be considered only as a rational and moral creatures of a God infinitely wise and good; for such a Being could not have suffered them to remain ignorant of his existence, and of the relations which they bore to Him [Deluc does not tell his reader how he knows all this, upon what *facts*?]. This reflection alone enabled the Theist to repel the arguments of a few sceptics against the reality of a revelation from God to man.

6. But a complete change was made in this state of things by geology. All those, who have formed geological systems, have endeavored to rest them upon facts, which relate to the history of the earth. This has called back the attention of men to that history, as it is contained in *Genesis*, and consequently to that of the human race, with which it is connected; and it was soon inferred, that if geology were contrary to *Genesis*, the latter must be fabulous. All attempts to elude so evident a conclusion would prove vain; and it must of necessity be admitted, that if geology, a science founded on facts, and on strictly logical deductions, having attained all the characters of truth, were in reality found to be in opposition to what *Genesis* relates of those physical events of the earth, which are intimately connected with the human race, the history of the latter would become vague and uncertain. This is a consideration which I offer to those, whose profession it is to teach and defend revealed religion. The weapons by which it has been attacked have been changed, and our [Deluc here aligns himself with those "whose profession it is to teach and defend revealed religion"!] modes of defence must be adapted to the arms of its assailants. They now attack it by geology; which therefore becomes a science as essential to theologians, as the study of the learned languages, or of those ancient arguments, which are already much neglected in the present times, and which must henceforth derive their chief support from the very science, through the medium of which, under the pretence of an appeal to facts, it is attempted to set them aside.

Why was Deluc so committed to the defense of the revealed religion? He tells us himself in the next paragraph (p. 4–5):

7. In proportion, therefore, to the influence which geological systems will necessarily have, must be the caution observed by the friends of mankind in their inquiries respecting this great object. No general inference, indeed, drawn from the physical sciences, could be more important to men than that in which *Genesis* was involved; for to consider that book as fabulous was to plunge them into a final uncertainty, with regard to what

is most concerned them to know, viz. their origin, their duties, and their destination: it was sapping the very foundation on which the great edifice of society has always rested: it was, in short, abandoning men to themselves; and those must have been little acquainted with them, who did not foresee the fatal consequences which would inevitably ensue. [55]

Deluc, the onetime progressive democrat of Geneva, was horrified by the savage consequences of the French Revolution, which he ascribed to its hostility to Christianity, was out to defend what he called "revealed religion" to save mankind from the "fatal consequences" of its loss. He saw the greatest threat to it in the developments in geology, at the time the only historical natural science that had begun to show the untenability of the Biblical narrative of the origin and evolution of the earth and of mankind. He realized that no religious rhetoric could defeat geology, so he resolved to attack those holding geology against the Bible by standing on their own ground. To do this, he took on the geology as espoused by Hutton and his exponent Playfair, because in its arguments and its data on which those arguments were based, he discerned the greatest danger to Christian creed. In the following paragraphs, I list the geological claims of Hutton (and Playfair as his mouthpiece) as cited by Deluc and Deluc's counter arguments. That listing shows the degree of modernity in Hutton's geology and the failure of Deluc in his crusade against it.

After having rightly pointed out that all "geological systems" so far proposed failed in their aims Deluc says that both those trying to support the mythology of the Bible and those against it were equally impotent. He then mentions Playfair's explication of the Huttonian geology in his *Illustrations*. Deluc is most impressed with Playfair as a careful observer and judge of what he sees and praises what Deluc himself considers to be Playfair's devotion to Bacon's method of doing science. Deluc greatly appreciates his book and will often refer to it in his own. In fact, his own is nothing but a long attempt at refuting everything Playfair and therefore Hutton claimed, as already pointed out by Deluc's contemporary Scipione Breislak (1811a, p. XIII).

Deluc begins with the age of the present terrestrial topography. He thinks that it was not created by the external agencies, but "are original characteristics of our continents, and that the action of external causes, so far from having produced these characters, has in fact a direct tendency to efface them" (p. 27). He says "the cavities of vallies and dales, and the abrupt cliffs of our coasts, are original characteristics of our continents" (p. 27). He presents no facts, as he would call it, to support these assertions. However, had he considered Antoine-Laurent de Lavoisier's (1743–1794; Fig. 8) detailed observations in Normandy (Lavoisier, 1789)[56], he would have seen that his remark that "the abrupt cliffs of our coasts, are original characteristics of our continents" was entirely indefensible. He seems totally unaware that Lavoisier's great teacher Guillaume-François Rouelle (1703–1770) had already distinguished littoral and pelagic sediments (Deluc could have read about Rouelle's ideas of submarine deposition in Desmarest's *Géographie-Physique* in the *Encyclopedie Méthodique*:

Figure 8. Antoine Laurent de Lavoisier. Lithograph after Jacques Louis David (1748–1825). Credit: Wellcome Collection. Attribution 4.0 International (CC BY 4.0).

Desmarest, An III (1794–1795), especially p. 415–422) and that Lavoisier had shown that both are currently being produced along the shores of Normandy, dispersing the sediments into the deep-sea. Lavoisier illustrated in detailed figures how the cliffs were being created and retreating by the erosive action of the sea and the debris resulting from their creation were dumped into the sea to be further dispersed by currents.

Neither, it seems, Deluc was aware of the Italian physician, botanist and geologist Giovanni Targioni Tozzetti's (1712–1784; Fig. 9) criticism in his *Relazioni d'alcuni Viaggi Fatti in Diverse Parti della Toscana* (=Relation of some voyages made in diverse parts of Tuscany: Tozzetti, 1752, v. 3, p. 407–412) of de Buffon's attempts to make valleys a product of marine currents in a sea that once supposedly covered our continents (de Buffon, 1749, p. 451–453). Tozzetti was a respected naturalist, not unknown to the Francophone world: de Buffon cited him on the Quaternary vertebrates in northern Italy in his *Les Époques de la Nature* and called him the "savant Naturaliste d'Italie" (de Buffon, 1778, p. 515). Tozzetti clearly showed that valleys were results of fluvial erosion by the streams flowing in them and that they clearly post-dated any lowering of the sea level (for an incomplete English translation, omitting certain passages and cutting off within the first paragraph on p. 12, see Mather and Mason, 1939, p. 74–75, and, reproduced from the prematurely truncated text in Mather and Mason, in Chorley et al., 1964, p. 16). The Venetian naturalist and traveler Alberto (actually Giovanni Battista) Fortis (1741–1803) wrote in the second volume of his *Mémoirs Pour Servir à l'Histoire Naturelle de l'Italie* that Deluc "did not have

Figure 9. Giovanni Targioni Tozzetti (1712–1783). Wellcome Library no. 9082i. Attribution 4.0 International (CC BY 4.0).

the time to consult this work, somewhat heavy and condensed, of the fine Tuscan naturalist" (Fortis, 1802, p. 50). Both Guettard (1770, memoires 5 and 6) and Desmarest (1779[57], 1806) recognized the potency of subaerial erosion on the continents, to which Deluc makes no reference. The Austrian linguist and naturalist and professor of German language in the University of Vienna, Johann Siegmund Valentin Popowitsch (1705–1774; Fig. 10) published a remarkable book in 1750 entitled *Untersuchungen vom Meere*,[58](=Investigations about the sea) in which he describes in great detail the power of subaerial erosion and transport by rivers and the origin of the present topography including the advance of the coastline in places where there is sediment aggradation (see his p. 178ff). To document the changes along the sea shore, Popowitsch uses reports concerning the burial of Roman constructions under sediment! He is extremely learned and cites both modern and antique authors concerning such changes. Of all such observation reports, there is not a word in Deluc. Thus, to oppose Hutton and Playfair, Deluc was going against his own counsel of honoring observations, but remaining true to his precept that (paragraph 7, p. 5), "No general inference, indeed, drawn from the physical sciences, could be more important to men than that in which *Genesis* was involved." If *Genesis* implies that the present topography was an original feature of the post-Diluvial world, then, for Deluc, no geological observation could possibly negate it.

Deluc next writes (p. 27, italics his) "I will shew that the *sediments deposited by rivers at their mouths, and the materials detached by the action of the sea from those shores, which were originally steep, and against which it beats, are all accumulated along the coasts*." Unfortunately for his thesis, Deluc shows nowhere in his book that the sediments cannot be dispersed away from the coasts. I quote below the remarkable observations by Polybius in *The Histories* about sediment dispersal in the Sea of Azov and the Black Sea, which Deluc ignores. Closer to his time, had he consulted the Italian soldier and polymath Count Luigi Ferdinando Marsili's (1658–1730; Fig. 11) famous *Histoire Physique de la Mer* (=Physical history of the sea; Marsilli [Marsili], 1725), the founding text of modern oceanography, he would have seen from the dredging data the Count reported from the Gulf of

Figure 10. Johann Siegmund Valentin Popowitsch. Engraving by E. Mansfeld. Public domain (BILD-PD-ALT).

Figure 11. Count Luigi Ferdinando Marsili. (G.B. Vai, by permission of Biblioteca Universitaria di Bologna.)

Lyon, where the profiles he reproduces on his 3rd Plate extend as far as 100 km to the shelf edge south of the Rhône delta (Fig. 12) displaying a variety of clastic sediments strewn on the shelf. No observations were available beyond the canyons cut by the Rhône during the Messinian Salinity Crisis, called the *abyme* by the Count, but the data Count Marsili reported are enough to show that Deluc's statement that "*the materials detached by the action of the sea from those shores, which were originally steep, and against which it beats, are all accumulated along the coasts*" was simply not true. Here was *fact* supporting Hutton's contention. Although Deluc was so insistent on having facts, he recklessly ignored them when he thought they negated the presumed veracity of Moses.

But Deluc did not even have to consult Marsili's big book. He could have simply looked at Lavoisier's 1789 paper I quoted above to see his own error. But Deluc's mind was closed to everything that might jeopardize his belief in the *Bible*.

Next, Deluc repeats the arguments made by many before him against Hutton's theory of heat induration of strata and from there moves on to say that bringing the indurated strata to the surface of the sea could not have been accomplished by the internal heat of the planet as assumed by Hutton. He is vehemently against the action of slow causes and is convinced that "the birth of our continents has been the effect of some great revolution upon our globe" (p. 29). Why does he think that? Because he sees widespread deformation in the rocky vest of our earth. He thinks this has happened by subsiding former continents and concentrating the seas in the basins thus formed by one sudden global catastrophe. Here Deluc still adheres to the sponge earth theory of Empedocles (ca. 494–434 BC), adopted by Plato (in *Phaedon*: see Şengör, 2003), Aristotle (in *Meteorologikon*: see Şengör, 2003), Kircher (1665) and Leibniz (1749a), among numerous others, to allow the subsidence of alleged former continents into empty cavities in the earth[59]. Had he read d'Alembert's (1749) and Euler's (1751) famous papers on the precession of the equinoxes and the nutation of the axis of the earth, however, he would have seen that his assumption of large empty cavities within the earth

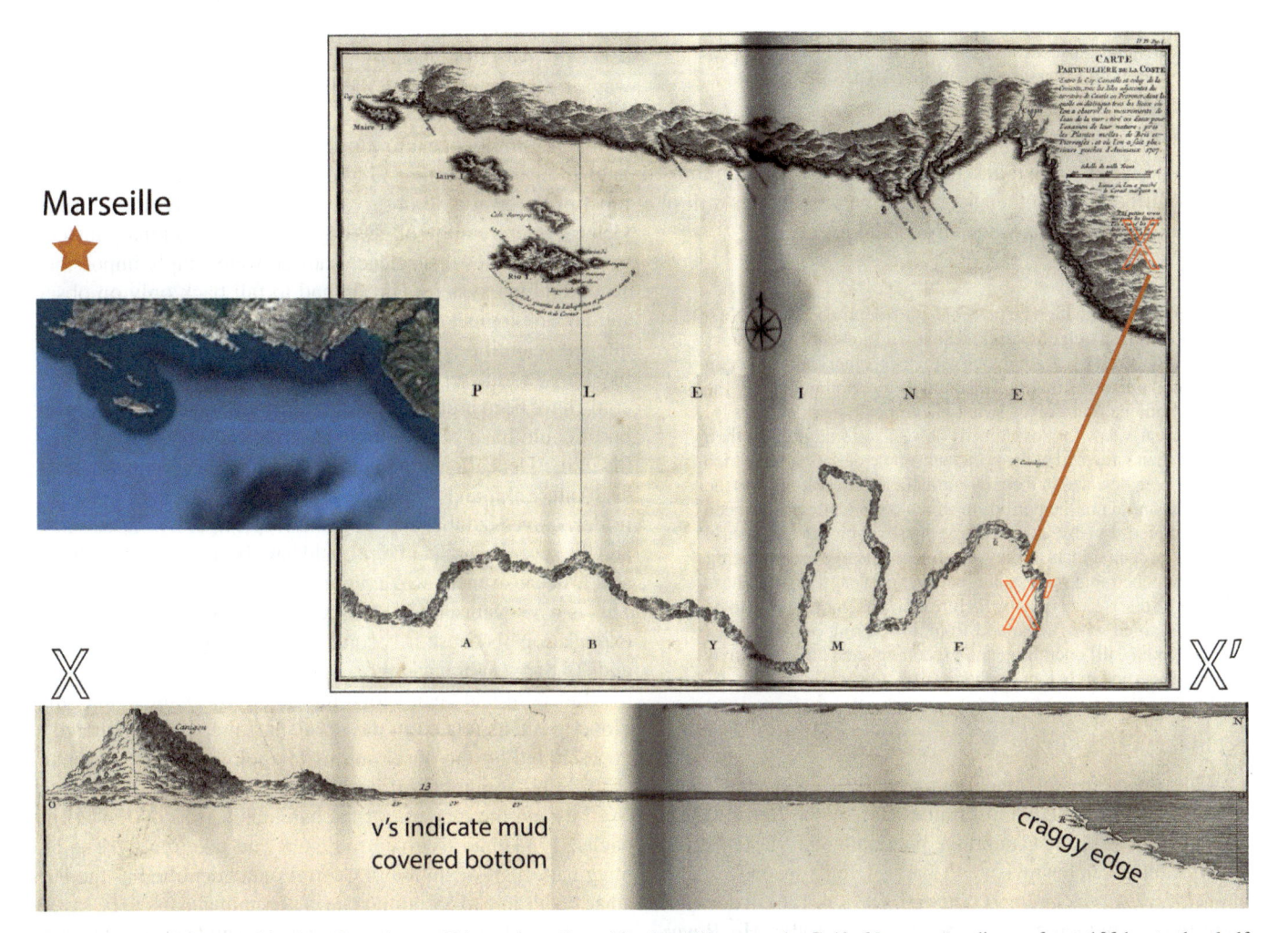

Figure 12. Marsili's map (his plate II) and one of his sections (from his plate III) across the Gulf of Lyon, extending as far as 100 km to the shelf edge south of the Rhône delta displaying clastic sediments on the shelf (from Marsilli [Marsili], 1725). The Google Earth cut-out shows the location just south of Marseille.

capable of receiving entire continents was physically impossible, i.e., total nonsense (see d'Alembert, 1749, especially chapter III discussing the possibility of a layered structure of the planet and Euler's opening arguments in his 1751 paper). In addition, the Astronomer Royal Nevil Maskelyne's (see Charles Hutton, 1778) and physicist and chemist Henry Cavendish's measurements of the earth's density (Cavendish, 1798) were available to him and they were in fact used against Deluc's gratuitous assumption of immense cavities in the earth by geologists (e.g., Breislak, 1811a, p. 69; Rees, 1819, unpaginated). Although this knowledge was accessible to him, Deluc chose to ignore it, which made it possible to uphold his Biblical Deluge interpretation.

Deluc next chastises Playfair for having "excluded the lights" resulting from certain well-determined origins[60].

> The only means of putting an end to the conflict of systems was to study them with redoubled attention; seeking first those, which might lead to immediate consequences with respect to their causes. It was thus only that the complication, so perplexing in these phenomena, could be unravelled; for by separately considering those proceeding from known causes, the parts of which the causes were as yet undiscovered, would appear divested of all necessary combinations. (p. 36)

From causes known and causes unknown, Deluc jumps, without further ado, to their identification with causes still in action and those that had allegedly ceased to act.

> If this discrimination be possible, it evidently becomes a first guide in research of causes, which will prevent many errors. Now, when we [i.e., together with his brother] had fully convinced ourselves that this distinction was pointed out by the phenomena themselves [what phenomena, he does not say], we clearly saw [not explained "how?"], on studying the theories then known, that the principal source of the errors, which had been detected in them by subsequent observation, was the confounding of the periods in which certain effects had been produced. For by attributing to causes which were seen in action, such effects as they were incapable of ever producing, an impenetrable veil was thrown over past causes; since these can be discovered only by their real effects, more surely to be ascertained when separated from those produced by causes, which are still operating, and producing such effects, as may be discerned to belong to them. (p. 37)

This was a frontal attack on the world view of those who were soon to be called "uniformitarians" by the Reverend Dr. William Whewell, the Cambridge polymath and Master of the Trinity College, in his review of Charles Lyell's second volume of his *Principles of Geology* (Whewell, 1832, p. 126). It does not seem to have bothered Deluc (neither did it bother His Reverence Dr. Whewell in either of his two reviews of Lyell's book![61]) that his identification of causes unknown with causes that had ceased to act, robbed him of the possibility of knowing what had happened before the causes now in operation commenced. Here a digression into philosophy of science is in order to be able to appreciate where Deluc stood with respect to Hutton along the scale of being scientific and being modern from the viewpoint of the uniformity of the earth processes both in time and in space (see footnote 54 above). The main aim of this digression is to clarify the methods the two antagonists employed to question the earth.

The first part of this question is particularly important for the historical sciences, such as geology and cosmology. I think there is now general agreement, among natural scientists at least, that the definition Popper (1935) gave for science is the most apposite: *science is a system of thought and attendant action in which conjectures can be falsified by observation statements*[62]. The more "unlikely" a conjecture is, the more "information content" it has, because it evidently contains much, that, if true, we would find surprising and therefore it potentially explains a large or an especially complicated part of the universe, which we previously did not know or did not understand or did not even suspect. David Deutsch's (2011, p. 24) "good explanations—hard to vary" in the sense of "hard to replace by any other explanation," is another way of expressing this. In the light of this, Hutton's theory of the earth was hard to vary as a whole with the knowledge then available, simply because it did not clash with any natural law known during his time and long afterwards and yet explained most of the available observations satisfactorily. By contrast, Deluc's interpretation of geology, especially his binary history, required miracles that science showed were simply impossible. To defend his system, Hutton had to fall back only on observations and logical consistency; by contrast, Deluc needed the alleged scientific authority of the *Bible*, which only believers like himself took seriously. The Biblical Flood he was defending could have been brought about by the Jewish (and Christian) god, but it could have been brought about equally feasibly by Zeus (as in the Deucalion story) or by the Sumerian pantheon of El, An, Anu, Enki, and Ninhursanga (as in the Ziusudra myth and its derivatives Izdubar and Gilgamesh epics), or by Snow White and the seven dwarfs, or there could have been no universal flood at all, because there was no objective evidence for it anyway. This is what Deutsch (2011, especially p. 19–24) called a "bad explanation" because it is so easy to vary (from the Sumerian gods to Snow White and the seven dwarfs).

Indeed it had been varied repeatedly by Deluc's time: the Flood myth, as related in the *Bible*, had already been altered[63], necessitated by the observations in zoology: Carl von Linné (1707–1778) in his *Oratio de Telluris Habitabilis Incremento* (Address on the increase of the habitable earth) of 1744 (Linnaeus, 1744) had to give up the ark and take recourse to an imaginary high mountain in Central Asia, identified as the Paradise, to be able to accommodate all the animal pairs to be saved plus all the climate zones in which they could survive, but still using an inspiration derived from the *Bible* itself (*Ezekiel*, 22, 24, 28: 13–14; for equating Jerusalem with the Paradise, see 47)[64], the Rabbinic tradition (e.g., in the *Babylonian Talmud: Berakod*, 40, 1; 45, 1) and perhaps even the early Syrian fathers such as St. Ephrem (306–373) in his *Hymns on Paradise* (15 hymns in his *Madrāshē*; see Brock, 1990, p. 78–79) plus perhaps such medieval scholastics as Guillaume d'Auvergne (1190–1249) in his *De Universo Creaturarum* (written around 1230–1236; *editio princeps*, in his collected works, is Nürnberg, 1496; corrected reprinting: Orléans, 1674) and Renaissance humanists as the Carthusian monk Gregorius Reisch (1470–1525; in his *Margarita Philosophica Totius Philosophiae Rationalis, Naturalis & Moralis Principia Dialogice Duodecim Libris Complectens* {Philosophical Pearl—Dialogues on the principles of all of rational, natural and moral philosophy in twelve books} of 1517[65], Freiburg i. Br.; book 7, chapter 45, where Reisch quotes Guillaume d'Auvergne—as Guilhelmo Parisiensis—from his book *De Universo* about the location of the Paradise on earth and Johannes Damascenus {675?–749}—presumably from his *De Orthodoxa Fide*, although Reisch simply says "Damascenus, book 2"— on its location in the orient and that it is higher than any land; see also Hoheisel, 1979, p. 63). But von Linné also had other sources of inspiration for his mountain: Nils Matsson Kiöping (1621–1680), the Swedish traveler in southern Asia, whom von Linné read and quoted on Ceylon (Frängsmyr, 1994, p. 120) was one such source. The Florentine Franciscan missionary, Papal legate and traveler Giovanni de Marignolli (1290–?later than 1357 and before 1359) also had heard about the Paradise mountain and its having escaped the Flood in Ceylon earlier in the 13th century (Yule, 1914[1966], p. 245). The island in the flood/mountain idea to save the animals that could not find a refuge in Noah's Ark was subsequently also used by Deluc (see Hübner, 2011, p. 34). But none of these changes or "reports" made the original Flood myth any better as a scientific explanation.

Cuvier documented animal extinctions without a shadow of doubt, which, however, created the problem of where to get the "newer animals." Cuvier never said the new animals had to have been created anew (Cuvier, 1812, p. 81–85 of the "Discours Préliminaire"; in the first independently published version of the *Discours*: Cuvier, 1813, p. 125–131; in the first independently published French version, which was much enriched by new footnotes: Cuvier, 1825, p. 129–139); that was done by his pupil Alcide Dessalines d'Orbigny (1802–1857), first in his magnificent expedition report on South America, where he ascribed the creation of the new animals to *toute-puissance créatrice* ("creative omnipotence": d'Orbigny, 1842, p. 274). Cuvier simply said that they had to have come from elsewhere, implying that the catastrophes he was positing to explain his extinctions could not have been universal. To come up with a universal flood, one had to interpret the geomorphological and sedimentary evidence in terms of a preconceived model, which was believed miraculous. Hutton's theory, on the other hand, was capable of explaining the same observations by employing mundane events that are taking place before our very eyes, without recourse to any fabulous cause.

In fact, the postulate that the past was no different from the present in terms of the behavior of the cosmos was *the* giant step that gave birth to science in ancient Greece. So far as I know, this was first emphasized by the Cambridge classical scholar Francis Macdonald Cornford (1874–1943; he was married to Darwin's granddaughter, the poet Frances Darwin):

> As Hesiod looked back in time from his own age and the life he knew and dealt with every day, past the earlier ages—the Heroic Age, the Silver Age—to the dominion of Cronos and the elder gods, and beyond that to the birth of the gods themselves from the mysterious marriage of Heaven and Earth, it must have seemed that the world became less and less like the common world of familiar experience. The events—the marriage and birth of the gods, the war of the Olympians and the Titans, the legend of Prometheus—were not events of the same order as what happened in Boeotia in Hesiod's time. We may get the same impression by thinking of the Book of Genesis—all the events from creation down to the call of Abraham. As we follow the story we gradually emerge into the world we know, and the superhuman figures dwindle down to human proportions. That is how the past had looked to everyone before the rise of Ionian science. It was an extraordinary feat of rational thinking, to dissipate the haze of myth from the origins of the world and of life. ***Anaximander's system pushes back to the very beginning the operation of ordinary forces such as we see at work in Nature every day. The formation of the world becomes a natural, not a supernatural, event.*** (Cornford, 1932, p. 19–20; bold italics are mine)

Exactly thirty years later, this momentous step and its consequences were described in what the *Times Literary Supplement* called a "brilliant essay on the origins of Greek thought" by the late laureate professor of Collège de France, Jean-Pierre Vernant (1914–2007). What Vernant wrote is so important for our understanding of what science is and the place it occupies in human thought and history that I quote the relevant passage from his book in full below, at the risk of repeating some of what Cornford says in the above quotation[66]:

> In the history of humankind, beginnings ordinarily elude us. But if the advent of philosophy in Greece marked by the decline of mythological thought and the beginning of rational understanding, we can fix the date and the place of birth of Greek reason—establish its civil status. It was at the beginning of the sixth century, in Ionian Miletus, that such men as Thales, Anaximander, and Anaximenes ushered in a new way of thinking about nature. They made it the object of a detached and systematic investigation (a *historia*) and offered a comprehensive view of it (a *theoria*[67]). The explanations they proposed for the origin of the world, its composition and structure, and all meteorological[68] phenomena were unencumbered by the dramatic machinery of earlier theogonies and cosmogonies. The figures of the great primordial powers were now obliterated. Gone were the supernatural agents whose adventures, struggles, and exploits formed the web of creation myths that traced the emergence of the world and the establishment of order; gone even any allusion to the gods that were linked to the forces of nature by the beliefs and

observances of the official religion. For the "natural philosophers" of Ionia a spirit of positivity[69] pervaded the whole existence from the outset. Nothing existed that was not nature, *physis*. The human, the divine, and the natural worlds made up a unified, homogeneous universe, all on the same plane; they were parts or aspects of one and the same *physis*, which everywhere brought into play the same powers and revealed the same vital force. The ways in which *physis* had come into being and been diversified and ordered were entirely accessible to human intelligence: **nature had not functioned "in the beginning" otherwise than it still functioned every day**, when fire dries a wet garment or when a sieve is shaken and the larger particles are separated from the rest. As there were but one *physis*, which excluded the very notion of anything supernatural, so there was but a single temporality. **The ancient and the primordial were stripped of their grandeur and mystery; they had the reassuring banality of familiar phenomena**. To mythological thought, daily experience was illuminated and given meaning by exemplary deeds performed by the gods "in the beginning." For the Ionians, the comparison was reversed. The primordial events, the forces that produced the cosmos, were conceived in the image of the facts that could be observed today, and could be explained in the same way. It was no longer the beginning that illuminated and transfigured the everyday; it was the everyday that made the beginning intelligible, by supplying models for an understanding of how the world had been shaped and set in order.

This intellectual revolution appears to have been so sudden and so radical that it has been considered inexplicable in terms of historical causality: one has spoken of a Greek miracle. All of a sudden, on the soil of Ionia, logos presumably broke free of myth, as the scale fell from the blind man's eyes. And the light of that reason, revealed once and for all, has never ceased to guide the progress of the human mind. (Vernant, 1982, p. 102–104; bold italics are mine)

From what Cornford and Vernant say, it is clear that uniformitarian thinking was no invention of Hutton or of anybody else during the scientific revolution and the Enlightenment: it had been there all along and defined the nature of "science," ever since it was invented by the Presocratics (Popper, 1958–1959 [1989]; 1998; this book contains the earlier essay with two addenda). In fact, the scientific revolution of the seventeenth, but especially the eighteenth, century came about by discarding the Aristotelian "binary space" (supra- and sub-Lunar) introduced by the post-Socratic religious thinking separating a divine heaven from a corruptible earth and the "binary history" dictated by almost all religions separating an imaginary time of gods and prophets full of miracles (see Eliade, 1961, p. 23), from a time of ordinary human beings and events, i.e., by returning to the way of thinking of the Presocratics!

Was it possible for Deluc to know these opinions of the Presocratics? Certainly, yes; at least second hand, as he was able to read none of the classical languages. Not only the very designation "presocratic"[70] had already been introduced by the polyglot German theologian and philosopher Johann August Eberhard (1739–1809) in his *Allgemeine Geschichte der Philosophie* (General history of philosophy; 1788, p. 47), but Eberhard called it also "scientific philosophy." He also claimed that it was Anaxagoras (fifth century BCE) who brought back the idea of "god" into the Ionian philosophy[71]. But, unfortunately, Deluc was also ignorant of the German language. Nevertheless, in English there was the two-volume book by the British Unitarian minister William Enfield (1741–1797), called *The History of Philosophy* (Enfield, 1792), based on the German cleric and historian of philosophy Johann Jakob Brucker's (1696–1770) multi-volume *Historia Critica Philosophiae* (published in Leipzig 1742–1744; second edition in 1766–1767). None of these early books on the history of Greek philosophy was much good by our current standards; they contained many serious errors of omission and commission, yet they were the best available at the time. As such, they nevertheless did make the correct point that Anaximander (and his pupils until Anaxagoras) clearly thought that the formative force in the universe was not a sentient being separate from the material of the universe (e.g., Eberhard, 1788, p. 49–50). However, even if Deluc had been aware of these historical accounts of philosophy, they would not have been able to make him reconsider his position vis-à-vis geology, but would have only served to increase the alarm in his pious mind about the safety of the Christian faith and he might have started a separate crusade against the historiography of philosophy.

In the light of what is said above, Deluc's geology (if one can call it that) is mythology, not science, whatever correct observations and inferences may be intercalated into it. Indeed, the German sedimentologist Karl Erich Andrée (1880–1959) wrote in 1938 (his p. 163) that if actualism is eliminated, "it would bring about the danger of endless phantasies resulting from a metaphysical look inwards, a tendency to mysticism, that would have little or nothing in common with true natural scientific research," a statement that pretty much describes the route Deluc had taken. Deutsch (2011, p. 22), justifiably calls myths "*unscientific*" because they constitute what he calls *bad* explanations. By contrast, what Hutton did was science, notwithstanding his errors. Deluc's binary history was mythology-based as he himself expressly says: it is based on, or invented to save, the narrative of the *Bible*. It is in this light that one must regard Deluc's attack on uniformitarianism and on Hutton: it was an attack of religion, i.e., myth, an *un*scientific claim, on science[72].

But let us return to Deluc's text: he next tells us that he and his brother had decided that the "production of the mass of our continents, in regard both to their composition and general form, as well as their existence above the level of the sea, should be ascribed to causes no longer in action on our globe" (p. 37). Why, on the basis of what observations, we are not told. He thus decided to study the causes now in operation: "that thus, by their being everywhere determined, our continents might be traced back to their original state." (ibid.) Deluc was convinced that he not only can reconstruct the state of the continents since their formation, but that he can even tell how much time since then had elapsed. With the appearance of the second volume of de Saussure's *Voyages dans les Alpes* in 1796, Deluc says, "It was then a veil, through which I had hitherto surveyed these monuments of our globe, had been suddenly withdrawn." (p. 41). What he discovered was that there was precipitation going on at the bottom of the sea and that the accumulated series had started with the precipitation of granite. He considers it a great error that Hutton thinks granite, porphyry, etc., igneous, because de Saussure's observations "proved" them to be sedimentary. In "no instance I found any exception to them; and I am persuaded that the great geological fact which they have established will soon be universally admitted" (p. 43). Again here we have assertions, but no observations, or any criticism of Hutton's observations of the cross-cutting relationships, or those on xenoliths making a sedimentary origin for granite impossible. Here Deluc calls the aqueous origin of granite a "fact" and claims that Hutton had been aware of it. Up to p. 109 Deluc repeats assertion after assertion that the continents are young, the coastal cliffs were not produced by the sea and that alluvial plains had nothing to do with rivers. He presents not one observation to support these assertions, while repeatedly saying that he will present them later and accusing Hutton (and Playfair) for not having taken into consideration his publications[73] in their claims of subaerial erosion both on the continents and along their steep shores. The usefulness of this part of Deluc's book from the viewpoint of the present book is that he quotes long passages from Playfair thus making his (and thus Hutton's) statements accessible to his own readers, among them, presumably, many objectors to Hutton's views.

Deluc then takes Playfair to task for his claim that rivers emptying into lakes excavate their own valleys. He says that waterfalls are obliterated not by headward erosion, but by accumulation of alluvium. Then there is a long discussion of why the Lake of Geneva (=Lac Léman) could not have been excavated originally as a river valley and, if it is so old as Playfair claims, why is it still empty. In the layers of the delta of the Rhône Deluc sees "chronometers" as in all other river mouths entering lakes. He then returns to the origin of valleys, this time using de Saussure's conclusions that the rocks south of Lake Geneva (the Préalpes) are highly deformed and fractured. He notes, correctly, that the two sides of the Arve valley between the mountain of La Môle (now known to belong to the Médianes Plastiques: Triassic, Liassic, Dogger forming a part of the Prealpine nappe pile) and the mountain of the Brezon (now considered a part of the Subalpine chains, tectonically underlying the Préalpes and belonging to a more external position in the Alpine edifice), do not correspond in structure and details of rock content. From this correct observation, Deluc jumps to the conclusion that the Arve Valley could not have been a result of river erosion but must have been produced by tectonic movements.

One recent revisionist, Martin Rudwick, puts great emphasis on Deluc's natural chronometers, consisting of delta deposits, turf growth and even soil formation as if they were something new in geology or suitable for the service into which Deluc pressed them (e.g., Rudwick, 2005, p. 156–157, 309–310; 2011). In his praise of Deluc, Rudwick overlooks earlier, similar chronometers used by other authors—who came up with vastly different time scales for the age of our continents from Deluc's—and which were all well known in Deluc's time because of the great fame of their authors. Deluc ignored them, because he was more keen to support the biblical mythology than to discover a geological truth. For instance, the "father of history," Herodotus (ca. 484–ca.425 BC) of Halicarnassus (Bodrum in present southwestern Turkey) gave an elaborate description of how he thought the river Nile must have created Egypt in the past ages in tens of thousands of years—as John Phillips also reminded his readers (1860, p. 138–139) followed in our times by Oldroyd (2009, p. 11). Herodotus wrote:

> The greater portion, then, of this country whereof I have spoken was (as the priests told me, and I myself formed the same judgement) land acquired by the Egyptians; all that lies between the ranges of mountains above Memphis [ancient capital of Lower Egypt, 20 km south of Cairo, on the left bank of the Nile, almost at the apex of the Delta] to which I have referred seemed to me to have been once a gulf of the sea, just as the country about Ilion [containing the alluvial plain and delta of Karamenderes river; ancient Skamander] and Teuthrania [on the delta of the Bakırçay river; ancient Kaikos] and Ephesus [on the delta of the Küçük Menderes river; ancient Kaistros] and the plain of the Maeander, to compare these small things with great [note that Herodotus was not describing what he thought a unique event, but the consequences of a well-known process, documented at his time on many river mouths, namely sedimentation and the siltation of narrow estuaries]. Four of the rivers that brought down the stuff to make these lands there is none worthy to be compared for greatness with one of the mouths of the Nile; and the Nile has five mouths. There are also other rivers, not so great as the Nile, that have wrought great effects; ... Now in Arabia, not far from Egypt, there is a gulf of the sea entering in from the sea called Red, of which the length and narrowness is such as I shall show: for length, it is forty days' voyage for a ship rowed by oars from its inner end to the wide sea; and for breadth, it is half a day's voyage at the widest. Every day the tide ebbs and flows therein. I hold that where now is Egypt, there was once another such gulf; one entered from the northern sea towards Aethiopia, and the other, the Arabian gulf of which I will speak, bore from the south towards Syria; the ends of these gulfs pierced into the country near each other, and but a little space of land divided them. Now if the Nile chooses to turn his waters into this Arabian gulf, what hinders that it be not silted up by this stream in twenty thousand years? Nay, I think that ten thousand would suffice for it. Is it then to be believed that in the ages before my birth a gulf even much greater than this could not be silted up by a river so great and so busy? (Herodotus, II. 10–11)

Herodotus also knew that the Delta was a more recent construction than the rest of Egypt:

> ... for we have seen that (as the Egyptians themselves say and as I myself judge) the Delta is alluvial land and but lately (so to say) come into being. (Herodotus, II, 15)

So here we see a geological chronology, in which first Egypt inland from the Delta had formed in a matter of ten or twenty thousand years. Then, in a more recent time, the Delta itself. Herodotus then tied these estimates to human history in support of the vast antiquity of the Egyptian people, much as Deluc tying his geology to the Hebrew mythology, masquerading as history, to argue that the present continental surfaces were very young. The year estimates Herodotus presented for the formation of the Egyptian land were not pulled out of thin air, but were based on what was known from similar settings in Mysia, Lydia, and Caria, forming the hinterland of Aiolia and Ionia along the western shores of Asia Minor (i.e., present-day western Turkey), where rivers such as Scamander (Kara Menderes), Kaikos (Bakırçay), Hermus (Gediz), Kaistros (Küçük Menderes), Meander (Büyük Menderes), were rapidly building out their deltas threatening the Greek ports (including Troy, Ephesus, and Miletus) at their mouths. This siltation process was clearly partly the basis of the universal regression hypothesis erected by Anaximander in the sixth century BCE. What Herodotus had written was no different from what Deluc wrote about the European rivers and lakes and it is equally wrong in terms of its temporal estimates counted in years. Aristotle (384–322 BC), in his *Meteorology*, gave a similar chronometer for the silting up of the Palus Maeotis (the Azov Sea):

> Furthermore there has been such a great increase of river silt on the shores of Lake Maeotis that the ships that ply there now for trade are far smaller in size than they used to be sixty years ago. And from this fact it is easy to deduce that, like most other lakes, this too was originally produced by rivers and eventually it must all become dry. (Aristotle, *Meteorologikon*, I, 14)

From many such observations, Aristotle concluded that the time indicated by geological processes we can observe is infinite! But much more detail concerning such geological inferences and speculations about the geological chronology of the Azov and the Black Seas we read in the Greek soldier and historian Polybius' (ca. 208–125 BCE) great work *The Histories*. A well-educated Greek aristocrat, Polybius toured most of the Mediterranean countries as a soldier in Roman service and developed a deep interest in geography, not only practical, as necessary for a general, but also theoretical, as befitting an intellectual. I quote him at length below, as he is less widely known as a physical geographer and geologist than as a soldier and worthy successor of Thucydides. His critical treatment of his historical sources has long been lauded, but few mentioned the similarly evaluative approach he had to geographical sources.

> There are two reasons why there is a constant outflow from lake Maeotis [Sea of Azov] and the Black Sea. The first, which is glaringly obvious, is that many streams flow into a basin of limited circumference, the water level constantly increases; if there were no outlets, the water would inevitably rise ever higher and occupy a larger area of the basin, but where there are outlets the extra, surplus water keeps overflowing into, and streams away through, these channels. The second reason is that, after heavy rainfalls, the rivers carry large quantities of all kinds of soil into the basins, and the silt forces the water to rise and then flow, on the same principle as before, through the channels. Since the depositing of silt and the inflow of water are unremitting and constant, the outflow of water through the mouths is also bound to be unremitting and constant. These are the true reasons why water flows out of the Black Sea. They are not based only on merchants' tales, but also on observation of laws of physics, and it is hard to imagine a more accurate method than that.
>
> So far so good. But there is no point in my stopping there, with a mere statement of facts, and leaving things undeveloped. This is what most writers do, but I want to give a detailed account, to make sure that I leave my readers in no doubt about the answers to any questions they may have. For it is a distinctive feature of our times that, since everywhere in the world can now be reached by land or sea, we no longer have to rely on poets and storytellers to fill the gaps in our knowledge, as our predecessors did in most cases. They give us in Heraclitus' phrase, no more than "unreliable witnesses" to disputed facts, but I must try to give my readers an account that carries conviction on its own merits.
>
> I maintain that the silting up of the Black Sea has been going on for a very long time, that the process is continuing now, and that therefore both lake Maeotis and the Black Sea will become entirely silted up, if the region stays topographically the same and the factors that cause the silting remain in force. For given infinite time and basins that are limited in volume, it follows that they will eventually be filled, even if silt barely trickles in. After all, it is a natural law that, if a finite quantity goes on and on increasing and decreasing—even if, let us suppose, the amounts involved are tiny—the process will necessarily come to an end at some point within the infinite extent of time. And when the amount carried is not trivial, when a great deal of soil is being carried in, the outcome I am talking about will obviously happen relatively soon, not just some time in the distant future.
>
> That this is actually what is happening is easy to see. Lake Maeotis, at any rate, has already become silted up; most of it is no more than five or seven fathoms deep, which means that large ships can sail there now only with a pilot to guide them. Originally, as all ancient authorities agree, it was a sea that was confluent with the Black Sea, but now it is a freshwater lake, since the sea water has been displaced by the silt and replaced by the incoming river water. The same will happen to the Black Sea as well, and is already happening, though the size of the basin makes it very hard for most observers to tell what is going on. But even so, a moment's thought will reveal the truth of what I am saying[74].
>
> Take the Danube, for example, with several mouths issuing into the Black Sea from Europe: the sediment that is carried down into the sea through these mouths has formed a 1000-stade-long sandbank out at sea, a day's journey from land. Ships that cross the open sea still accidentally run aground there at night, on the "Breasts" as sailors call the shoals. But why does the sediment not form shoals close to land? Why is it pushed far out to sea? The reason must be that for a while the river's currents are the dominant force and push their way through the sea. As long as that is happening,

earth and whatever else is caught up in the currents necessarily continues to be pushed out to sea, without just stopping and settling; but when the sea has enough depth and volume to cancel the force of the streams, then, by the laws of physics, the sediment will of course stop moving, fall to the bottom, and settle. That is why the sediment carried by large, turbulent rivers forms shoals way out at sea, with the inshore seabed retaining its depth, while the sediment carried by smaller, gentler streams forms sandbanks by their mouths.

There is especially good evidence for this during heavy rainfall, when insignificant streams gain enough impetus to overcome the waves at their mouths and push sediment out into the sea to a distance that is proportionate in each case to the force with which the streams flow in. It is foolish to be skeptical about the size of the sandbank formed by the Danube, or in general about the vast number of rocks and logs, and the vast quantity of earth, that issue from rivers into the sea. We often see with our own eyes how rapidly an insignificant stream, one that flows only in winter, can scoop out a bed and cut a swathe through high ground, and deposit so much wood, earth, and stones that sometimes places are altered beyond recognition.

It makes little sense, then, to doubt that large rivers with a strong, year-round flow can have the effect I have been attributing to them and eventually will fill up the Black Sea. This is not probable, but a logical necessity. The future is indicated by the fact that just as lake Maeotis is less salty than the Black Sea, so the Black Sea is distinctly less salty than the Mediterranean. This proves that when an amount of time has passed that is proportionate to the time it took to fill up lake Maeotis, in the same ratio as the size of the Maeotis basin to the Black Sea basin, the Black Sea will be a shallow, freshwater lake, like Maeotis. In fact, this presumably happens at a faster rate, because the rivers that flow into the Black Sea are proportionately larger and more numerous.

I hope to have said enough to convince the sceptics that the Black Sea is now silting up and will continue to silt up, until, for all its size, it turns into a shallow lake. Above all, I hope to have countered the false and fanciful yarns of seafaring traders: we should not be condemned by our ignorance to believe everything we hear, like children. Where we have certain traces of truth, we can use them to deduce the truth or falsity of the stories we hear. (Polybius, *The Histories*, book IV, chapters 39–42[75])

Has Deluc ever read these amazing lines? Since an English translation of *The Histories* by the Cambridge-educated English translator and historian Christopher Watson (died 1581) had been available since 1568, he at least had the possibility to read them. Not only is a completely different time-frame of sedimentary phenomena at river mouths narrated here (recall Polybius' reference to a different geographical state of the Sea of Azov during the time of the "ancient authorities" who had written about it), but also a completely different theory of geomorphological evolution of coastlines near big rivers is presented, one not only different from Deluc's ideas, but very close to Hutton's thoughts. Since Polybius was a favorite reading material for the classics education, Hutton may very well have read him.

Thus, many well-known authors even in antiquity were after a sort of chronometer or another to date earth movements well before Deluc. In the Middle Ages, Master Jean Buridan (1300–1358), twice the Rector of the University of Paris, estimated in his *Questiones super tres primos libros metheororum et super majorem partem quarti a magistro* (Questions to the master on the first three books of *Meteorology* and major part of the fourth; Duhem, 1958, p. 293, footnote 1), a book devoted to the topics treated in Aristotle's *Meteorologikon,* the age of the earth from erosion and sedimentation rates to be on the order of millions of years! He thought that the high continents were high, because they were light and the ocean floors were depressed because they were heavy. He then imagined that ongoing erosion of mountains would further lighten the inhabited world (Buridan's world consisted of one major continent occupying one hemisphere, and the ocean the other) and the debris carried into the ocean would further load its floor and depress it. If this process continued for a very long time (Buridan had no scruples about thinking in terms of thousands of millions of years!), one can imagine that the original sea floor would eventually approach the center of the earth and finally emerge from the other hemisphere (though turned inside out) as new mountains. His pupil Albert of Saxony (ca. 1320–1390) very much followed his master in his view of the earth's behavior (Duhem, 1906[1984]); this is important, because Albert's writings went through a number of editions toward the end of the fifteenth and in the beginning of the sixteenth century (see esp. Duhem, 1906[1984], p. 334–338), whereas Buridan's writings remained unedited until the twentieth century! Thus the men of science from the Renaissance onwards, Leonardo da Vinci among them, have learnt of Buridan's theory of the earth through his illustrious student's publications. Had Deluc been able to read Latin or German he could have learnt more such chronometers from Leibniz' *Protogaea* (1749a) or from Scheidt's German translation of it (Leibniz (1749b).

The question thus naturally arose by the eighteenth century as to which of all these time estimates should have been believed. Hutton knew that none of them offered any secure foundation, because sedimentation and erosion rates were very variable and both lakes and rivers were ephemeral and could give no reliable information as to the age of the emergence of the continental surfaces, as it had become well known by the time of the late eighteenth century. Hutton certainly knew it at least from Pallas' descriptions of Central Asia, which we know he had read. Therefore, even then, De Luc's alleged chronometers, although valuable to make very rough estimates about the antiquity of small lakes and some deltas, were meaningless as indicators of the time of the "last catastrophe," because they were based on ephemeral objects, whose origins were independent of those of the last marine regression from continental surfaces. In fact, the lakes of Aboukir and Edko in northern Egypt had formed during Deluc's lifetime between 1780 and 1801 as a result of irruptions of the Mediterranean onto the Delta and this was reported by Napoleon's engineers (du Bois-Aymé, 1812, p. 15–16). Not having a handle on anything seriously quantitative, the best the geologists at the time could hope for was to secure a relative chronology. This is what Hutton recommended them to do and it remained the best that geology could do until isotopic dating methods became available in the twentieth century. Deluc's "natural chronometer" from Lake Geneva, had he really been able to use it (he would have needed deep drilling and even that would not have yielded the age of the lake as Lyell, 1830, p. 223, already pointed out[76]), would have shown him that the Rhône Delta is twice as old as the age of the Earth, let alone the time of the Deluge, as allowed by *Genesis*. But let us take the initial assumption of the veracity of the *Bible* out of his reasoning and look at what he believed the chronology that the infilling of Lake Geneva yielded him. With the knowledge available to the polymaths of the late eighteenth century, one could still point to the fossil lakes in Europe and Asia and say that the chronology that de Luc deduced from Lake Geneva was simply irrelevant to the question of the age of the continents now extant, because lakes come and go on continental surfaces, as work in Egypt and Asia had then already shown (e.g., Pallas, 1776; Patrin, 1783; Renovantz, 1788; du Bois-Aymé, 1812; from the references he makes, we know that Deluc was at least familiar with the works of Pallas and Patrin: p. 302 and 357).

What is even more remarkable is that the ephemeral nature of the immense Aral[77] Sea in Central Asia was already known in the Middle Ages! Its absence on Ptolemy's (ca. 100–ca. 170 CE) maps had been often ascribed to a mistake of the great Alexandrian astronomer and geographer by our modern classicists. However, when the Franciscan friar William of Rubruck's (ca.1220–ca. 1293) travel report to the court of Möngke Khan (1209–1259), grandson of Genghis Khan and the fourth Khagan of the vast Mongol Empire, was first published in Richard Hakluyt's (ca. 1552–1616) much enlarged second edition of his *The Principall Navigations, Voiages, Traffiques and Discoveries of the English Nation*, v. 1, in 1598, in London[78] (see Jackson and Morgan, 1990), anybody could have surmised that the Aral Sea simply did not exist when Rubruk passed very close to what should have been its northern shores (he may very well have passed through it and never noticed that there had been a vast lake there earlier!) and that Ptolemy's maps probably reflected the truth. Krivonogov (2009, p. 1146) pointed out that "The court historian Khafizi Abru[79] wrote in 1417 that Khorezmian Lake (the ancient name of the Aral Sea) did not exist any longer, and the Jeikhun River (Amu Darya = Oxus) had taken a new course, flowing to the Khazarian (Caspian) Sea." When the English merchant Anthony Jenkinson (1529–1611) visited the area in 1588, the Amu Darya was still flowing into the Caspian Sea (his account was also published in Hakluyt's second edition). Jenkinson's information was used in Abraham Ortelius' posthumously published 1606 atlas in English, printed in London, as we know from the frequent references to a three-volume edition of Hakluyt's famous book in the atlas. There is no Aral Sea shown in the map entitled *Persici sive Sophorum Regni Typus* (=Wise Kingdom or Persia) between p. 109 and 110 and the Amu Darya clearly flows into the Caspian Sea. The *Amu Lacus* (=Amu Lake) indicated in that map astride the ancient course of the Amu Darya to the south-southwest of the Aral Sea (now known as the Uzboy Channel), is the swamp region of Sarikamish (=Yellow Reed). Thus, the Aral Sea must have been reborn between the second century[80], when Ptolemy published his famous atlas (Γεωγραφιχή Υφήγησις: *Geographike Uphegesis*: Manual of Geography) and disappeared again by the thirteenth century to be reborn yet again after the fifteenth century[81]! At least Ptolemy's maps (in various editions and printings in Deluc's time), William of Rubruk's and Anthony Jenkinson's travel reports and Ortelius' Atlas were available to Deluc in the libraries of England to show him the ephemeral nature of even the largest lakes; they would have documented for him that they not only disappear but are reborn even in recent times, so that they can yield no information as to the age of the continental surfaces. But it seems that Deluc was totally unaware of them or of the Aral Sea problem.

However, the most objectionable aspect of Deluc's so-called natural chronometers was that they all were made to serve to authenticate the time scale Deluc deduced from the *Genesis* myth and not to establish a scientific truth. His insistence that continental surfaces are "new" is not a scientific deduction, but a Christian legacy since John Philoponus (ca. 490–570) argued against Aristotle's idea of the eternity of the world in his *De Aeternitate Mundi contra Aristotelem* (Philoponus, after 529; see also 1987) and such medieval authors as William of Auvergne repeated and enlarged his arguments defending the "newness of the world" (Teske, 1995 and references in it to other medieval authors of both Christian and Islamic and Jewish faiths with similar views and arguments). In reality, none of Deluc's chronometers had any possibility of measuring calendar time and Phillips (1860, p. 140–142) very nicely shows why that is so. Deluc pretended they did for the reason I just mentioned. As Lord Bertrand Russell once said of Socrates' conduct, "this is treachery to truth and the greatest of philosophical sins."

The introductory "Discours Préliminaire" of Cuvier's *Recherches sur les Ossemens Fossiles de Quadrupèdes* brings us to the problem of past sea levels and Deluc's view of them. Cuvier's famous *Discours* had its first separate publication in 1813 in English translation made by the Scottish surgeon and writer Robert Kerr (1757–1813), a student of Joseph Black and the translator also of Lavoisier's groundbreaking *Traité Élémentaire de Chimie*. The translation of Cuvier's text was embellished by "mineralogical notes and an account of Cuvier's geological discoveries" written by Robert Jameson (1774–1854), Regius Professor of Natural History in the University of Edinburgh and a former pupil of Abraham Gottlob Werner in Freiberg. Among the "mineralogical notes," note H is a lengthy quotation from Playfair's *Illustrations* "On the diminution of the waters of the ocean" describing the depression of sea level in the Baltic and stating that other observations elsewhere "make the *gradual depression*, not only of the *Baltic*, but of the whole *Northern Ocean*, a matter of certainty." Jameson follows this quotation with another lengthy quotation from Deluc disputing the observations that had led to the idea that the sea level around the Baltic (and indeed all around the Scandinavian Peninsula) had been dropping[82]. In the face of the robust observations since the publication of the answers to Swedish chemist and geologist and Fellow of the Royal Society of London Urban Hiärne's (1641–1724; Fig. 13) questionnaire and his synthesis of them (see Hiärne, 1702, p. 99), Deluc's statements are very surprising and were clearly motivated by his desire to uphold his own geological theory, conceived to support the mythology of the *Bible*, that no sea-level change had happened since the present continental surfaces had formed (see also his *An Elementary Treatise on Geology*, in which he repeatedly states that "Since the birth of our continents, the level of the sea has not changed, either absolutely, or in relation to the land," e.g., Deluc, 1809a, p. 347; in the French original, Deluc, 1809b, p. 363, a statement flatly contradicted by all observation statements known at the time: see Şengör, 2003)[83].

Another issue Deluc disagreed with Playfair (and thus Hutton) was the problem of erratic blocks, "the blocks of granite, and other kinds of stone, which are found disseminated over the surface of our continents." (p. 260). Playfair, following Hutton, thought that they were probably carried to where they are by ice (see Davies, 1968), but Deluc cannot accept that (p. 270–271), because heaps of rock are seen away from the Alpine glaciers that resemble those at the snouts of glaciers. Instead of making the obvious inference made by Hutton and Playfair that such heaps of rock probably indicate the earlier extent of the glaciers, Deluc thinks them as remnants of the great catastrophe that was caused by the subsidence of the continents that now underlie our oceans. To support this interpretation, he even presents a rather labored theory of soil formation ascribing different origins to the soils on the Po Plain and to those in the Alpine slopes to its north, calling the humus-rich agricultural soils *terreau* (a soil preferred for gardens and pots) and the Alpine soils that also contain humus just ordinary soil (p. 329–330).

This is not the place to analyze Deluc's work on geology[84]. I cited it at length, because of his detailed criticism of Hutton's views in the first decade of the nineteenth century and to emphasize the nature of that criticism. Deluc's book, written to discredit Hutton's ideas, was indeed useful to show what those ideas were and how seriously they were taken then, through the numerous quotations from Playfair's *Illustrations* and other contemporary literature; but Deluc's rebuttal carried no weight whatever in the eyes of the geologists at the time. Contrary to what Rudwick (2005, 2011) repeatedly asserts, this was not because Deluc was a religious man (so were many of his opponents), but because he adjusted his "observations" to results favoring the *Genesis* myth, thus straying away from scientific objectivity[85]. He openly said, discussing Kirwan's ideas in an appendix to his book here under discussion, "our object in geology is the same—that of pointing out the errors of the systems which are in opposition to the Book of *Genesis*, the only authentic source of the knowledge possessed by men respecting their own origin, and that of the universe" (p. 367; italics Deluc's). As Gillispie (1951, p. 49) so aptly pointed out "Deluc expressed an early and a limited attempt to come to terms with nature, although on his

Figure 13. Urban Hiärne. Credit: Portrait of U. Hjaerne. Credit: Wellcome Collection. Attribution 4.0 International (CC BY 4.0).

own conditions." Rudwick (2005, p. 496) himself pointed out that Brongniart could not take Deluc's objections to his and Cuvier's interpretation of the Paris Basin stratigraphy seriously, because—again according to Rudwick—Deluc was unwilling to accept anything outside his own interpretations (isn't that what Schmieder, 1807, p. VIII, also says?!). No doubt this was because he was fearful for Christianity, not for geological truth. Even the Catholics, who otherwise welcomed Deluc's defense of the *Bible*, thought that Deluc's geology was "subjective," because of his Protestant commitment (Hübner, 2010, p. 167). But even the Christian theologians of the late eighteenth century, following in the footsteps of Thomas Hobbes (1588–1679), Benedict Spinoza (1632–1677) and Richard Simon (1638–1712) and even some Medieval and Renaissance predecessors (see, for example, the papers in Niewöhner and Pluta, 1999), had given up the idea that Moses was the actual inspired author of the *Pentateuch* (Reventlow, 2001; Becker, 2015, p. 43). It had by then become clear that the *Old Testament* was a hodgepodge of texts that originated at widely different times and places[86] and contained double transmission of certain texts, clear contradictions within the same text (especially relevant for Deluc would have been the contradictions in the Flood myth in *Genesis*, 6–9) and different designations used for God, such as Elohim (although grammatically plural, used as singular![87]) and Yahweh[88], in different books (see Becker, 2015, p. 43; see also Reventlow, 2001, p. 214). These theological developments alone would have been sufficient to pull the rug from under Deluc, but he was obstinately oblivious to all such difficulties for his views.

Scientifically, already in the eighteenth century, geologists Horace-Bénédict de Saussure (1740–1799) and Dieudonné Sylvain Guy Tancrède de Gratet de Dolomieu (1750–1801) and the physician and zoologist Johann Friedrich Blumenbach (1752–1840) could not agree with Deluc, simply because they thought Deluc's system clashed with observations (Hübner, 2010, p. 205). Already the young Alexander von Humboldt (1790, p. 121–122) complained about the fact that in the Unkel area along the Rhine River between Koblenz and Bonn, Deluc saw what he wanted to see in the basalts there, not what was really there—so, even Deluc's observations were suspect. Neither Louis Albert Necker (1786–1861), nor the great Alpine geologist and one of the early pioneers of experimental geology, Jean Alphonse Favre (1815–1890), both Deluc's compatriots, mentioned him in their geology lectures as one of the founding figures of geology, or even among the early traveling geologists, despite protestations from Deluc's nephew, who unjustifiably ascribed these omissions to jealousy (Hübner, 2010, p. 28)!

At the end of the first decade of the nineteenth century we have thus an entire book devoted to Hutton's geology with a view to combatting it. It underlined the importance of Hutton's interpretations, but its author Deluc's explicit and furious alarm that they may seriously harm Christianity and his efforts to bend geological observations to support the *Genesis* myth, i.e., an unscientific assertion, largely destroyed his credibility. After the seventeenth century, as Lord Russell nicely put it, "Any one might still believe that the heavens exist to declare the glory of God, but no one could let this belief intervene in an astronomical calculation. The world might have a purpose, but purposes could no longer enter into scientific explanations" (Russell, 1946, p. 559). Deluc had simply violated these simple rules, whereas most of his Christian critics did not. Élie de Beaumont, for example, who did not take Deluc's theory of mountain-building by subsidence seriously, still talked about the Biblical Deluge even in the middle of the nineteenth century, but—although himself an ardent Catholic—not as God's punishment, but as a natural phenomenon the record of which was found in ancient documents, including the Jewish *Bible*. Thus, the assessment that Charles Coulston Gillispie gives of Deluc's geological work, following the authoritative assessment by one of the greatest geologists of the nineteenth century, Karl Alfred von Zittel (1899, p. 109), that "it is not necessary to consider any more closely the views of this assiduous but fleeting observer and prolific author of phantastic writings; Deluc's publications have been deservedly forgotten and his strong attacks on Hutton and Playfair became pointless," was a much more apt evaluation than Rudwick's own apologetics in the name of the *philosophe Chrétien* Deluc (e.g., Rudwick, 2005, 2011)[89]. Oldroyd (2009), in his otherwise fine paper, wrote that he did not think that Gillispie's evaluation of Deluc as a geologist was impartial and dispassionate, because of Gillispie's metaphysical position as a "freethinker." However, he provides no detail as to what aspect of Deluc's geology he thinks Gillispie misrepresented. It is a pity that Oldroyd died before he could tell us what he thought of a liberal Protestant theologian's son von Zittel's statement, which was harsher than Gillispie's.

Geologist and agriculturalist, Sir George Steuart Mackenzie, 7th Baronet, (1780–1848; Fig. 14) traveled to Iceland in the company of Sir Henry Holland, Bart. (1788–1873), and Mr. (later Dr.; the "father of nephrology" after whom the Bright's disease was named) Richard Bright (1789–1858) in 1810. Sir George described his journey, with the help of his friends[90], in his *Travels in the Island of Iceland during the Summer of the Year MDCCCX,* that had two editions: 1811 and 1812. Dean (1992, p. 55–56) mentions the first edition and summarizes Sir George's view of Hutton's theory, although it is unclear to me what he meant by "Thus, in defending Hutton, Mackenzie jettisoned a great part of his theory" (Dean, 1992, p. 56). Let me here quote, from the second edition of the same book, which hardly had changed from the first (except for new additions to the map of newer details of the great Laki eruption of 1783), because only six months had elapsed between the publication dates of the two editions. The purpose of the trip was mainly geological:

> The importance of the study of Mineralogy has of late years acquired, and the intimate connection which it is now acknowledged to hold, with all legitimate views of geological science, rendered the examination of Iceland particularly desirable. (Mackenzie, 1812, p. viii)

Figure 14. Sir George Steuart Mackenzie, 7th Baronet. Portrait by Sir Henry Raeburn. Creative Commons Attribution 4.0 International License.

... mineralogical research was the principal object of the voyage.... (p. x)

In Chapter IX, devoted to mineralogy, Sir George points out that the novelty of many of the geological objects they encountered made description very difficult because of the poverty of the existing scientific nomenclature. He complains, in addition, that the existing nomenclature is too theory-laden:

We cannot help, however, expressing a wish, that the language of mineralogy were divested of theory. In so far as it relates to his own system, the language of Werner is admirably contrived; but few, it is believed, will acknowledge the right of the proposer of a theory to alter the whole language of science, expressly for the purpose of adapting it to that theory. Neither the disciples of Dr. Hutton, nor Werner, have any title to assume such authority; and there ought to be a language, independent of all theory, which might be used in all discussions, as being universally understood. Such a language does not at present exist.

...The language of Werner having excited a good deal of attention, is perhaps better understood than any other; and though we by no means approve of it, some of its terms shall be used. Were a descriptive language employed, and mineralogists left to guess at what was meant, our labours would infallibly prove fruitless; and therefore it is considered better, in the present instance, to make use of a faulty language, which can be explained, than one in any respect ambiguous. When any of Werner's expressions which may involve theory are used, they must be understood in no other than a descriptive sense. (p. 353–354)

Having thus taken care of the terminology he wishes to use, Sir George goes on to tell his readers what he intends to describe:

In the present state of geology, nothing can be of greater importance than to ascertain with accuracy what are the results of heat acting on bodies under strong compression; since by means of that knowledge, we are enabled to compare the ordinary deductions from Dr. Hutton's principles, with the phenomena of nature, and bring to the test of actual observation the merits of a system which promises fair to put us in possession of a most simple and beautiful view of the mineral kingdom.

Dr. Hutton has clearly pointed out the reason why calcareous spar may occur in the productions of subterraneous heat, while it could not exist in those of open fire, by ascribing the presence of carbonic acid to the effects of powerful compression. Since Dr. Hutton's death, Sir James Hall has confirmed, by actual experiment, the truth of these theoretical deductions. The rocks of Etna, to which Dolomieu ascribed a submarine origin, present a result in the works of nature by which these principles are brought to a test. But that philosopher, having no such ideas in contemplation, could derive no such advantage from the facts before him. (p. 357)

Let me interrupt Sir George's text here to observe what he had just written: he points out that Dolomieu, not having at his disposal Hutton's theory, could not derive the same satisfactory results from his observations as Sir George and his friends could from theirs. Dolomieu was thus considered disadvantaged in not having Hutton's theory in his intellectual apparatus. This is the only example I know in the entire Huttonian literature that explicitly spells out the usefulness of Hutton's theory in guiding observations *and* interpretations in reaching results deemed best in accord with Nature. On his p. 373, in footnote † Sir George presents a wonderful defense of theoretical science[91]. But now let us continue following his narrative:

His natural sagacity, however, induced him, from other circumstances, to conclude that the rocks of Etna, which contained calcareous spar and zeolite, had actually been in a state of fusion, and that their vesicles had been filled by submersion in the sea. Doing every justice to so admirable an observer, we may be allowed, availing ourselves of the elucidation of the subject by Dr. Hutton, to derive still greater light from the phenomena which present themselves in Iceland; ... (p. 357–358)

Sir George goes on to emphasize the identity of basalt with lava seen in Iceland. He then compares the Wernerian with the Huttonian interpretation of volcanoes:

The chemical analyses, by the much lamented Dr. Kennedy [Robert Kennedy, 1774?–1803], have proved their perfect similarity in composition; and there now remains no doubt that the materials of which both consist, are exactly the same. But the supporters of the two great theories which divide geologists differ materially in their accounts of the production of lava. The followers of Werner maintain that lava is melted greenstone; and this supposition is not grounded on the result of the chemical analysis of the two substances, but is a necessary acknowledgement arising out of the Wernerian theory of volcanoes, which assumes the heat to be occasioned by the burning of beds of coal. The theory of Hutton is not so limited; for whatever variety there may be in lavas, the matter composing them is supposed to have been produced as well from materials in the bowels of the earth, independent of any rock formation, as from the destruction of rocks; to which last circumstance the Wernerians confine themselves. They have given no satisfactory account of the mode in which beds of coal may be set on fire, or how the combustion is to be kept up, when a sufficient mass of fuel is provided. The presence of water they very properly consider indispensable for producing an eruption; but a variety of other causes are required to combine and act in regular succession. The Huttonians are not under the necessity of going in search of any accidental causes. They infer the existence of an internal source of heat; but how, or where the heat is produced, or maintained, is not of any importance to the fundamental principles of their system. Philosophers may speculate respecting the existence of a central source of heat, the casual effects of electricity, or the inflammation of the metals of the earths and alkalis; but though it was absolutely impossible to prove that any of these causes was the true one, it is surely as reasonable to infer the existence of internal heat from the phenomena of volcanoes, as to believe in the Wernerian rising and falling of the waters of the globe, without the evidence of any analogous fact whatever. Much ingenuity has been exercised in combating and defending what is really not necessary for the support of the Huttonian theory; and it is by no means fair to interweave detached speculations on the possibility or probability of the existence of a central source of heat, with a theory which has the widely extended phenomena of volcanoes to refer to, in proof of the existence of subterranean fire. The point, whether lava is, or is not, melted greenstone, is really not worth disputing. (p. 359–360)

In the rest of this chapter, Sir George describes various examples of volcanic manifestations, recognizes the tectonic nature of the Þingvallavatn and surrounding area and gives us an eye-witness account, most likely of the Allmannagjá structure (Sir George identifies it, without naming it, simply as the one trough they investigated of the two they saw), ascribing its origin to the subsidence of its floor. The whole is presented within the framework of the Huttonian theory with occasional criticism of the interpretations of the opposing camp:

Nothing seems to bear more strongly against the Neptunian system than the alternation of what, in its language, are called mechanical and chemical deposits. It is nothing uncommon, in this country [i.e., Scotland], to see highly crystallised greenstone resting on sandstone; and, in Iceland, we see it placed above tuffa, which has been formed by mechanical exertions, far more violent than those required to form sandstone. To suppose (which is necessary in such cases) that, when the waters of the globe rose at one time, they were fit only to act mechanically; while afterwards, they were in every respect adapted to act the part of a chemical solvent; appears to be a violent trespass on the steady, simple laws of nature. (p. 383)

No wonder, Sir George's unpublished "Huttonian theater piece," *Helga, or the Rival Minstrels, a Tragedy Founded on an Icelandic Saga, or History* was greeted by considerable clamor during the last two acts, as the Wernerians, who apparently filled the theater, found their contents exceptionable (see Wawn, 2000, p. 53; also see below).

Dean mentions on his p. 161 and 225 Captain Basil Hall's (1788–1844; second son of Sir James Hall) observations in South Africa on granitic intrusions as drawn up by John Playfair in the seventh volume of the *Transactions of the Royal Society of Edinburgh*. In that short paper there is no mention of Hutton explicitly, but Playfair wrote that Hall's observations "to be an *instantia crucis*, with respect to the two theories concerning the formation of rocks" (Hall and Playfair, 1815, p. 278). Dean unfortunately does not cite the surgeon and naturalist Clarke Abel's (1780–1826; Fig. 15) book relating the observations he made during Lord Amherst's (William Pitt Amherst, First Earl Amherst, Earl of Arrakan, 1773–1857) Embassy to the Court of Pekin in 1816 and 1817. There, Abel describes basaltic dykes cutting a granite basement from a "small island separated from Hong-kong [sic] by a channel not more than a hundred yards wide" (Abel, 1818, p. 61)[92]. Regrettably he was able to spend only two hours on the island. By contrast, the Embassy spent from 27 May to 11 June at the Cape Colony and there Abel was able to make extensive geological observations parallel with those earlier ones by Hall described in Hall and Playfair (1815), to which he devoted an entire chapter in his book (Abel, 1818, p. 285–312).

Abel's geological account of his South African observations is so long and detailed that it is impossible to quote it here in full. Suffice it to say that he walked, presumably from Simon's Town north to the broad valley between the Table Mountain and the Lion's Head south of Cape Town. The area is underlain by parts of a Pan-African orogenic belt segment consisting of the Malmsbury Supergroup including mainly clastic sedimentary rocks metamorphosed to greenschist facies. The deposition of the clastic sediments had begun 575 million years ago and they were deformed and intruded by the Cape Granite Suite some 540 million years ago, during the final assembly of Gondwana-Land.

What attracted Abel's attention, and aroused his theoretical interest, was the intrusive relationships he observed:

Like the granite forming the base of the Lion's Head, they are frequently traversed by large veins of quartz. Beyond these, and near the margin of the sea, I observed immense ridges of rock apparently of different colours, mingled together.

Figure 15. Clarke Abel. Lithograph by M. Gauci after P.W. Wilkin. Credit: Wellcome Collection. Attribution 4.0 International (CC BY 4.0).

I hastened to visit them, in the hope of meeting with an example of those *instantiæ crucis* on which different geological theories are supposed to turn. (Abel, 1818, p. 288)

And what were these "different geological theories"? Abel explicitly spelled them out on p. 305 as being those of Werner and Hutton. He thought that unconformable sedimentary contacts were compatible with Werner's, while intrusive contacts spoke for Hutton's theory. He wrote:

Another consequence of the fact that I have cited appears to be, that the mountains at the Cape of Good Hope exhibit phenomena illustrative of certain positions both of the Huttonian and Wernerian theories, but only to be entirely explained by the agency of both the elements on which the respective systems are founded. (Abel, 1818, p. 305–306).

What, according to Abel, was compatible with Werner's theory, namely the sedimentary contact on metamorphic rocks, was also compatible with Hutton's theory, but intrusive igneous dykes were not compatible with Werner's, although he did not belabor that point in the passage quoted.

A remarkable detail in Abel's narrative is that he uses Playfair's usage (in Hall and Playfair, 1815) of Bacon's term *instantia crucis* and is keen to test the two competing theories by finding such crucial instances. I must here quote his very careful observations and clear reasoning at length, speaking decisively in favor of Hutton's interpretation of field relationships of igneous rocks:

The theory of the igneous origin of granite adopted by Captain Hall and Mr. Playfair to explain the phenomena visible in Table Mountain, very happily meets the facts, and must appear, I apprehend, to one standing on the spot where they occur, as incontestable.[93] They are indeed of that nature which strikes the conviction at the first glance. Carry to the spot one who never heard of geological theories, and ask him what he infers from the appearances before him, and he will exclaim "The white rock has broken the black in pieces." Whatever may be the true explanation of these appearances, this I apprehend must be the first impression that affects the mind of any person visiting Table Mountain. A more deliberate investigation, by a more cautious observer, would also, I think, lead to similar conclusions. If he were one desirous to explain all geological phenomena by the agency of water, and be therefore disposed to consider the mixture of schistus and granite as resulting from contemporaneous formation, he would hesitate over the imbedded masses of the former uncontaminated by their matrix, and would be still more perplexed by the fact that they can be separated without mixture. He would in vain seek in them those instances of the wedging of one rock into the other, which has been supposed confirmative of this opinion.[94] In looking at the principal line of junction between the granite and the schistus, he would see no gradual gradation of one into the other, but a distinct, though interrupted line of separation. This interruption he would find to be occasioned by veins of granite passing directly from the principal mass of granite into the schistus; in other words, the subjacent rock shooting into the superincumbent one[95]. Below this line he would indeed find an intimate mixture of the two rocks, occasioning a compound of a lighter color than the schistus; such an appearance as might in itself be explained on the supposition of coeval formation. But he would meet a fact exceedingly unfavorable to this conclusion on a closer view; he would see here, as in the granite, two pieces of schistus of the natural dark color imbedded in the mixed mass, and if he subjected them to the most accurate measurement, he would find their edge and angles corresponding. If he should leave the Table Mountain and visit Green Point, where the mixture of the two rocks is more intimate, he would still have to contend with facts scarcely explicable on the Neptunian theory. The appearance most likely to arrest his attention, would be the intimate intermingling of the two rocks in the large field, a portion of which is represented in page 289, and which he would probably consider, on a first view, not only explicable on the theory of aqueous formation, but as favorable to it. Here he could see that intermingling, and wedging, and gradation of one rock into the other, which his opinions require. But this conclusion would perhaps give way to a wider view of the phenomena around him. The leading feature stamped on all the facts at Green Point is exceeding commotion at the period of the mixture of the two formations. To conceive that they were deposited from a fluid in a state of rest, seems to me impossible for any one crediting the evidence of sense. Although it might perhaps be said, that their intimate mixture was the consequence of the agitation of the fluid while they were crystallizing. But supposing a Neptunian to have formed this inference, he must, I apprehend, yield it to one of the conditions of his own theory, and one of the laws of crystallization[96]. Bodies of a perfect crystalline structure can only be formed as a chemical deposit from a fluid in a state of rest. Rocks of an earthy fracture are formed from a fluid more or less agitated. What then are the characters of the rocks at the point of junction? Certainly not those which result from a deposit more or less mechanical. Their fracture is highly crystalline, and more so in proportion to the vicinity of the granite.

On the other hand, one who adopts the theory of the igneous origin of granite would find no difficulty in explaining all the phenomena which present themselves; he would consider them, indeed, as a beautiful illustration and a powerful confirmation of his doctrine. These instances, so puzzling to a Neptunist, of the detached fragments of the rock which overlies, and its penetration by veins from below, and the crystalline fracture of the two rocks when compounded, would appear to him necessary to the verification of his opinion. The granite in fusion bursting through a superincumbent rock should split it into an infinite variety of fissures, and fill them, like melted metal poured into a mould, and should dislodge and insulate fragments, is an inference too obvious to be much dwelt upon in this place. (Abel, 1818, p. 298–301)

I have not seen anywhere else, so early, such a clearly described and detailed comparison between the Wernerian and Huttonian theories concerning a critical aspect of them on the basis of personal observations. The text that comes closest is that by the French traveler Louis Simond, quoted below while reviewing the French reactions to Hutton's theory.

Only a year later, in 1819, the 36th volume of the Welsh non-conformist priest Abraham Rees' (1743–1825; Fig. 16) monumental *The Cyclopædia; or, Universal Dictionary of Arts, Sciences, and Literature*—much maligned at the time because of its alleged irreligiousity—was published (volumes 1–39 were published as a set in the same year, but one must be careful in ascribing any priorities on the basis of that set, because parts and fascicules had been published earlier; volumes 40–45 were published later), in which, under the heading "System of Geology," Hutton's theory is given an exhaustive review in four-and-a-half pages of small print. It is surprising that Dean (1992) did not consider it. The author of the article "System of Geology," seems very well-informed about geology in general and concludes by writing:

> Whatever changes the Huttonian system must undergo to adapt it to a more advanced state of the science, or whatever may be its ultimate fate, it cannot be denied that it is distinguished by characters of grandeur and simplicity. When we consider how little was generally known of geology at the time this system was published; how extensively it applies to the various phenomena which subsequent observations have discovered; what a probable, at least plausible, solution it offers to many difficulties in the science; and what confirmation it has received from chemical experiments; we must regard it as one of the happiest efforts of speculative philosophy. (Rees, 1819, unpaginated)

Under the same heading of "System of Geology," the author also reviewed Werner's and Deluc's work. His account of Werner's work ends by listing the objections to it that could not be overcome. He wrote "The objections made to many parts of the system of Werner, as originally announced by his followers, it would be difficult, if not impossible to remove;…" (Rees, 1819, unpaginated). Rees then says that the system may be improved to make it more conformable to modern geology and ends up by acknowledging that Werner's work contributed substantially to the advancement of geology, especially mineralogy. He thought Deluc's work was better than Werner's because he at least thought of a means of disposing the waters of the Deluge, but, the author hastens to add, Deluc's solution is made impossible by the density of the earth.

Another publication that Dean does not mention and that contains summaries of Hutton's and Werner's theories before the publication of Lyell's *Principles* in 1830 is the famous book entitled *A Treatise on a Section of the Strata from Newcastle-upon-Tyne to the Mountain of Cross Fell in Cumberland* by the mining agent Westgarth Forster (1772–1835). Forster's book was first published in 1809 under the title *A Treatise on a Section of the Strata, commencing near Newcastle upon Tyne, and concluding on the West Side of the Mountain of Cross-Fell. With remarks on Mineral Veins in General, and Engraved Figures of Some of the different Species of those Productions to which are added Tables of the Strata in Yorkshire and Derbyshire the whole intended to Amuse the Mineralogist and to Assist the Miner in his Professional Researches*; a second edition appeared in 1821 which carries the title first cited above and which is the best-known version of the book. A posthumous edition was published by the Reverend W. Nall in 1883, which was however, heavily edited. The editor, the Reverend W. Nall, deleted the comparative accounts of the theories of Werner and Hutton with the justification that

> The controversy between the Huttonians and the Wernerians respecting the origin of the rocks composing the crust of the earth, which raged during the closing years of the last and the opening years of the present century, is now as dead as the controversies of the ancients concerning the shape of the earth. The statement of the two theories which Forster inserted in the second edition of the "Strata," has, therefore, been left out of the present edition. (Nall in Forster, 1883, p. v–vi)

Forster reviews the two theories, because (what follows is verbatim after Millar, in Williams, 1810, v. II, p. 560):

Figure 16. Abraham Rees. Credit: Stipple engraving by J. Thomson, 1820, after J. Opie. Wellcome Collection. Attribution 4.0 International (CC BY 4.0).

It will not be improper to insert here, an account of the two late Theories, which have divided the opinion of some of the Geologists of this country, and indeed have excited a good deal of the warmth, and some degree of asperity, of controversial discussion. The theories alluded to, are those of HUTTON and WERNER, as stated in Dr. Miller's [sic] edition of Williams's Mineral Kingdom, Vol. II. [97](Forster, 1821, p. 143)

Forster repeats, again verbatim, what Millar wrote in Williams' second edition (1810, v. II, p. 560–569; see Forster's p. 144–153), but then adds the following which hardly follows from what Millar wrote (but is in keeping with the spirit of Williams' first edition):

The theory of Hutton has gradually sunk into disrepute in proportion as geological facts and observations have been multiplied and extensive; and it is not improbable but even the beautiful theory of Werner may share a similar fate; as some parts of it have met with considerable and powerful opposition (Forster, 1821, p. 153.)

Forster's statements are prescient to the extent that Werner's theory did indeed fall into oblivion, but what he says of Hutton's theory was certainly not true as the later developments in the British Empire (and in the rest of the civilized world) showed. The verbatim quotations from the Millar edition of Williams's book and the incongruous comments on Hutton and Werner based on Williams' first edition can now be understood in the light of Forbes' (2015) remarkable documentation that Forster's book was cobbled together, as Forbes picturesquely expresses it, almost entirely from plagiarized sections from others' books, principally from those by John Whitehurst (1713–1788) and John Williams (but see a contrary view in http://westgarthforster.blogspot.com/2015/06/where-was-westgarth-forsters-section-of.html). As I pointed out in Şengör (1991), one can discern two traditions in geology well into the last quarter of the nineteenth century: a physiographers' and a miners' tradition. The first tradition embraced Hutton, while the second Werner, until the Wernerian was largely dragged away from it by the advances in geology in the second quarter of the nineteenth century. Nall's justification of the elimination of Forster's account of two theories implies, by the allusion he makes to controversies on the shape of the earth, the triumph of one of them—and we know today which one the triumphant theory was.

Another publication that mentions both Hutton and Werner is the theological tract by Granville Penn (1761–1844), the grandson of William Penn (1644–1718), the first Quaker theologian and founder of the colony of the Province of Pennsylvania. The present state of Pennsylvania takes its name from him and thus from the name of Granville Penn's family, although the founder died in England, penniless. The only importance of Penn's book, from the viewpoint of the present book, is that he regarded both Hutton's and Werner's views as regrettable novelties that had risen against the authority of the *Bible*. His, the first edition of which came out in 1822, is really an all-out attack on geology for disagreeing with the Biblical narrative. Just like Andrew Ure later, Penn ridicules the notion that the history of the earth can be known by geological investigations and repeatedly asserts the veracity of the Biblical account of the history of the earth. Here is a sample of his reasoning:

The Neptunian has established the fact against Hutton, that secondary formations are of aqueous production; by showing, that the perfect preservation of sea-shells in inland soils could not have taken place, if the revolution which transported them thither had been effected by fire; for the shells, being calcareous, must have been dissolved, and mingled with the general mass. But he would infer from thence, that primary formations must likewise have been of aqueous production; which is more than his premises can yield. He has refuted Hutton, indeed, in the one argument, but he has left him as strong as ever in the other: and yet not a whit stronger than himself; for the force of their arguments is so nearly poised and balanced, that they neutralize each other. The result is, that that there will remain for ever a ground for hypothetical contest between the two; and, therefore, as there exists no accessory weight of *truth* to determine the scale *definitively* on either side, the just conclusion is, that *both* are equally erroneous with respect to *fact*; consequently, that "the crystalline character stamped upon the primitive mineral masses," was not stamped by either of the secondary causes assigned; but that it was impressed by the *first, Creating Cause*, who anticipated the effects of each, in giving to those masses *"the properties which most conduced to the end for which He formed them."*

A principal and obvious "end" of those "properties," as we have seen, was the solidity and durability resulting from the grain and texture of their composition; so that the granite summits, traversed by Hannibal 2000 years ago, are identically the same which we now witness; and we are sure that they stood identically the same, twice 2000 years before him. Whatever may be the destructive and wasting power of the atmospheric agents upon some bodies, it is null with respect to these, and therefore idle to take account of it in geology; and it is only resorted to, to aid a limping system. We discern a manifest *"end,"* likewise, in their *"sizes"* and their *"figures;"* for, the altitude of the *former*, is owing the accumulation of supplies for the rivers which are to irrigate the globe; and, to prolongations and inclinations of the *latter*, are owing the conduct and direction of the rivers which actually irrigate it. And how is it possible to contemplate the unchangeable arrangement, by which all these perfect means conduce to their several perfect ends, without "rendering immediately to GOD, the things which are GOD's! (Penn, 1822, p. 96–98, italics Penn's)

This is repeated verbatim in the second edition (Penn, 1825, p. 115).

Penn is, however, ready to resort to geological observations, to Deluc's natural chronometers, "which the earth exhibits; and which remain perpetual vouchers for the veracity of the Mosaical *chronology*…" (p. 246, italics Penn's), when they suit him,

although he severely criticizes Deluc for having sought a compromise between geology and the Bible! Penn's excellent knowledge of the *Bible* is remarkable, but his profound ignorance of the geology of his own times is no less remarkable, as the quotation above illustrates. Hutton is mentioned only once in Penn's book (see the quotation above) and Werner four times (p. 125 and 298–299), both only to be attacked for not following the *Bible* to the letter. The second edition of Penn's book (Penn, 1825), in two volumes, takes cognizance of the appearance of new and important studies on geology. In his "Introduction to the present edition" he lists the following books he says he took into consideration:
1. The first volume (actually all published) of the *Geology of England and Wales* by Conybeare and Phillips.
2. *Reliquiæ Diluvianæ* of William Buckland.
3. *Ossemens fossils* by Cuvier, the new edition.
4. *Geognostical Essay on the Superposition of Rocks in both Hemispheres* by Alexander von Humboldt.
5. The Edinburgh and Quarterly reviews of *Reliquiæ Diluvianæ*.
6. *Geological Disquisition* introduced by the Reverend G.S. Faber into his Treatise on *Three Dispensations*.
7. Reviews of the first edition in the *Eclectic Review, Journal of Science* and *British Critic*.

However, despite his new reading, Penn found nothing to change in his original narrative and arguments presented in the first edition. He staunchly holds on to the interpretation of the Hebrew word יום (*yom* = day) as standing for the time interval between sunup to sundown (thus repudiating the attempts to interpret it allegorically to represent epochs of indeterminate duration) and finds von Humboldt's account of global stratigraphy too theoretical and grounded in what he calls "mineral geology," which he finds inadmissible: "M. Humboldt's 'Geognostical Essay' has interwoven itself a theory as fanciful, as arbitrary, and as injurious to the attainment of fundamental Truth, as any of those which have drawn down upon them this condemnation: yet, it is the most recent and avowed effort of the Mineral Geology to supersede the authority of the Mosaical, which, however, by contrary and unexpected consequence, it pointedly confirms it every stage." (Penn, 1825, v.1, p. 349). Penn was either unaware, or simply incredulous, of the geological knowledge of his time and thought that whatever geology finds out had to be judged against the narrative of the *Bible*, taken as the ultimate truth.

Penn's last salvo against what he called "mineral geologists" was his little book, cast in the form of a dialogue between a certain Mrs R. (really representing Penn's voice), a woman knowledgeable about the geological literature, but clearly a supporter of Penn's own views, and two children, Edward and Christina, entitled *Conversations on Geology; Comprising a Familiar Explanation of the Huttonian and Wernerian Systems; the Mosaic Geology as Explained by Mr. Granville Penn; and the Late Discoveries of Professor Buckland, Humboldt, Dr. Macculloch and others*, published in 1828. This book is much more scientific in nature in that it truthfully describes not only the opinions of Hutton and Werner (actually more of Hutton than Werner), but rejects both because, he says, both try to deduce the origin of the earth and its inhabitants from what we can observe today. He thinks this is inadmissible, because it negates the Biblical narrative that these entities were created by divine fiat at once! However, in this book, the reader obtains the impression that Hutton's views were more generally accepted by geologists than anybody else's and Penn combats it simply because they are incompatible with the mythology of the *Bible*.

Dean (1992, p. 190) refers to George Julius Poulett Scrope's (1797–1876) approval of Sir James Hall's experiments (but only one of his three page references is correct, namely p. 111 in Scrope, 1825) and quotes the *Monthly Review* saying that his concepts were "borrowed from Hutton" and that Scrope is "an advocate of that [i.e., Hutton's] theory"[98] although Scrope rejected the hypothesis of a central fire. An unexpected omission in Dean's book is a reference to the magnificent first edition of Scrope's sumptuous *Memoir on the Geology of Central France; including the Volcanic Formations of Auvergne, the Velay and the Vivarais* published in 1827 (but had been handed to the publisher already in 1822). Although Hutton is not mentioned in it by name, the entire book (Scrope called it "an appendix, or *pièce justificative*" for his 1825 book: Scrope, 1827, p. x) is a rebuttal of the Wernerian position and a demonstration of Hutton's views.

The penultimate publication before the appearance of Lyell's *Principles* that I would like to mention here is Andrew Ure's *A New System of Geology in Which the Great Revolutions of the Earth and Animated Nature are Reconciled at once to Modern Science and Sacred History* (Ure, 1829). Andrew Ure (1778–1857; Fig. 17) was an extraordinary person: he was an accomplished chemist, a competent anatomist, a knowledgeable man in a number of technical subjects and a fine lecturer. His books on chemistry enjoyed an international success. He became a member of the Royal Society of London in 1822. His teaching in chemistry and mechanics in the Andersonian Institution (which became the University of Strathclyde in 1964) inspired the foundation of a number of mechanical institutions in the UK and the École des Arts et Métiers in Paris. His geology book betrays an immense background also in global geology. Yet, Ure was a biblical literalist! In the introduction of his book, he mentioned "our two famous geologists Werner and Hutton" (p. xviii), "the worshippers of Water and Fire" (p. xxi), and discussed and dismissed their theories. His criticism of both ultimately came down to their ignoring the Mosaic account of Genesis, which Ure thought had to be adhered to rigidly. Interestingly, he also spurned the geological writings of Kirwan and Deluc (p. xv), because he thought they did violence to both religion and science. His only substantial criticism of Hutton was that the crystalline rocks such as gneisses and micaschists contained no fossils of former organisms. Werner, on the other hand, Ure thought proposed an entirely impossible system that ignored both physics and chemistry. The only thing on which he agreed with Deluc was the importance of Bacon's inductive methodology.

Ure wrote that even a discussion of Hutton's theory would cause an unnecessary fatigue (p. xxviii):

Figure 17. Andrew Ure, F.R.S. Engraving, original size 15.9×11.3 cm. Public domain (PD-US-expired).

I have no desire to fatigue my readers with a detailed examination of the theory propounded by Hutton, and embellished by Playfair. Its defects and inconsistencies, and indeed its whole hypothetical tenor are now so notorious, that no practical naturalist of eminence will venture to adopt its conclusions. My sole object here, was to unveil its vain spirit of theoretic cosmogony; to exhibit its efforts to build a Babel-tower that should make a name, and enfranchise it from the control of a creating and directing Providence. The world, according to Hutton, shows no trace of a beginning, or of an end; but has been the theatre of an indefinite series of transformations in time past, and will continue to be so in time to come. The mountains of a former earth were worn down and diffused over the bottom of a former ocean. There they were exposed to the agglutinating power of subjacent internal fires; and after due induration, were heaved up by the explosive violence of the same force, into the inclined or nearly vertical position, in which the great mountain strata now stand.

The man who wrote this scathing evaluation of Hutton's interpretations then offered the results of his own studies in the following itemized fashion:

1. That a great portion of the present dry lands, more particularly the secondary strata which are replete with sea shells of the most delicate texture, distributed entire in regular beds, have lain for a long period at the bottom of the primeval ocean.

2. That within the schistose crust of the globe, explosive materials exist, which have given evidence of their convulsive and disruptive powers in all its terraqueous regions in every age of the world from the protrusion of the primordial dry land, till the present day.

3. That the ocean at whose bottom many of our present earthly strata were deposited, has not been lessened by dissipation of its waters into celestial space or by their absorption into the bowels of the earth.

4. That, therefore, its channel must have been changed by transference of, a great portion at least, of its waters from their ancient to their present basin; an effect referrible to volcanic agency, which has operated by sinking old lands, and upheaving the new. (p. 435–436)

Ure was also of the opinion that both granite and trap (i.e., basalt) were igneous rocks, respectively intrusive and extrusive, and that granite was responsible for upheaving mountains. He wrote:

The primitive envelope of the globe, seems to have originally consisted of concentric strata of gneiss, mica-slate, with partial layers of semi-crystalline limestone; for such, with a few inconsiderable exceptions, constitute its rocky crust, and are spread over all its regions. These coats, however, no longer lie in layers concentric with the spheroid, but are thrown up into nearly vertical planes, and transpierced in many points by towering masses of granite and porphyry.

On the primordial spheroid covered with its illimitable ocean, these stratiform coats lay in horizontal planes; but with the gathering together of the waters, on the emergence of the land, they were heaved up abruptly into near vertical tables, in which they now universally stand. This remarkable position corresponds to the eruptive violence that caused it. From the shoulders and flanks of the stupendous granite peaks, mantles of gneiss and mica-slate depend in magnificent drapery. These schistose coverings are arranged near the summit in folds almost upright, which lower down, become sloped off with clay-slate and limestone, into a gentle declivity. The coats of gneiss are often contorted into the most singular flexures of rock scenery, demonstrating a certain pliancy of texture at the instant of erection; resulting either from the moisture out of which it rose, or the softening influence of subterranean fire.

The wood-engraving at the bottom of this page [Fig. 18 herein], taken from D'Aubuisson, represents the usual structure of the primitive mountains. It is impossible for an unbiased mind to contemplate this sketch, given by a Wernerian geologist, without seeing its entire conformity with the eruptive mode of formation. (p. 91–92)

Ure even reported granites cutting older granites, on the basis of MacCulloch's observations, to emphasize that there were granites of different ages (p. 116). He employs a very poetic language to emphasize the great uplifting power of the granite:

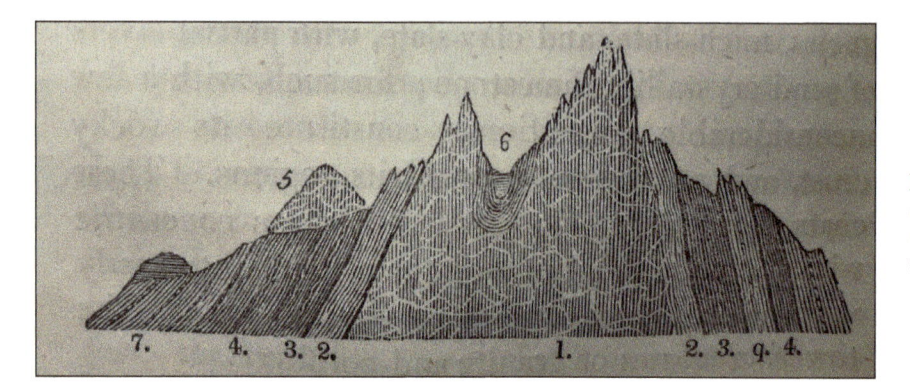

Figure 18. D'Aubuisson's figure, reproduced by Ure, "representing the usual structure of the primitive mountains" (Ure, 1829, p. 92).

In the early epochas of the antediluvian world, soon after the granitic atlas had uplifted the primitive mountains, and before the extensive series of mineral beds ... were deposited beneath the ocean, its waters resting on the nearly concentric, or slightly broken zones of gneiss and mica slate, necessarily lay in closer proximity with the interior fires, than any at subsequent period. Hence two important consequences: 1. From thinness of the solid crust, the smallest chink or fissure in it, would be an immediate focus of submarine explosion, accompanied and followed by a commensurate comminution and dispersion of solid rocks and organic deposits through the agitated waters. 2. The ocean would then attain its maximum temperature; a pitch certainly far higher than at present.... (p. 429)

Anybody reading Ure's statements cited above and numerous similar ones in his book, would be justified in wondering where he thought he differed from Hutton. Ure agreed with essentially all of Hutton's interpretations, except one: according to Ure all the phenomena cited above had happened by divine fiat and that the history recounted in the *Genesis* myth is an expression of unimpeachable truth. He could not conceive the structures he described to be products of slow evolution, because if they had been so, then there would be no moments of intervention of the divine power and indeed no need for a creative and governing God. This, Ure was unwilling to accept. He realized that Hutton's geology, albeit when he compared it with the former "theories of the earth," was found to be far superior to that of Werner and that he could agree with most of Hutton's deductions, but he could not accept a world without God's constant meddling. Although much better than Deluc's geology, Ure's also remained a mythology, because he tried to tie everything to God's creation and constant intervention and to see Moses as the truthful inspired historian. Largely because of this attitude and because his book contained nothing original except the expression of his religious zeal, Ure's name quickly disappeared from the geological intercourse during the nineteenth century, despite his great respectability in other fields of science and technology.

The last publication I consider here is the small book by the famous English chemist William Thomas Brande, F.R.S. (1788–1866; Fig. 19) based on his lectures delivered at the Royal Institution in London. Dean (1992, p. 172–173) mentions Brande's book based on his 1816 lectures in the same place and published a year later (Brande, 1817). I have before me also a newer version, published in 1829, incorporating evidently newer material derived from other lectures and not only that of 1816. In it, Brande again presents both the Wernerian and the Huttonian theories (p. 20–26), but even here, his antipathy to Wernerian teaching is evident:

There is one disadvantage and difficulty attending an attempt to expound Werner's doctrines,—which is, that we are obliged to take them at second hand, since he has not published any connected view of them himself; and moreover, the Professor and his scholars have generally affected a mysterious phraseology which is very difficult to construe into common sense, or intelligible terms; and which is sometimes so harsh and

Figure 19. William Thomas Brande, F.R.S., by the portrait artist, the Royal Academician, Henry William Pickersgill (1830). Public domain (PD-US-expired).

uncouth as to border upon the ridiculous, when, at least, we attempt to put it into an English dress [these protests almost echo Sir George Mackenzie's complaints]. The darkness which he has thrown round his doctrines seems, indeed, often as if it were expressly intended to keep them from the eyes of the vulgar and uninitiated, and may be compared to that mystic and symbolical language, in which the alchemists delighted to veil the accounts of their researches, and which after all were no great things, when by dint of much labour and study they were deciphered and done into plain and legible terms. (Brande, 1829, p. 21)

Then comes a damning verdict:

... in a word, that it is a system which might pass for the invention of an age when sound philosophy had not yet alighted on earth, nor taught man that he is but the minister and interpreter of nature, and can neither extend his power nor his knowledge a hair's breadth beyond his experience and observation of the present order of things. (Brande, 1829, p. 24)

Notice the much more lenient judgement of Hutton's views:

Though the details of these theoretical views there is very much that is fantastic and improbable, it must be allowed that there is also much that is consistent with the known agency of bodies, and which is even directly borne out and verified by the actual result of experimental investigation. Indeed, the progress of modern chemistry has disburdened the Huttonian doctrines of some of their heaviest inconsistencies. (Brande, 1829, p. 26)

Brande returns to Hutton's opinions when discussing the origin of basalts and metallic veins and finds them the best available (see his sections VIII and IX), although when it was the turn of the geomorphology to be explained, he retained a diluvial cause as the chief agent of sculpting the land.

Brande's final judgement on Hutton's theory is an emphatic, but not comprehensive, endorsement and a clear, previsional, negation of the claims of the recent revisionists of the history of geology, who represent Hutton's theory as an old-fashioned member of the chain of theories of the earth since the seventeenth century:

It is, however, unnecessary that I should here recapitulate the Plutonian and Neptunian evidence. The former, as modified by Dr. Hutton and his commentators, is that which most plausibly accounts for the phenomena, is least at variance with facts, is least hypothetical, and best entitled to the appellation of a theory of the earth;.... (Brande, 1829, p. 233)

Perhaps I should end this account of Hutton's pre-Lyellian reputation in the British Empire with an amusing episode in Edinburgh, related in the memoires of Sir Henry Holland (1788–1873)

of Icelandic fame mentioned above, the one-time president of the Royal Institution of Great Britain and physician ordinary to Queen Victoria. Sir Henry was in Edinburgh between 1806 and 1811 as part of his medical education. He was much impressed with the intellectual milieu he encountered there, although still "partially dissevered ... by the political feelings still strong in Scotland at this time" (Holland, 1872, p. 80). After having briefly mentioned the rivalry between the Whigs and the Tories, Sir Henry writes:

A minor and more whimsical cause of disruption in the Edinburgh society was the controversy going on between the Huttonians and Wernerians, as they were then called—the respective advocates of fire and water, as agents concerned in moulding the crust of the earth. The basaltic and other rocks clustering round Edinburgh furnished ample local material for discussion. No compromise of combined or successive agency, such as reason might suggest, was admitted into this scientific dispute, which grew angry enough to show itself even within the walls of a theatre. A play written by an ardent Huttonian, though graced with a prologue by Walter Scott and an epilogue by Mackenzie (the author of the "Man of Feeling")[99], was condemned the first night—as many persons alleged, by a *packed house* of the Neptunian school. It may be that the play itself was more concerned in the fate that befell it. (Holland, 1872, p. 81, italics Holland's)

There is, in this recollection, the hint that Werner's neptunism was already dying in Edinburgh, despite the presence of its supporters enough to fill a theater for angry encounters between Huttonian and Wernerian groups! Sir Henry Holland certainly knew the background of the hostility to this particular play at that particular time: it resulted mainly from Sir George Steuart Mackenzie's enthusiastic Huttonian account of his observations in Iceland during the trip together with Sir Henry and Mr. Bright, as I quoted above and particularly the last two acts of his unpublished[100], but performed play *Helga, or The Rival Minstrels, A Tragedy, Founded on an Icelandic Saga, or History* found offensive by the Wernerians. Wawn says the play had the shortest imaginable run in the Theatre Royal in Edinburgh (Wawn, 2000, p. 52).

In pre-Lyellian days, the religious sentiment was far stronger in the British Empire than it was in continental Europe and its colonial dependencies (and it remained so for the rest of the nineteenth century), except for the Spanish Empire and its successor states. This was evidently because of the robustness of the monarchical and feudal system in Great Britain than elsewhere in Europe, with the exception of the Russian and Spanish empires. But, unlike the Russians and the Spaniards, the Britons managed to democratize their society without destroying their old institutions and, among those, religion was a fundamental pillar. In England, for example, the person of the monarch sat (and still does so) at the top of the Church of England as "Supreme Governor of the Church of England," i.e., as *pontifex maximus* and *rex sacrorum*, combining (although mostly symbolically) the powers of both. Any attack on religion was thus seen as an attack on the state and the society it embodies, even if one was not in England, but in other UK countries all ruled by the monarch, who was (and still is) also the monarch of England[101], whereas, on the continent, the French Revolution forever broke the undisputed authority of religion, except in the Catholic countries north of the Mediterranean. Therefore, the religious fervor directed against Hutton's ideas in the UK countries is not surprising, although even his religious critics acknowledged the novelty of his theory. Those not scared by its religious implications realized that it was the herald of a new geology even before Lyell became the midwife of the geology based on Hutton's theory of the functioning of our planet.

After Lyell and before Geikie

As for the post-Lyellian/pre-Geikian period in the British Empire, Dean competently reviews the prominent reactions to Hutton's geological ideas. In the following I deal only with those that he did not mention.

In the same year as the publication of the second volume of Lyell's *Principles*, the well-known popular writer of children's books for purposes of education, the creator of "little Harry" (in her *Harry Beaufoy, or The Pupil of Nature*, published in 1821) Maria Hack (1777–1844) published her *Geological Sketches and Glimpses of the Ancient Earth*. This wonderful little book, although written for the edification of the youth, "is not designed exclusively for young persons, but for all to whom the subject is new, and who have not inclination or opportunity for studying it scientifically. The references will show that the works of those who are considered the best authorities, have furnished the materials. As to the mode of arranging them, the adoption of colloquial intercourse seems to afford the greatest freedom and variety of illustration" (Hack, 1832, p. iv).

The book again features Harry, somewhat grown up (he is now fifteen) and "taller," in conversation with his mother Mrs. Beaufoy, during a morning walk they take through Nature near their home. They first come upon a nightingale and then see a team of horses pulling a heavy wagon. This reminds Harry's mother of the presence of a chalk quarry nearby and they decide to explore it. So begins their geological conversation. After having seen a few outcrops and studied strata of different colors, Mrs. Beaufoy explains to Harry that some strata are horizontal, some are inclined and some even vertical. She tells Harry that she cannot tell why perpendicular strata are called vertical but suspects it may be to emphasize that they may have attained that position "by a force acting *from beneath* the strata" (Hack, 1832, p. 59, italics Hack's). Harry expresses surprise and asks "what could that force be?" In answering this question Mrs. Beaufoy gets into a description of the prevailing theories of earth behavior:

> There have been various opinions as to the causes employed to produce the arrangement observed in the strata of the earth. Some philosophers think it may be accounted for by successive depositions from *water*; these have been called *Neptunists*.

> Others, ascribing the formation of terrestrial bodies to the action of *subterraneous fire*, are termed *Plutonists* or *Vulcanists*. (Hack, 1832, p. 59, italics Hack's).

Harry asks his mother which theory she thinks is right. Mrs. Beaufoy responds by saying that she really is not qualified to pass judgement, but suspects that the truth may lie somewhere in between. Upon Harry's request to learn more about this "third system," Mrs. Beaufoy responds as follows:

> It has been called, from Dr. Hutton, (who thought that the structure of the earth may be most satisfactorily explained, by supposing that it has been subjected to the *successive* agency of water and fire,) *The Huttonian Theory*. (Hack, 1832, p. 60–61; italics hers)

Hack here references Playfair's biography of Hutton and says that the truth of Hutton's theory has not yet been established and it is one among many guesses. But then, while discussing the origin of granite (with reference to Playfair), she comes again to Hutton. While talking about granite veins intruding country rock, she says:

> To prove the existence of such veins, was a subject of great anxiety to Dr. Hutton, while he was endeavouring to establish the truth of his theory respecting the formation of this rock. He wished very much to find a clear example of granite in a melted state, having been injected among other rocks, in the manner I have described. Resolved to try the truth of his theory by this test, he set out on a journey to the Grampian mountains, intending carefully to examine the places in which it was probable that the junction of granite-rocks with those lying immediately above them, might exhibit the appearance he desired. In Glen Tilt he was so fortunate as to find the object of his search. Veins of red granite are there seen branching out from the principal mass, and penetrating the adjacent limestone and other rocks, which appear so different from the granite, not only in colour, but in structure, being *stratified*, (which granite, you know, is not) that they afford a very striking example of the fact which Dr. Hutton desired to establish. When he saw this undeniable confirmation of his theory, he was so delighted that he could not restrain the expression of his joy and exultation; and his guides, who probably had no conception that any thing relating to the formation of a rock could excite such emotions, were led into a strange mistake: they made sure that the gentleman they were attending had discovered a vein or silver or gold. (Hack, 1832, p. 71–72; italics hers; here Hack gives a reference to Lyell's *Principles*)

When Harry says he could not understand why Dr. Hutton was so delighted, Mrs. Beaufoy replies:

> Do you not perceive that the discovery Hutton made in Glen Tilt confirmed in several particulars the truth of his theory? (p. 72).

In the next chapter, Mrs. Beaufoy explains to Harry Dr. Hutton's theory of sediment induration by heat. When Harry thinks it silly, his mother proceeds to tell him of the experiments made that corroborate Dr. Hutton's opinion. Let me note here that Hack, a one-time Quaker who later joined the Anglican Church, herself a strict believer in the Biblical myths, expressed her sympathies in Mrs. Beaufoy's name: in French it means "beautiful faith"! But, remarkably, although there is sporadic talk of God's creation of our world in her book, she managed, in an admirable way, to keep her narrative of geology almost free of religious interference.

This is how much Mrs. Beaufoy explains to her son about Hutton's ideas. Although she did mention the controversy between the neptunists and the plutonists, she seemed to favor Hutton's theory which she termed the "third system," allowing a rôle both to water and to fire in the generation of terrestrial processes. Werner's name is not once mentioned in Hack's book.

Only two years after Hack's popular book, the chemist and medical doctor Henry Samuel Boase (1799–1883), the secretary of the Royal Geological Society of Cornwall, published his *A Treatise on Primary Geology*, a book Dean refers to very briefly as the "last book of importance to oppose Huttonian plutonism" (Dean, 1992, p. 229). Boase's criticism of Hutton's theory was not, by any means, a total rejection, but a modification, still along the lines of what one might call Huttonian geology. Boase thought that the primeval crust of the earth was not entirely of granitic composition, but it also had schists. He further protested against the presumption that the earliest stages in the evolution of our planet may be interpreted using the measure of the current geological phenomena prevailing on the earth. This is another one of those "correct guesses" of nineteenth century geologists for entirely wrong reasons. But this is not why Boase's book is important for the purpose of the present book. He pointed out that Hutton's theory was "the prevailing geological theory" when he was writing his own book, i.e., just after Lyell's *Principles* was first published! This is what he wrote:

> In short, *the prevailing geological theory is that of Hutton;* modified indeed, and improved on, so as to adapt it to the numerous and important facts that have been collected since his day. One of the arguments which he triumphantly advanced against his opponents, the Wernerians, was founded, on the penetration of the stratified rocks by granitic veins, proceeding from the main mass of this rock; and likewise, on the occurrence of angular pieces of slate, enveloped in the granite; and other apparent signs of confusion and violence, which are generally found at the junction of the stratified and unstratified rocks. These facts are still esteemed, as affording an impregnable position for the Plutonic doctrine; and, though the followers of Werner have gradually submitted, yet the Huttonians are continually adding outworks to this strong-hold, in hopes of increasing its security. (Boase, 1834, p. 217, italics mine).

From here Boase proceeds to his criticism which amounts to saying that granite is not the only original component of the earth's crust.

In 1838, the English physician and paleontologist, the famous discoverer of the Iguanodon[102], Gideon Algernon Mantell (1790–1852; Fig. 20), published his two-volume *Wonders of Geology*, which proved to be an extremely popular book, running through many subsequent editions. While discussing the effects of high temperature on rocks, Mantell first references Sir James Hall's experiments and then writes:

> We shall hereafter find, in accordance with the beautiful and philosophical theory of Dr. Hutton, that all the strata have been more or less modified by heat, acting under great pressure and at various depths; and that the present position and direction of the materials composing the crust of the globe, have been produced by the same agency.[103] The Huttonian theory, indeed, offers a most satisfactory explanation of a great proportion of geological phenomena, enabling us to solve many of the most difficult problems in the science; and it is but an act of justice to the memory of an illustrious philosopher, and of his able illustrator, Professor Playfair, to state that this theory, corrected and elucidated by the light which modern discoveries have shed upon the physical history of our planet, is now embraced by the most distinguished geologists. (Mantell, 1838a, p. 79–80)

The book proved very popular with the third edition coming out already in 1839, which was also published in the United

Figure 20. Portrait of Gideon Algernon Mantell by Davey after Sentier. Credit: Wellcome Collection. Attribution 4.0 International (CC BY 4.0).

States as the first American printing. This passage occurs in all the subsequent editions, including those that came out after Mantell's death (e.g., in the 7th edition, 1857, p. 104)

In the second volume, after having referred to Fournet's statement that all the primary rocks are simply sedimentary deposits metamorphosed by igneous action[104], he continues:

> I will only add that this opinion is but a modification of that long since expressed by our illustrious countryman, Hutton, that granite rocks are consolidated and altered sediments which have accumulated at the bottom of the ocean. (Mantell, 1838b, p. 650–651)

The Scottish naturalist and one-time conservator of the Museum of the Royal College of Surgeons of Edinburgh, William MacGillivray (1796–1852)[105] published in 1840, the first edition of his small *A Manual of Geology* as a part of the series of his Natural History manuals. In this book, there is no direct reference to Hutton, although the geological theory adopted is entirely Huttonian, despite the fact that MacGillivray started his professional life as an assistant to Robert Jameson, the Wernerian (MacGillivray, 1912). By the time a second edition was called for, MacGillivray had become, in 1841, professor of natural history in the Marischal College and the University in Aberdeen, where he remained until his death. During his tenure in Aberdeen, MacGillivray published a second edition of his geological manual (MacGillivray, 1844). It is little different from the first edition, but in this newer edition, he mentioned Hutton and Playfair in relation to the formation of metallic veins (his p. 260). He is noncommittal as to the theories of vein origin, but no other authors are mentioned in his account. Before I take leave of MacGillivray, I wish to remind my readers that he was principally an ornithologist and his geological training was confined to the time he spent as assistant and secretary to Robert Jameson.

George Fleming Richardson (1796?–1848), geologist, lecturer, and one-time employee of the British Museum, who ended his life by suicide as a result of pecuniary embarrassments, published in 1842 his *Geology for Beginners*, which Mantell, his former employer, called a "scandalous piracy"[106]. This led to a controversy with Mantell that adversely affected the sale of the book and is believed to have contributed to Richardson's economic difficulties (Torrens, 2004b). However, although it may not have corresponded to the author's expectations, the book did sell well and had an enlarged second edition already in the following year (Richardson, 1843). The demand was such that after Richardson's death, it was edited by Thomas Wright under the new title *An Introduction to Geology, and its associated sciences Mineralogy, Fossil Botany, and Palæontology* and published in 1851 with a reprint to follow in 1855.

Richardson finds the essence of Hutton's theory in the sentence "no single substance in nature is either permanent or primary" (p. 41). He then elaborates on the details of the theory:

> The most distinguished opponent of Werner in this country, or, we should rather say, the complete refuter and destroyer of his system, was Dr. Hutton, a Scottish physician, whose opinions on all essential points, were the very reverse of those of the philosopher of Freyberg, substituting for the limited views of the professor of mineralogy, the more extensive generalizations of the observer of universal nature; and for data and decisions formed from one of the natural sciences, facts and conclusions derived from nearly all.
>
> As the substance of his theory, he taught the following important and deeply instructive truths, which there is some difficulty in expressing in a condensed form.
>
> His first lesson was, that no geological phenomena afford any proof of the beginning of things.
>
> The oldest rocks are merely derivative compounds of the ruins of rocks which existed before them, and which were destroyed, chiefly by the slow action of atmospheric causes; while their detritus, borne by rivers to the ocean, and loosely strewed over its bed became consolidated by heat, and subsequently upheaved and fractured.
>
> The metamorphic rocks were originally sedimentary deposits, similar to the secondary, but altered by the long continued action of heat.
>
> Granite has crystallized from a state of igneous fusion, under conditions of heat and pressure. In other words, it has been melted by fire, at great depths in the earth, and has cooled under pressure, so vast as to have prevented the gaseous portions of its elements from escaping, and has thus assumed its present crystalline texture.
>
> Such is the imperfect outline of the celebrated system of Dr. Hutton; a system which has not only supplanted that of Werner, but has formed the foundation of the researches and writings of our most enlightened observers, and is justly regarded as the basis of all sound Geology at the present day. (Richardson, 1842, p. 41–42.)

With the addition of a word or two, this passage is repeated verbatim in the second edition (Richardson, 1843, p. 69). When Richardson first wrote these lines, Sir Archibald Geikie was only seven years of age.

After the lines I just reproduced, Richardson proceeds to corroborate them. Among the corroborating observations he cites Sir James Hall's experiments on marble and points out that they produced "the very result presumed to have occurred in nature, by the hypothesis of Dr. Hutton" (Richardson, 1842, p. 43; 1843, p. 70). To support the idea of the fluidity of granite during its intrusion, he cites Hutton's own observations in Glen Tilt.

Richardson returns to Hutton's views while describing the "Cambrian and Cumbrian Systems" in his chapter 20. He concludes his introductory statement to "Metamorphic rocks; mica-schist and gneiss systems" with the words:

> The splendid conception of Dr. Hutton, described in a previous page [the reference here is to his p. 42, cited above], by

which the whole of the primary rocks stratified and unstratified are considered to be merely sedimentary deposits, variously acted on by heat, affords the only means of reconciling, in a satisfactory manner, the common origin of these rocks, with the varied and dissimilar aspects which they now present. (Richardson, 1842, p. 476)

The "Cambrian and Cumbrian Systems" occur in chapter 21 in the second edition, and, in it, the statement above occurs on p. 560. In the second edition, Richardson added the following to his description of "The Gneiss System" (p. 562):

On the other hand, it may be urged that the graduation of gneiss into mica-schist, and various slaty rocks, and of these last into rocks which are unquestionably of aqueous origin, confirm the probability of the whole being of the common origin to which Dr. Hutton has assigned them, and owing their present dissimilarity of texture and aspect to the dissimilar degrees of heat to which they have been exposed.

After Richardson's tragic end, the popularity of his introductory book did not diminish and the distinguished Scottish geologist and physician Thomas Wright (1809–1884), F.R.S. and winner of the Wollaston Medal of the Geological Society in London, edited a new edition, which was published in 1851. Although Wright introduced many changes to bring the book up-to-date, the introductory chapter in the new edition was preserved verbatim, including the statements that Werner's errors and his entire system had long exploded and that modern geology was based on Hutton's theory (Richardson, 1851, p. 36, 38–39). Richardson's posthumous book maintained its popularity in its revised form and was reprinted in 1855.

The author of popular books on geology, including a remarkable textbook, the professor of geology in the Addiscombe Military Seminary (formally Royal India Miliary College after it was taken over by the British government) belonging to the East India Company in Surrey (now the London district of Croydon), David Thomas Ansted, F.R.S. (1814–1880) brought out in 1844 his two-volume *Geology, Introductory, Descriptive, & Practical*. It is a book addressed to the practical men and contains no history of geology. However, in the second volume, where he discusses the origin of mineral veins, he mentions Hutton's ideas, although he says:

The Huttonian hypothesis, that the contents of veins were in all cases injected from below in a state of igneous fusion, is scarcely more probable or better founded than the rival theory of the Saxon Geologist. (p. 275)

He instead prefers that the mineral veins were filled by means of "electricity." From his text, however, it is clear that at his time, Hutton's ideas were the reigning ones, which, in the case of mineralised veins, he finds unsatisfactory.

The noted English geologist Alexander Henry Green (1832–1896; Fig. 21), one of the early defenders of the power of subaerial erosion by rivers in the British Empire at a time when papers in England were being refused publication for mentioning that "improbable notion" (Miall, 1896)! After having become professor at the new Yorkshire College of Science (now the University of Leeds), Green wrote a textbook that became popular enough to go through four editions (1876, 1877, 1882, 1898). Here I consider only the first three editions, as the fourth falls outside the temporal limits I chose for the present book. In the first chapter of the first edition, entitled "The Aim and Scope of Geology, with a Sketch of its Rise and Progress" Green pointed out that "Hutton labored successfully to show that the forces now in action are fully competent to form rocks, and to bring about a large portion of the changes, which we learn from Geology must have passed over the earth's surface" (Green, 1876, p. 4; also the same pages in 1877 and 1882). Green thought that Hutton did not consider whether there were times in the past of our planet when things were substantially different. He added, significantly, that Hutton "may be fairly looked upon as the veritable father of the science" (Green, 1876, p. 4). In the second edition he wrote more emphatically: "we may go further and call him the founder of Geology as a whole" (Green, 1877, p. 4; 1882, p. 3). Green continued: "Hutton, like most great men, was in advance of his age; his teaching fell dead till it was revived and illustrated by Lyell; but it is now universally recognized as the principle on which we must base all speculations relating to that part of the science of which he treated" (p. 4–5). To this passage, Green added a footnote after the word "dead" containing the quotation I gave above from Forster's book to illustrate how completely Hutton's theory had been forgotten or discredited. (These passages occur on the same pages in the second and third editions.) I quote this footnote to show that in 1876 thinking Hutton's theory outdated was viewed as odd, although Green was certainly not right in thinking that Hutton's theory had been so completely forgotten as the numerous quotations in the present book shows!

Figure 21. Alexander Henry Green (from Miall, 1896).

Continuing his praise for Hutton's ideas, on p. 115 (on p. 116 in the second edition, on p. 207 in the third edition) Green wrote, after having emphasized the importance of subaerial agencies in eroding the land, "It is now however very generally recognized that they perform by far the larger part of the denudation that is going on before our eyes, and they have at least had their true place granted them in the roll of denuding forces, a place Hutton long ago pointed out that they were entitled." On p. 116 (in the second edition, on p. 117; in the third edition on p. 207), Green emphasized, with a direct quotation from Hutton, that

> Hutton was the first clearly to enunciate the laws of denudation, which he had learned by observation; the summary of one of his chapters is worth quoting: "Whether we examine the mountain or the plain; whether we consider the degradation of rocks or of the softer strata of the earth; whether we contemplate Nature and the operation of time upon the shores of the sea or in the middle of the continent, in fertile countries or in barren deserts, we shall find the evidence of general dissolution of on the surface of the earth, and of decay among the hard and solid bodies of the globe."

After having quoted a parallel passage from Playfair's *Illustrations*, Green returns to Hutton on the same page (116 in the first, 117 in the second edition, on p. 208 in the third edition):

> Another useful lesson may be learned from this bit of the history of Geology. It is now nearly eighty years since the Theory of the Earth was published, and it is only quite lately that geologists have come to recognize the truth of its teaching. So slow are men, even when the right road is pointed out to them, to leave a groove which they have been for a long time following.

Then, in a footnote, Green added, "In connection with the subject of Denudation, the student will do well to read, Theory of the Earth, Part I, chap. I sec. iv, Part II. Chaps. iii to vii." In the third enlarged and penultimate edition of his textbook, Green (1882) repeated what I quote above (p. 4 and 4–5 and 116 and 117 in the earlier two editions) on his p. 3 and 207–208 exactly, including the footnote. But then, in other parts of the book, he added further references to Hutton: on p. 573 he reiterates that "The truth that the present inequalities of the surface are mainly due to denudation was first clearly seized upon by Hutton." On p. 585–586 he presents a history of the idea of subaerial denudation, in which he laments the late recognition of the importance of this process and proceeds to say that "The whole truth was first thoroughly seized upon by two of the master minds of the science—by Hutton in 1795 and by Scrope in 1826."

Green thought that Hutton had discovered unconformities. Although not true, I quote Green's assertion here just to emphasize the perception of some of the geologists in the UK in the eighties of the nineteenth century: "The existence and meaning of unconformity were recognized first by the master mind of Hutton" (p. 517, footnote). In reality, Hutton had merely reminded the geological community of what Steno had discovered already in 1669, but it was a useful reminder, because, under the influence of Werner, Steno's interpretation of *unconformities as witnesses of rock deformation* had begun to be forgotten (on Werner's culpability in this amnesia, see Élie de Beaumont, 1832, p. 344, the last paragraph of his long footnote).

In 1880, the author of fictional and non-fictional popular books, Arthur Nicols (who was a fellow both of the Geological Society in London and the Royal Geographical Society) published his small *Chapters from the Physical History of the Earth—An Introduction to Geology and Palæontology*, in the introduction of which he wrote

> The names of Werner and Hutton can never be remembered but with gratitude; for the very fierceness of the controversy aroused by their conflicting doctrines showed that nothing could be done without patient observation, and extreme caution in generalization while the known facts were so few and isolated. (Nicols, 1880, p. 8)

In 1884 and 1885, two English geologists, Robert Etheridge (1819–1903) and Harry Govier Seeley (1839–1909), the man who first divided the dinosaurs into the two orders of Saurischia and Ornithiscia in 1888, revised and greatly expanded the last edition of John Phillips' (1800–1874; Fig. 22) *Manual of Geology: Practical and Theoretical* (1855). This book was first published as a 68-page article in the 15th volume of the seventh edition of the *Encyclopædia Britannica* under the article Mineralogy, in which Hutton's name is mentioned once in connection with the

Figure 22. John Phillips. Photograph by H.J. Whitlock.

origin of mineralized veins (Phillips, 1842, p. 235). However, four years earlier, Phillips had published, with the same publisher, a much longer version in independent book form, but still advertised as an article in the seventh edition of the *Encyclopædia Britannica* (1838). Hutton is mentioned also in this book, in the same context as in the later-published shorter article (p. 273). In 1845, Phillips published a 279-page article in the sixth volume of the *Encyclopædia Metropolitana* (Phillips and Daubeny, 1845) It was very similar in content to his 1838 book. Finally, the *Encyclopædia Metropolitana* text was readjusted to become a textbook with emphasis on English geology, Daubeny's tract on volcanoes in the *Encyclopædia* (p. 711–749) was excised, and Phillips' revision of his part in the *Encyclopædia* was issued as a separate book in 1855. In the *Encyclopædia* article (on p. 534) and in the subsequent book (on p. 14) there is mention of Huttonian and Wernerian theories, but first, to tell the reader that the author will not prejudge either theory or any other hypothesis (he had already discussed Werner at length, but not mentioned Hutton!) and, secondly, to discuss the idea of uniformitarianism.

Phillips' discussion of uniformitarianism is most informative and modern. Having described the stratified rocks in earlier pages of the *Encyclopædia Metropolitana*, he comes to discuss the processes that made them (p. 701; the text cited below occurs in the 1855 book in chapter XV, p. 466–467 with italics and bold-face type added):

> Having now concluded our descriptions of the strata and aqueous products recognized in the crust of the globe, and also traced the effects of subsequent extraordinary inundations upon the surface, arising from local changes of level or general internal convulsions, it remains to be seen whether the causes now in action in the modern economy of nature are of the same *kind* as those which were formerly concerned in producing arrangements and disarrangements observed in the crust of the globe.
>
> This is the true cardinal point of theory. According as the one or the other conclusion on this point be adopted, we may attempt to explain the ancient phenomena by modern laws of nature, and thus connect the present and the past, the extinct and the existing history of our planet into one system of progressive change, according to the school of Hutton, Playfair, and Lyell; or suppose that in the chaotic infancy of our planet, laws peculiar to that period prevailed, and properties of matter were unfolded then which never show themselves at present; and that the ancient rocks and organic bodies belong to a wholly distinct set of causes, were the produce of a peculiar creative impulse, no longer permitted to operate on the finished and man-inhabited planet. The Wernerian cosmogony bears very much this aspect.
>
> But though, put thus in direct opposition, the rival hypotheses appear to have no point of union, we find, in fact, that between the opinion of Hutton, who considers creative nature to be perpetually in progress—the same to-day, yesterday, and for ever—and the dogma of Werner, that the world was made by a certain settled sequence of events to which nothing similar now happens, every variety of theory is adopted and defended.

> We may, however, with rigid accuracy and much convenience, rank them in three classes.
>
> 1. The favorers of Hutton's and Lyell's views, who maintain that the causes now in action to change the level and alter the relations of the masses of matter in and near the crust of our globe, are those which have ever been in action, identical in kind, and equal in degree, in all times past, and which may be expected to continue the same, in kind and degree, through the future.
>
> 2. **Views of the English School on this Subject**—The general school of English geologists, who has always maintained, and labored to prove, that the causes operating on the surface and in the interior of the earth have remained through all times past unchanged in kind, and are still operating with the same tendencies as they always did, but often on smaller areas, and with less effect. This view of the subject has a double aspect. English geologists have generally believed that as volcanos were supposed to become languid through want of fuel, the *circumstances* under which the modern operations of water and fire are manifested in the general economy of nature, approach more nearly a state of equilibrium or saturation, and therefore afford no opportunity for the same extraordinary display of energy as in ancient times; but since the relative periods of the great convulsions which have elevated chains of mountains, and given new boundaries to the ocean have been investigated upon sound principles, the mind has become gradually familiarized to another notion, and habituated to contemplate long periods of ordinary and regular action of natural causes interrupted by transient, local, or general convulsions. According to this modification of the hypothesis, the present is a period of ordinary and regular action, succeeding upon an epoch of violent disturbance[107].
>
> 3. The old notion of despairing speculators in cosmogony, who found it easier to cut the Gordian knot, by flatly denying the analogy of modern and ancient operations, and either referring the whole beautiful order of the ancient works of nature which they could not comprehend, to a momentary fiat of Deity, or to the rude and prolonged confusion of elements in chaos.

Phillips refers to the third class of interpretation as a "mere dream of indolence and deficient observation; for we have already proved that the stratified rocks are certainly analogous in all points to the products of modern waters, and that the unstratified rocks clearly prove their special origin from fire." (ibid.). On p. 799 of the *Encyclopædia* and p. 630 of the book, Phillips says that Werner "with far less moderation than Hutton wished to begin at the beginning." He thus dismisses any Wernerism in the strongest possible terms and presents what he calls the view of the English school as a modified form of Hutton's theory. The modification was clearly imported from France and soon proved faulty.

On p. 772 of the *Encyclopædia* and p. 537 of the book, Phillips points out that mineral veins postdate their country rock and says this is allowed by both the Wernerian and the Huttonian hypotheses. On p. 800 of the *Encyclopædia* and p. 632–633 of the book, Phillips mentions Hutton's interpretation of limestone as a clastic rock and disagrees with it and on p. 764 of the

Encyclopædia and p. 517 of the 1855 book, he describes the Salisbury Crag near Edinburgh and agrees with Hutton's magmatic interpretation of the "greenstones" seen there.

When it comes to tectonics, Phillips approves of Hutton's ideas, but, as the last quotation showed, he thought the revised "English interpretation" of Hutton's theory would still be Hutton's theory in tectonics also. He wrote (I annotate his paragraph between square brackets):

> There is, perhaps, no point of theoretical Geology more certainly established than that, in any given small area of the surface of the Globe, long periods of ordinary action of natural causes have been several times interrupted by epochs of extraordinary disturbance; that the relation of the level of the sea and land has remained for a long time the same, or very gradually changed [von Buch, 1810(1870) and many others as far back as Urban Hiärne] and afterwards been altered by internal convulsions [so far, these are derived from Leopold von Buch's publications on mountain-building: for mechanism, see: von Buch, 1818–1819 (1877), 1824 (1877); for speed, see: 1827 (1977)]. It is also admitted that this law has an extensive though less than exact application; that the periods of ordinary and crises of extraordinary action were respectively contemporaneous over very large regions of the Globe [Cuvier, 1812, 1825], and even with respect to some of the cases admit of general application [this is pure Élie de Beaumont, 1829–1830, 1830, 1831, 1833]. It appears, also, that the nature of the strata deposited differs more or less according to the several successive periods [a curious Wernerian remnant], and that the races of organic remains, in several important cases, are subject to contemporaneous crises [Cuvier, 1812; 1825; Élie de Beaumont, 1829–1830]. On this evidence, joined to some theoretical considerations, is founded the modern admission of the doctrine of alternating periods of convulsion and repose; a doctrine which was held by ancient Philosophers, revived by Leibnitz [sic] and Hutton, and illustrated by Cuvier and De Beaumont. Perhaps this view of the subject was never more clearly expressed than by Leibnitz [sic], whose just sense of philosophy of Geology has been lately placed in a strong light by Mr. Conybeare[108]. His view is, that the powerful agencies exerted a displacing and altering the solid crust which gradually thickened over the ignited nucleus, have many times renewed the face of the young Globe by the eruption of concreted igneous rocks from below, and the deposition of stratified rocks by water above; and the Globe was more and more diversified with mountains and valleys, and subjected to various physical conditions [these last two sentences are perfectly Huttonian, if divorced from the catastrophist statements above]; *donec quiescentibus causis, atque æquilibratis, consistentior emergeret rerum status* [109]. (Phillips and Daubeny, 1845, p. 798)

This passage occurs on p. 628 and 629 in the 1855 book. Despite his approval, Phillips did not understand the critical implication of Hutton's theory, already emphasized by Lyell, that there are, or never have been, global events on earth; that geological events are local and contingent; they may affect a whole continent, but never the entire planet. The planetary evolution was without interruptions.

Phillips' position is easy to understand. His geological training took place under the tutelage of his maternal uncle William "Strata" Smith (1769–1839), who was a stratigrapher who invented his trade in the little-deformed areas of southern England (see the copiously and brilliantly illustrated magnificent book by Wigley et al., 2018). Later, combined with Cuvier's biostratigraphy, the stratigraphers came to think that major units defined by characteristic fossils to be universal, in a way reviving the concept of Werner's "universal formations." Only the presence of universal events can make the existence of such neatly separated universal strata possible[110]. Although criticized by Alexander von Humboldt already in 1823 (see his p. 41–42)—because such a view ignored the great variations in geographical conditions on the surface of the globe at any given time—the idea of universal correlatability became one of the foundations of geology and as such formed one of the two bases (the other being Leopold von Buch's theory of catastrophic mountain formation by magma-propelled uplift) of Élie de Beaumont's theory of globally synchronous episodic orogeny. As we saw above, Lyell objected to it on the grounds that rocks (i.e., strata) cannot be assumed to represent instants in time; they embrace finite time intervals the duration of which cannot be assumed, but must be measured. Lyell's judicious objection was ignored not only by Phillips and most of his contemporaries, but even in the later twentieth century, as we saw above (see Şengör, 2016).

The revision and expansion of Phillips' *Manual* by Seeley and Etheridge so enlarged the book that it now had to be divided into two volumes. The first, subtitled *Part I Physical Geology and Palæontology* was revised by Seeley. By 1885, there remained little of the old disagreements between neptunists and plutonists or actualists and catastrophists, so that Seeley deleted all the passages pertaining to them and with them Hutton's name also became excised from the old text. However, Seeley wrote on the last page of the revised version, "In earlier times it was necessary, as a preliminary, to show how the present order of Nature is analogous to that exhibited in the strata, and that work as well done by Hutton and Lyell" (Phillips, 1885a, p. 529). Nothing contrasts with the importance of this revealing theoretical work Werner's defunct theory than Seeley's following words:

> A peculiar set of opinions concerning the formation of the earth has been honoured by the title of the Wernerian theory; and the pupils of Werner, who had found proof of the truth of his practical rules and inferences, may be readily pardoned for the determination which they manifested to uphold Werner's hypothetical notions. But if we wish to ascertain the real value of the benefits which researches of Werner have conferred upon geology, we must forget his theory, and view only the data which he collected for its foundation. (Phillips, 1885a, p. 4)

Seeley too thought that Hutton and his follower Lyell had laid down the critically important stone into the foundation of geology, namely the idea of the identity of past and present processes. The second volume, edited by Robert Etheridge, is devoted entirely to stratigraphy and historical geology, and Hutton's name does not appear in it (Phillips, 1885b).

Sir Joseph Prestwitch, F.R.S. (1812–1896), professor of geology in Oxford, published his beautifully illustrated two-volume *Geology—Chemical, Physical, and Stratigraphical* in 1886, full of statements about opinions and observations that one usually thinks belong to a later time. In his chapter I of the first volume, entitled "Objects and Methods of Geology," he remarked that

> Towards the end of the last century the structure of the rock-masses of the globe began to be more systematically studied and better understood. Werner, Hutton, Dolomieu, Saussure and others showed that the earth consisted partly of sedimentary stratified rocks, and partly of unstratified rocks. (Prestwitch, 1886, p. 1)

On p. 398 of the same volume, Prestwitch refers to "The well-known experiments of Sir James Hall" having proven that heat and pressure could make marble out of limestone, in agreement with Hutton's views. Sir James' experiments here are referenced not as historical predecessors, but as part of the current argument.

With the revision of Phillips' textbook and the publication of Sir Joseph's, we are slightly more than a decade away from Sir Archibald's 1897 book.

Appendix to Hutton in the British Empire: Hutton in the *Memoirs of the Wernerian Natural History Society*

It would not be without interest to examine the rise of Huttonism in the short-lived journal of the rival camp, the *Memoirs of the Wernerian Natural History Society*, published in seven volumes between the years 1811 and 1838 (the second volume was published in two separate volumes, so the total number of tomes comes to eight). In the earlier years, most of the geological authors of the *Memoirs* were decidedly anti-Hutton, some others simply questioning some of the "Huttonian" conclusions. In v. 1, Robert Jameson published a paper entitled "Mineralogical queries, proposed by Professor Jameson." On p. 123 he refers to Playfair: "Mr. Professor Playfair, in his eloquent work, the 'Illustrations of Huttonian Theory,' maintains that they are primitive: to me they appear to be transition." (The reference here is to the inclined slaty strata near Plymouth that we now know to be of Devonian age; thus, here, Jameson was right). In James Ogilby's paper "On the Transition Greenstone of Fassney" (1811a) Hutton and Playfair are attacked on p. 130, for misidentifying a "sienitic greenstone" (according to Jameson) near Fassneyburn for granite. Here again, since the rocks referred to are essentially alkali basalts and dolerites, in places differentiated creating teschenites and lugarites (Francis, 1991), Ogilby was justified in following Jameson's identification. When one thinks that the main mineral in these rocks is nepheline, which was first identified as a mineral species in 1801 by Haüy, it is easy to forgive the misidentification of the Huttonians, who might have thought the nephelines to be feldspars. But that the rocks lack quartz in the context of granite misidentification was important and that did not escape Ogilby's notice, who asked "We may ask, Where was Granite ever seen without Quartz, and united with hornblende, and in such a situation?" (p. 130).

In his long and entirely descriptive paper entitled "On the mineralogy and local scenery of certain districts in the Highlands of Scotland," the Scottish churchman, mathematician and naturalist Thomas MacKnight (1762–1836) describes the geology of the Glen Tilt with no explicit reference to Hutton, but he is clearly aware of his work, as he presents Hutton's hypothesis in some detail. He disagrees with Hutton's interpretation, because he thinks that the fill of the dykes and the surrounding rocks seem contemporary on the basis of the presence of veins and nodules of feldspar terminating within the dyke fill and the angular masses of hornblende, which, he confidently asserts, "could not possibly have been injected by external force" (1811, p. 367). He does not consider the fact that his observations are equally (actually more) compatible with Hutton's interpretation.

In his "On the veins that occur in the newest flœtz-trap formation of East Lothian," James Ogilby (1811b), wrote that his observations established the universality of this formation as against the claims of the Huttonians:

> I had occasion to make what I conceive to be an important remark in the present state of the science, namely, that much much more than one-half of the extensive tract of country which I explored, was covered with that assemblage of rocks of the flœtz-trap,—a fact which considerably weakens, if not entirely destroys, an objection of the Huttonians against the universality of this formation, from its supposed non-occurrence in this country. It is now not only a fact, that this formation is extensively distributed in this vicinity; but also that a series exists here perfectly analogous to those at Scheibenberg, upon which the Wernerian theory of aqueous deposition was particularly founded. (p. 472)

Here Ogilby unknowingly tries to correlate Carboniferous dolerites in Scotland with Tertiary basalts in Germany; but even in those days, before our present stratigraphy became known, many German geologists such as Johann Karl Wilhelm Voigt (1752–1821), himself a Werner student, would have taken exception to this interpretation by Ogilby.

Thus, in the first volume of the *Memoirs of the Wernerian Natural History Society*, covering the years 1808–1810, Hutton's ideas were indeed acknowledged, but rejected.

The second volume already announces a change in the wind: the Infantry Lieutenant-Colonel Ninian Imrie of Denmuir (ca. 1750–1820), a former correspondent of James Hutton, wrote, in his, "A geological account of the southern district of Stirlingshire, commonly called the Campsie Hills, with a few remarks relative to the two prevailing theories as to geology, and some examples given illustrative of these remarks" (1814) that "Those geologists, who, ... assert, that columnar forms are not produced by volcanic operation, must surely have never visited a real volcanic region. Proofs of this form having taken place in consequence of volcanic agency, are to be seen around the sites of almost every volcano where lava has erupted" (p. 43). Then a little farther ahead: "I must here repeat what I have already observed in another part of this memoir, that those who hold obsidian and pumice not to be of igneous origin, most certainly can never have trod true volcanic ground" (p. 46). However, although he clearly defended Hutton's igneous interpretation of the traps and other volcanic rocks, the realistic soldier warned against too enthusiastic an adherence to any one theory:

> The above observations relative to some of the facts and tenets connected with the two theories in geology, which are so much at variance with each other, are certainly in a great degree foreign to the greater part of the subject which this paper was originally intended to embrace. But having, in the first part of this memoir, carefully avoided mixing *description* with *theory*, and as a phenomenon occurs in the district described, which is one of the controverted points, I have presumed to lay before this Society a few of the facts which relate to both of the theories, and which consist of those alone which have fallen under my own observations when visiting various parts of the globe; but, in doing so, I am afraid, that according to the opinion of some, and perhaps according to my own, I have been led into a digression, which I have made too multifarious, and considerably too long; for which I here beg leave to apologize.
>
> I, however, cannot here refrain from adding one more observation relative to an extreme bad effect which a violent support of theory in geology leads to, and which I am certain, must have been observed and regretted by all impartial geologists. Where the too keen and extravagant support of theory has crept in, and where prejudice has taken root, we must bid adieu to all candid geological description. Not that I here mean to allege, that all describers of geological scenery, under these influences, would intentionally mislead; but many of them stray into error without they themselves being aware of their giving false description. Their minds are warped without their knowing it; and their jaundiced eyes see all objects around them yellow, and they describe accordingly. Reasoning and argument may then be dropt; as the strongly prejudiced mind is not to be convinced, even by a clear demonstration of truth. (p. 48–49)

In the second part of v. 2, the Reverend Dr. James Grierson published his "Mineralogical observations in Galloway" (1818). After having reviewed some of the older "theories of the earth" he wrote:

> Of all the theorists of this sort, or intellectual adventurers, as we may call them, the late Dr. Hutton of Edinburgh, was certainly the boldest, and the celebrated Werner of Freyberg, is, I think, the most modest. The former took him to account, not only for the present appearance of the earth, but for all the revolutions it had undergone in time past, as well as for all these it would undergo in time to come. He assumed as a cause, an agent of whose existence he could give no satisfactory proof, and of whose existence many have thought they could demonstrate the impossibility.
>
> Werner, on the contrary, took induction for his guide. He did not first form a theory, and then examine the mineral kingdom for its confirmation; but he examined this kingdom, and thence drew a theory. (p. 375–376)

Grierson then proceeds to describe how Werner approached geology. Although he admits that Werner's geological theory was wrong, he thinks this arose "not from the false and erroneous principles on which he proceeds, but from the difficulty of the subject, and the yet imperfect state of our induction. His opponents, themselves, have in effect been compelled to admit this, and have acknowledged, that 'though he has not put us right, he has put us on the way of being right,' or has shewn us the proper mode of investigation in this subject,..." (p. 377). But, at the end of his paper, Grierson's conclusion about the age of the Galloway granite is that of Hutton, and not that of Werner: "From the general appearance of the granite, and its relations to the neighboring rocks, it appears to belong to one of the newer formations" (p. 391).

In the third volume, there is no mention of Hutton, but there is a criticism of Cuvier's non-actualistic theory, also by Grierson (1821): "It is his opinion, however, that other causes besides those which we at present observe to operate in nature, must have then existed. But this is evidently giving up the question; it evidently amounts to the assertion that the laws of the physical world were formerly different from what they now are. And if so, then I am afraid we never can understand how these laws operated. It is the same thing as to tell us, that petrifactions were formed, we know not how. For a law of nature different from any that we at present know, is, I apprehend, to all intents and purposes, the same thing to us as no law at all. On this principle, therefore, we must for ever despair of being able to account for petrifactions; or to refer them to any law of nature now known to exist" (p. 165–166).

In the first part of the fourth volume, Hutton's theory begins to be mentioned in favorable terms and is contrasted with Werner's ideas, which are held to be incorrect. Jameson's German Huguenot student Ami (Amédée) Boué (1794–1881) described the geognosy of his country in a letter to his teacher, in which he defended the igneous nature of trap. Jameson, to his credit, published it under the title "On the Geognosy of Germany, with observations on the igneous origin of trap" (Boué, 1822), in which we read the following passages:

> Indeed, I can assert, that the Erzgebirge[111] contains many interesting facts, and distinct appearances, which might

be adduced in support of the Huttonian theory. Were I not apprehensive of being considered rash, I would mention the appearance of porphyry elevating itself, with various true igneous *accidents*[112], from below, and out of the claystone and gneiss rocks. It is also ascertained, I can say, that the beds of porphyry described in the gneiss of that country, are true veins, belonging, as would appear, nearly all to the beginning of the Flœtz Period[113]; in which there yet appears, as in Zinnwald, a granite, with ores of zinc, and other minerals, which have been erroneously named the very oldest ones. That the metalliferous veins are intimately united with the appearances of these crystalline igneous rocks, I cannot doubt; but, as may naturally be supposed, some of the minerals contained in the veins have got there from above, or have been formed in the aqueous way. The great metalliferous deposits in veins appeared to me to form a kind of network. Certainly nearly all that Werner has said about them is true; but his explanation, by filling up entirely from above, is no longer admissible. When we consider such vast bodies of rock impregnated with ores, as the auriferous transition-porphyries of Kremnitz and of Transylvania; and when we reflect that all the rich mines of Hungaria and of Transylvania are in porphyry masses, excepting a single one in greywacke (Verespatak); and when we entertain a strong suspicion that these are igneous products, we will not long be puzzled to comprehend the phenomena. You will probably oppose to the igneous hypothesis the Mercury-mines in the coal-formation; but these also seem to have been produced in the same way, as results from observations I lately made in the Bavarian Rhine provinces. The ores are there contained in small veins in porphyry, or in rocks in contact with such products; and these coaly or arenacenus rocks are almost always indurated or altered in a thousand various ways; but I shall discuss this subject at another time.

In the period subsequent to the Old Red Sandstone, the Basalts have protruded from below, probably at various irregular periods, even before the formation of chalk; for the cones, hills, and veins of basalts, so well known around the Meissner, Eisenach, the Rhonegebirge, and Göttingen, seem to indicate such an age. But this is a point that requires very minute investigation.

Germany possesses a great variety of basaltic deposits, most of which are analogous to those of Ireland, and to the most of those in Scotland. I say most of those in Scotland, because I am now inclined to classify the few hills around Edinburgh, Arthur's Seat, Salisbury-craigs, and such like deposits, with those of Eisenach. These are certainly in Germany the oldest, and they comprehend some conical hills and veins in different parts.

The other basalts in Germany can be divided into those which have been formed under water, like the preceding, and those which have flown [sic] in the open air. The first, like the basalts in Ireland and Scotland, posterior to the chalk formation, form conical or massive hills, a kind of platforms, or high plains very little inclined, and veins or dikes. The cones or hills are principally formed of various porphyritic clinkstones, which take, as in the Mittelgebirge and Rhongebirge, the place of the trachyte of other great deposits. The group of Mezen, in Auvergne, presents the same fact, although it was probably formed above the water. The plateaux, or nappes, present the same variety of basalt and tuffs as Ireland and Scotland, and some points in Auvergne, and also the same calcareous and zeolitic substances.

Basaltic veins are very frequent in Germany, and are found in almost all the formations of that country, with all the *accidents*, as in Scotland. The Quadersandstein of the north of Bohemia, and the Shell Limestone (Muschel-kalkstein) of the western part of Germany, present beautiful examples of these appearances.

The second less numerous class of igneous rocks which appear to have flown [sic] in the open air, occur not only in hills with craters, and with scoriæ, but also in currents. In the south of Germany, in Hungary, and Transylvania, this class is exceedingly well exemplified; for there the trachytes form great and high districts, more or less surrounded by or associated with basalts; for example, near Feldbach, in Styria. In Hungary and Transylvania, they are accompanied also with vitreous rocks, pumice, and great masses of re-agglutinated trachytic, or pumice-rocks, which show, by sometimes containing shells of the Parisian formation, the recent age of these deposits. Such is particularly the nature of these formations in Hungary, where they rest at Chemnitz and Cremnitz upon the transition metalliferous porphyries, and form, as it were, four or five great islands in the middle of that immense basin.

In the other parts of Germany there exist no trachytes, but only basaltic lavas, with scoriæ and craters, or indications of them. Thus there is a very beautiful crater near the Pferde Kopf, in the Rhongebirge; distinct lava-streams are observed in the Vogelsgebirge; at Eger, there are true volcanic scoriæ, and indications of a crater; near Hof, upon the borders of Moravia and Silesia, the Raudenberg is a great heap of red scoriæ, like the Puy de Graveneire, in Auvergne, or in Vivarais; and there is a portion of a crater and small streams of basaltic lava; lastly, even in the Riesengebirge, there is a crater and streams of lava.

I shall now conclude this long letter, by enumerating the characters which appear to me most distinctive of the two kinds of igneous rocks, those formed under, and those formed above the water.

1. The igneous rocks formed under water, at least those posterior to the chalk formation, do not rise into hills of so great a height as those formed above the surface; and, in general, the first class of rocks must have certainly, in all periods, had more difficulty in attaining the same height as the second.

2. The first class produce veins or dikes more easily, and in greater number, than the second.

3. When the first class of rocks form a kind of coulee or stream, these streams seem generally not to unite the length and the small breadth of the streams (coulees) above the water.

4. The rocks of the first class are generally more compact than those of the second.

5. The basalts of the first class are often intimately united with basaltic tuffs, and the porphyries with some kinds of felspathic breccia; an appearance which is almost entirely unknown in the basalts produced above the surface of the water, because in them the small pieces which form the tuffs had been ejected by the volcanoes under the form of rapilli [sic].

6. Rocks with the vitreous character abound much more in the igneous rocks formed above water, than in those formed under it.

7. The igneous rocks formed under water, contain many substances, produced by infiltration, unknown in the other class of rocks, and more frequently also substances produced by sublimation.

8. The basalts formed under water show imbedded, very often, pieces of the neighbouring rock, which are more or less indurated or altered. Beautiful examples of this I observed in the basaltic cone of Dosenberg, near Warburg, where the rock is full of pieces of the Shell Limestone (Muschelkalkstein); and in the small clinkstone-cone near Banow, upon the borders of Hungary and Moravia, the rock contains great and small masses of clay and sandstone, so much indurated and altered, that they are like the rock of Portrush in Ireland.

9. The neighbouring rocks are rarely altered near the lavas: on the other hand, near the basalts formed under water, these same rocks are very often subjected to various indurations, alterations, and penetrations of igneous gaseous matters. (p. 103–108)

This is the first full-fledged explicitly Huttonian manifesto published in the *Memoirs of the Wernerian Natural History Society* under Jameson's editorship and, as I said above, it is to his great credit that he published it, and, it seems, without touching a syllable (he did not even correct some of the slight slips in English of his Huguenot German student). Boué had come some ways toward Hutton since his *Essai Géologique sur l'Écosse* published only two years earlier.

In the part II of the same volume, Jameson gives an interesting example of how Wernerians and Huttonians could disagree on the interpretation of the very same outcrop. In his *Notes on the geognosy of the Crif-Fell, Kirkbean, and the Needle's Eye in Galloway*, Jameson concluded (concerning the inclusions and microdiorite dykes and aplite veins in the Emsian Criffel-Dalbeattie Pluton; Jameson spelled Criffel as Crif-Fell) that:

The rocks of the Needle's Eye and the neighbourhood, afforded to the active and enterprising mind of Sir James Hall proofs in favour of the Huttonian theory of the Earth; to me they were interesting illustrations of the doctrine of contemporaneous formation. (p. 547)

Needless to say, in this instance it was Sir James who was right.

Part I of the volume 5 of the *Memoirs of the Wernerian Natural History Society* has no papers on geology. In Part II there are seven geological papers. The first is a description of a part of the Deccan Traps in the middle part of the vast volcanic plateau south of the Tapi River by H.W. Vaysey (1826a). After having described the plateau and the lie of the tabular rocks, he writes: "A Wernerian would not hesitate in pronouncing them to be of the newest flœtz trap formation; a Huttonian would call them overlying rocks; and a modern geologist would pronounce that they owed their origin to submarine volcanoes" (p. 290). He concludes that the traps are of volcanic origin: "I have numerous collateral proofs of the intrusion of the trap-rocks, in this district, amongst the gneiss, which do not allow me to doubt of their volcanic origin" (p. 296). The second paper in this part is also by Vaysey and is devoted to a short description of the trap rocks in and around Nagpoor. Vaysey wrote (1826b, p. 299): "I shall not here enter into a more minute detail of these appearances, but shall content myself with observing, that the most satisfactory explanation of these phenomena is derived from that theory which ascribes to the trap-rocks an igneous origin, under pressure of a great body of water." Vaysey does not identify here the theory as that of Hutton, but at the end of his paper mentions an experiment he made by fusing a large piece of Seetabuldee basalt. This was clearly inspired by Sir James Hall's experiments as Vaysey's earlier emphasis on great pressure implies.

In the next geological paper, the Reverend Grierson reviews some "theories of the earth" (he begins with Burnet and proceeds through Woodward, Whiston, Leibniz, Descartes, and de Buffon) finding them "purely imaginative" and proposes to move to a "more sober cast" (Grierson, 1826, p. 404). But he finds "some of these, even of these soberer ones, hold opinions little, if at all, less strange than certain of those we have already enumerated." (ibid.) Among those he lists Delamétherie, Dolomieu, and de Marschall. He finally comes to Hutton:

Such was the creed of the late celebrated Dr. Hutton of this place, as illustrated by the late, no less celebrated Professor Playfair,—a gentleman, this last, remarkable for his many and valuable accomplishments, but for none more than for the elegance and felicity with which he always expressed himself when he wrote on subjects of natural science. His master in geology (I mean Hutton) had sublime and extensive views, but he could not clearly exhibit them to the minds of others [114]. No man ever stood more in need of an illustrator, and no man ever found a more admirable one than he did in the person of his pupil. But what proof have we that a central fire exists? And what proof have we that the materials carried down by the rivulets and rivers ultimately deposited in the bottom of the ocean? I can perceive none,—not even of this latter position. Everywhere do we see holms, and carses, and bars, and deltas, thus formed at the mouths of rivers, but no evidence can I perceive that the materials go farther. The first principles, therefore, of his theory are, like those of the others at which we have been looking, in my opinion, purely gratuitous. (p. 406–407)

He then contends that Werner "was the first to introduce into this subject what may be called the truly inductive method" (p. 407). He hastens to add that not everything Werner and his students advanced in the science of mineralogy was perfect. He does not think that Werner's great contribution lies in elucidating how minerals form, but in his methods of recognizing them easily in the field. He then proceeds to a panegyric of Werner's methods. Similar to what we have already seen above in Deluc's case, he does not seem well informed in geology; he even thinks that the term geognosy was introduced by Werner, clearly being unaware of Füchsel's priority ("viri scientia geognostica imbuti," i.e., "inspired by geognostical knowledge": Füchsel, 1761, p. 209; see Gumprecht, 1850). Because of what he thinks to be a more inductive methodology he finds the term geognosy preferable to geology, which he thinks expresses Werner's more "proper" compass:

> Why speculate, and fancy, and suppose, and take for granted, and bring our science into contempt, by what are called Theories of the Earth? Werner, in the comprehensiveness of his mind, seems to have felt this impropriety, and, in order to avoid it, introduced the term Geognosy, by which he meant to distinguish the legitimate and useful department of the science of Mineralogy, or rather, of Geology, from the purely fanciful and useless part of it, which consists in pretending to explain how the Earth was at first formed, and how the successive changes it may have, as a whole, from time to time, undergone, took place. Geology he leaves to express this, which it had too often understood to do (though Geogony is the more appropriate term); and Geognosy is understood by him to indicate the distinguishing, or classifying, on the great scale, the various substances which constitute the exterior parts or crust of the globe, that is, all that portion of it of which we ever can come to the certain knowledge. (p. 409–410).

Grierson ends his paper with a renewed attack on the uselessness of speculating about a central fire in the earth. His was the last paper to offer a partisan defense of Wernerism as against Huttonism in the *Memoirs of the Wernerian Natural History Society*. Later papers used Wernerian stratigraphical terminology indeed, but no longer his theoretical views and the final geological paper before the journal went out of business explicitly expounded Huttonian views.

Neither Blackadder's (1826), nor Drummond's (1826) descriptive papers mention anything theoretical. Witham in his *Notice in regard to the Trap Rocks in the Mountain Districts of the West and North-west of the Counties of York, Durham, Westmoreland, Cumberland, and Northumberland* (1826), regards both Hutton's and Werner's theories as stimulating new observations and hence useful. In his description of the traps he favors neither of the volcanic or the neptunian theories.

Finally, Haidinger (1826), in his paper on the drawing of crystal shapes in true perspective according to Mohs' method, simply mentions the great superiority of Haüy's mineralogy to that of Werner (his p. 486).

The three geological papers (one is actually an article on physical geography) of the volume six have no mention of either Hutton or Werner, or their theories. Only the paper on the fossil quadrupeds found in the Kirkdale cave, written by the polyglot and prolific Reverend George Young (1777–1848), an ardent Deluge theorist (although he had been a favorite student of John Playfair in the University of Edinburgh), offers an alternative Deluge theory to that of Buckland (Young, 1832).

In the seventh and final volume of the *Memoirs of the Wernerian Natural History Society*, there are only two long papers and only one is about geology, describing the Lothians. In that paper, geological theory is entirely Huttonian and only some of Werner's terms such as "Transition" or "Secondary" have remained. Not only the igneous rocks are interpreted as such, but even the structural geology is explained in Huttonian terms: "The junctional appearances exhibited in East Lothian are interesting in the history of Geological Science. From examining them, Hutton, Hall and Playfair formed those opinions which are now invariably held in regard to such phenomena" (Cunningham, 1838, p. 35).

Although the Wernerian Society lasted exactly two decades longer, it held no meetings between 1850 and 1856 and its journal ceased publication in 1838, and geologically, on a Huttonian tone. During its short life from 1811 to 1838, the papers it published evolved from being purely Wernerian to almost purely Huttonian (what survived of Wernerism was only its stratigraphical terminology). The authors of its latter volumes would have been shocked, had they heard that some historians of geology, two hundred years later, would think that Hutton's contributions to geology were limited or that had it not been for Sir Archibald Geikie, Hutton would have been entirely forgotten!

FRANCE

Before Lyell

Nobody outside the British Empire gave such close critical attention before Lyell to what Hutton wrote as the French scientists. Already in *An III* of the French Republic (22 September 1794–22 September 1795), the French geographer and geologist Nicolas Desmarest (1725–1815; Fig. 23) began publishing his four-volume *Géographie Physique* as part of the immense *Encyclopédie Méthodique* (more than 200 volumes published between 1782 and 1832) commenced by the influential French publisher Charles-Joseph Panckoucke (1736–1798), in the second part of the first volume of which (published in 1798; see Taylor, 2013, p. 42, footnote 8) he discussed Hutton's 1788 paper in an article of 50 pages! Regrettably, Desmarest was able to bring his book, which appeared in the form of a dictionary, only to the letter N. A fifth volume was published posthumously in 1828 by the soldier-naturalist-geographer Jean-Baptiste-Geneviève-Marcellin Bory de Saint-Vincent (1778–1846), the physician Guillaume-Tell Doin (1794–1854), and the geographer and naturalist Jean-Jacques-Nicolas Huot (1790–1845) that completed it to Z and added supplements concerning the earlier published items (i.e., supplements to articles covering the letters A through N; Taylor, 2013, p. 40, footnote 2, assures us that it was Huot who did much of the writing for the volume, including the entire "avertissement" and the "supplement"). These supplements are almost exclusively properly geographical except the articles on geological formations and fossils. In the former is the mention that granite is a plutonic rock as much as trap is (Bory de Saint-Vincent et al., 1828, p. 911).

Figure 23. Nicolas Desmarest, engraving by Ambroise Tardieu. Credit: Wellcome Collection. Attribution 4.0 International (CC BY 4.0).

Desmarest had earlier written the article *Géographie physique* in the epoch-making *Encyclopédie* of Diderot and d'Alembert (Desmarest, 1757), which was full of hints concerning his future geological thinking. One telling example is the following sentence:

> Nature is often revealed by a deviation betraying its secret in broad daylight; but one draws advantage of such irregularities only insofar as one is aware of what, in such or such circumstance, is the uniform course of nature, and that one can unravel whether these deviations influence the essence or they are only accessory. (Desmarest, 1757, p. 615)

It is hardly surprising that a man harboring such thoughts would be especially interested in what Hutton had to say. Moreover, Victor Eyles (1970) put forward two observations supporting the conjecture that Desmarest and Hutton knew each other personally, having met in Paris while Hutton was there between 1747 and 1749 and that they remained in contact by mail thereafter. Eyles suspects that Hutton probably sent Desmarest a copy of his abstract of 1785 (see Eyles, 1970, p. xix).

The abstract of Hutton's paper, published in 1785, had already been translated into French, before Desmarest wrote anything about Hutton's views, by the Spanish medical doctor (originally Italian, born in Rome and traveled at the Spanish king's expense in Italy, England, France, Spain, and the Netherlands, for purposes of studying university medical instruction and hospitals: Kenneth L. Taylor, written communication, 29 January 2019), author and translator José Iberti, whose translations played such an important rôle in importing European ideas of the Enlightenment into Spain (see Hutton, 1793; Eyles, 1970, his endnote 16, also mentions this translation)[115]. Iberti added four pages of his own critical comments to the end of his translation (Delamétherie, 1816, p. 280–282, reviews these comments under the title "Système géologique d'Iberti" and shows that Iberti's critique and alternative theory to Hutton's is untenable simply because fluids must maintain an equipotential surface—Delamétherie calls this "lois de l'équilibre des liquids"). Iberti was unhappy with Hutton's theory that heat caused the consolidation of the strata under the sea. He was remarkably well versed in geology and certainly knew his sedimentary rocks: he pointed out that the calcareous and siliceous fossils would have been obliterated by fusion caused by elevated temperatures. He gave petrified wood as an example of silicification of organic substances without fusion. However, Iberti believed that seawater may dissolve all sorts of substances, which may crystallize and some of them may give rise to violent reactions leading to eruptions across the superjacent strata. Iberti thought perhaps metals offered better support for Hutton's fusion theory, but then, he asks, why the metals could not form under the sea, since we do not see the production of metals by means of fire anywhere. He admitted the presence of what he called "subterranean volcanoes" and observed that their products, "lavas, basalts, whinstone, toadstone, traps" do not form regular beds everywhere but exist only where volcanoes are found. Although he agreed that lavas, basalts, whinstones, toadstones, and traps were all volcanic products, he could not understand why such igneous rocks did not everywhere produce the same contact phenomena. Neither was Iberti terribly sympathetic to the idea that magma-driven uplift formed mountains. In this matter his objection proved less well aimed (but, as late as 1886, the American geologist T. Sterry Hunt, had ideas that are surprisingly similar; see below). But it is indeed remarkable that this learned Italian physician found it important enough to translate into French and publish Hutton's abstract in the middle of the *Terreur* in Paris with such prescient criticism!

In *An III*, Desmarest published a much more extensive discussion in the first volume of his *Géographie-Physique* concerning Hutton's work, based on Hutton's 1788 paper. In his preface, Desmarest says that originally he had intended to discuss no "theory of the earth," because he had felt that they "were entirely opposed to the principles of physical geography" but then changed his mind, because he considered them like fables in relation to history and decided to follow "the plan of those writers who commence their history with a brief mention of the times of the heroes." He decided to review only the theories of those writers who had been deceased, with one exception: Peter Simon Pallas (1741–1811), because of "the extent and importance of his observations in a large part of the earth" (Desmarest, *An III* (1794–1795) p. 1). But then, he made yet another exception:

Although in the preceding notices we have confined ourselves to review the works of deceased savants, we believe we must deviate from this position with respect to those of that Scottish savant, because the English naturalists[116] have dealt with them extensively, both as supporters and as opposers. One cannot really regard the new views that Dr. Hutton developed and the new questions that he discussed with indifference. They are of interest only to those who, by their researches, have occupied themselves to enrich the natural history of the globe. (Desmarest, *An III*, p. 732)

What follows in Desmarest's book is a detailed, but critical 50-page review of Hutton's paper, with long quotations out of it. Just like Dr. Iberti, Desmarest cannot reconcile himself to the idea that sedimentary layers were consolidated into rock strata by heat. Neither does he like the idea that granite is a product of solidification out of magma, but he likes the idea that internal processes uplifted mountains and that mountain-building took a very long time.

Desmarest's long and detailed article about Hutton's ideas is important in its emphasis that what Hutton was saying was *new* and worth discussing because of the importance of the unresolved problems he tackled and that they are addressed to serious scientists. Desmarest was critical of substantial parts of Hutton's theory, but nevertheless greeted it as a fresh breath of air.

At about the same time, the geologist and one-time botanist and physiologist Jean-Claude Delamétherie (1743–1817; Fig. 24) published the three-volume first edition of his *Théorie de la Terre* (*An III*: 1795), in which the motor of geological evolution was thought to be crystallization. In the third volume, Delamétherie provided a summary of Hutton's theory (Delamétherie, 1795, p. 400–401) using his words from the 1785 abstract in Iberti's translation (Hutton, 1793) and proceeded to combat it from the viewpoint of his own theory (p. 401–404). He thinks that it had been "proven" that the "primitive" rocks such as granite, porphyry, and gneiss had crystallized out of water. "Neither can it be doubted,", he says, that "the secondary strata such as coals, bitumen, coquinas ... were produced in water." He asks what can be the origin of the internal heat that can fuse all these substances. Finally, he claims that no subterranean explosions are known to have uplifted the mountainous masses. The implication is that he believes he thus falsified Hutton's theory. But, he continues, how to answer Hutton's question as to the solubility of these same substances in water? Here he runs into trouble and his defense is most extraordinary:

Undoubtedly, this is a very difficult question to resolve. But the fact is certain and we cannot doubt it because we cannot assign a cause to it. (Delamétherie, 1795, p. 402)

What he cannot doubt is that all rocks precipitated out of water, because his theory of crystallization demands it. He recognizes two difficulties:

1. What were the solvents during the process of crystallization?
2. Every one of these substances require an immense amount of water to dissolve them.

He concludes at least a hundred times a mass of water is required to dissolve things like quartz, feldspar, tourmaline, hornblende, mica ... Then the question becomes, what happened to that immense quantity of water? He thinks he has a number of satisfactory answers to that question: He repeats that such substances as gypsum, fluors (he means various compounds of fluorine), heavy spars (barite, siderite and calcite are meant), calcareous stones, feldspars, quartz, tourmaline, micas, schorl (black tourmaline and quartz together, in places accompanied by muscovite forming usually vein fillings; the name was known even before 1400 in Saxony, but current usage restricts it to black tourmaline alone) ... were once in a state of dissolution. He says, with emphasis, "here is a certain fact" (p. 404). He then points out that all acids are highly soluble in water and therefore they exist in great quantities in waters. These acids, he says, can dissolve various mineral substances to different degrees and therefore not all such substances needed to be taken into solution at the same time. He identifies the "fixed air" (later shown to be carbon dioxide) as the principal agent creating an acid that dissolves most mineral substances. He therefore concludes that such waters in primitive times may have taken much mineral matter into solution.

The second edition of Delamétherie's book came out in *An V* (1797) and in it he dealt with Hutton's theory almost in the same way as he did in the first, but with some interesting changes: first, he introduces it with a complimentary statement lacking in the first edition:

He [Hutton] published a beautiful work on the theory of the earth, in which he maintains that the exterior layers of the earth have been due to the action of fire. (Delamétherie, 1797, p. 484)

Figure 24. Jean-Claude Delamétherie (from Delamétherie, 1797. v. 1, frontispiece).

Secondly, although he maintains that all rocks crystallized out of water, but, now he says one can recognize on them effects of later heating. He repeats Hutton's question as to how all the mineral substances could have been taken into solution by water. But this time he simply responds by saying that "these are without doubt, difficult questions to answer" (p. 486) without presenting any of the earlier chemical arguments. In his later lectures on geology, which he delivered in Collège de France upon Cuvier's request, he further retreated from his original position but still objected to Hutton, because he could not understand the source of the postulated interior heat of the planet (see below).

In 1784[117], the French geologist Barthélemy Faujas de Saint-Fond (1741–1819; Fig. 25) paid a visit to Hutton in Edinburgh in Hutton's home. Faujas had earlier published his magnificent account of the volcanoes of central France, in which he declared basalt as the primordial magmatic rock (Faujas de Saint-Fond, 1778, p. vj). Faujas greatly enjoyed his conversations with Hutton, which is hardly surprising as his own interpretation of geology was in full accord with Hutton's. He had already reported the occurrence of magmatic dykes in limestones using also the remarkable observations by Horace-Bénédict de Saussure (1740–1799) in the Venetian volcanic province of medial Eocene to Oligocene age in Valdagno, west-northwest of Venice and by Dieudonné Sylvain Guy Tancrède Gratet de Dolomieu (1750–1801) in the Lisbon Volcanic Complex of late Cretaceous to early Eocene age, where magmatic rocks intrude Cretaceous limestones. De Saussure, when he carefully examined the contact between the magmatic and the sedimentary rocks, saw that the latter "had somewhat suffered and in places a little calcinated" (Faujas de Saint-Fond, 1778, p. 333; for Dolomieu's letter to Faujas containing his observations, see p. 446–448). Little wonder that Faujas enjoyed his talk with Hutton[118]:

> Dr. James Hutton is, perhaps, the only individual in Edinburgh who has placed in his cabinet some minerals and a number of agates chiefly found in Scotland; but I observed that he had not been much concerned in collecting the different matrices which contained them and which serve to complement the natural history of these stones. I therefore experienced much more pleasure in conversing with this modest philosopher than in examining his collection, which presented me with nothing new, since I had seen and studied almost all the specimens in his collection upon a large scale and in the places where nature had deposited them.
>
> Dr. Hutton was at this time busily employed in writing a work on the theory of the earth in the tranquility of his cabinet[119] (Faujas de Saint-Fond, 1797, v. 2, p. 266)

Elsewhere in the same work Hutton is again mentioned in relation to his description in his *Theory of the Earth*, of the prismatic ore of Dunbar (Faujas de Saint-Fond, 1797, v. 1, p. 224, footnote 1).

Six years after the publication of his travel book, Faujas published a two-volume essay on geology, the first volume of

Figure 25. Barthélemy Faujas de Saint-Fond (engraved by Ambrose Tardieu from an original communicated by M. Brard).

which came out in 1803 and the second (in two separate volumes) in 1809. In the first volume, Hutton is mentioned among the scientific luminaries of Scotland.

> Scotland seems to be the classical land for a long time, which prides itself in producing the most distinguished men in the letters, in the sciences and in the arts. [William] Cullen [1710–1790], [Joseph] Black [1728–1799], [Alexander] Monro [*secundus*, 1733–1817], [James] Hutton [1726–1797], Hockard[120], [James] Anderson [1739–1808], Benjamin Wat[121] [sic], Lord Dundonald [Archibald Cochrane, 9th Earl of Dundonald; 1748–1831], [Robert] Knox [?–1812] and so many others honour and commend her in medicine, chemistry, anatomy, natural history, mineralogy and useful arts; David Hume [1711–1776], [Adam] Smith [1723–1790], [William] Robertson [1721–1793] have ennobled history, philosophy and literature. (Faujas de Saint-Fond, 1803, p. 24–25)

In the first part of the second volume, Faujas mentions Hutton again in relation to his description of the graphic granite:

> Dr. Hutton, who described and illustrated the graphic granite of Scotland, compares the forms of this latter with the runic characters. (Faujas de Saint-Fond, 1809a, part I, p. 179)

Despite Faujas' great respect for Hutton as a scientist and his knowledge of Hutton's 1788 paper, there is no mention in this book of Hutton's overall theory. The reason is simple: the resolutely empiricist Faujas discusses nowhere in his book any theoretical question, but gives detailed descriptions of fossils, minerals and rocks. Even where he criticizes Werner because of his neptunistic interpretation of the basalts of the High Meissner in the present federal state of Hessen in Germany (see Faust, 1789), he does so by giving a minute description of his own observations there to document their volcanic nature. (Faujas de Saint-Fond, 1803, p. 450–451 and 1809b, part II, p. 417–418, footnote 1).

Hutton's ideas were discussed in France during the last decade of the eighteenth century also by those who did not cite his name. For example, Taylor (1971) made, I think, a convincing case of Dolomieu criticizing Hutton's theory of the earth without citing him, disputing in particular the consolidation of sediments under the sea, the adequacy of the presently acting causes to explain the geological record and the assumption of a very high age for the earth.

Cuvier, in his celebrated "Discours Préliminaire" of the *Recherches sur les Ossemens Fossiles de Quadrupèdes,* reviewed Hutton's theory of the earth among others of a newer date (Cuvier, 1812, p. 30), on the basis of Playfair's book (1802), and asked himself how was it that so many people, observing the same things on the basis of the same principles could produce such contradictory views. His answer was, essentially, that they all had neglected stratigraphy. Werner had made an admirable attempt at doing stratigraphy, but had foundered at the end, because he had neglected the message of the fossils. That is where, Cuvier believed, his contribution came in by showing that the employment of fossils was a great help in correlating strata and establishing a relative chronology. He refused, however, to get involved in etiological arguments. He did think the succession of faunas he established happened because of near-global catastrophes of very short duration simply on the nature of the preserved fossils (for example, mammoth cadavers with still edible flesh and fur!), but only twice expressed his opinion informally to his friends that either crustal ruptures or a cometary fly-by of the kind Laplace had imagined could have caused the sort of catastrophes he had in mind[122]. Nobody at the time could possibly have imagined how close the great scientist had come to our modern interpretations (regrettably, for reasons we now consider incorrect)!

The first separate edition of the *Discours* in French appeared in 1825 (called the third French edition, because the *Ossemens Fossiles* had a second, enlarged edition in 1821), after it had already been translated into English and German and Cuvier gave in to the demands that a separate edition should also be available in French. In his preface ("Avertissement") he says he took into account the developments since the earlier editions (as parts of the *Recherches*), but he did so only those in zoology and paleontology, not in general geology. What he said about Hutton in 1812, appeared unchanged in the 1825 book (see Cuvier, 1825, p. 51). The staunchly positivist Cuvier was not interested in theoretical geology.

The Benedictine professor of literature in the former military school of Sorèze in southwestern France (which had replaced the old Benedictine Abbey and the College there in 1776), who later became the director of the École Normale and an officer of the Imperial University, César-Auguste Basset (1760–1828), translated Playfair's *Illustrations of the Huttonian Theory of the Earth* (1802), together with the Edinburgh surgeon and chemist John Murray's (1778–1820) *Comparative View of the Huttonian and Neptunian Systems of Geology in Answer to the Illustrations of the Huttonian Theory of the Earth by Professor Playfair* (1802), in 1815, because, he thought, at this time, the three Scottish authors appeared to be the most successful in their pursuit of geology. He dedicated his translation to the students of the famous School of Mines, saying

> As the science of geology is closely related to all the mathematical and physical knowledge forming the subject of your studies, I believe that I would be useful to you by translating into our own tongue two new theories of the earth that are based on researches and observations made with great care. (Basset, 1815, p. iii)

Basset here points out that he is translating two new theories. But in his own introduction he expands this point, first by discussing what is meant by the terms hypothesis, system, and theory. He says that with hypothesis alone not much can be accomplished and that systems are often portrayed by their critics as "dreams." Basset does not think so. He points out how complex the natural world is and that systems constructed by authors function as guides for further research. He lists the systems of Linné, de Buffon, Jussieu[123], Lagrange, Delaplace [sic], Lavoisier, Berthollet [sic], Cuvier, Haüy, Werner, and Klaproth and points out that their errors had led to questioning by their successors and caused progress. Therefore, he hopes that by presenting Hutton's views and those of one contemporary critic, he would encourage the students to think for themselves on geology by weighing the available evidence. Although he says he will not disclose his own opinion, not even in the notes he appended to his translations, he betrays his sympathies by adding, at the end of his book, an account of his trip to the island of Staffa in Scotland and describing the magnificent basaltic columns there. The only plates his book has are views from Staffa and Hutton's septarian nodule, as Murray's book has no figures.

In 1800, Cuvier was appointed to the chair of Natural History in the Collège de France replacing de Buffon's close collaborator Louis Jean-Marie Daubenton (1716–1799). Jean-Claude Delamétherie had been hoping to obtain that position. Cuvier, already a very busy man, realized that he could not cover the entire spectrum of natural history, asked Delamétherie to give the geology lectures and gave him first the one-third, but then the two-thirds of his remuneration in the Collège. To this generous act of Cuvier do we owe Delamétherie's book, entitled *Leçons de Géologie données au Collège de France* (=Lessons on Geology Given at the Collège de France), comprising three volumes like the first edition of his earlier *Théorie de la Terre* and containing the material of his lectures. In the third volume he reviews Hutton's theory of the earth and the experiments undertaken by Sir James Hall in support of it (Delamétherie, 1816, p. 177–185). Delamétherie briefly presents Hutton's theory on p. 177, then reviews in great detail Hall's experiments. On p. 184–185, he gives his own assessment. His difficulty in understanding the source of the heat postulated by Hutton and Hall continues undiminished. He says that the layered rocks are all assumed to have been laid down in the sea, but points to those containing plant fossils showing that they were terrestrial deposits. Delamétherie's greatest objection pertained to Hall's theory of subterranean currents creating uplifts and depressions. He asked, where a rock layer five hundred *toises* (one *toise* was 1.949 m during the eighteenth century France and exactly 2 m between 1812 and 1840) thick could be found to be shaped into mountains and valleys.

The French traveler Louis Simond (1767–1831; Fig. 26) from Lyon, who married first an English woman and then a Genevan lady and spent 20 years in America (mainly to escape the horrors of the French Revolution and its aftermath) finally ending up becoming a Genevan citizen (the Louis-Simond Square in that city is named after him), published the journals of a trip to Great Britain in 1815. In the second volume, there is a long description of Hutton's theory, mainly on the basis of Playfair's *Illustrations*. Simond's description and defense of Hutton's theory is so detailed, so learned, and so ingenious, in places going beyond Playfair and even criticizing Huttonians for not realizing the full potential of Hutton's theory, that I quote it below in full. Moreover, Simond's book was translated by himself into his mother tongue and published in 1816 in Paris by Treuttel under the title *Voyage d'un Français en Angleterre Pendant les Années 1810 et 1811: Avec des Observations sur l'État Politique et Moral, les Arts et la Littérature de ce Pays, et sur les Mœurs et les Usages de ses Habitans* and into German by Ludwig Schlosser, published in 1818 in Leipzig by Brockhaus as *Reise eines Gallo-Amerikaners (m. Simond's) durch Großbritannien in den Jahren 1810–1811*. Simond's book must have contributed immensely to the spread of Hutton's ideas. In fact, I feel almost certain that its French version must have inspired Élie de Beaumont to his tectonic ideas (see below).

Figure 26. Louis Simond. Public domain (PD-US-expired).

Edinburgh, January 1, 1811.

From metaphysics, which are rather out of fashion, the learned Scotch have lately passed to geology,—from mind to stones,—subjects, perhaps, alike impenetrable. The system of the earth, which made its appearance here some years ago, under the title of Illustrations of the Huttonion Theory of the Earth, by John Playfair, deserves, in many respects, the high celebrity it soon acquired. An inventor, who did not possess the talent of writing [Hutton is here meant], could not be happier in a commentator. Mr Playfair writes as well as Buffon, but with more solidity, more prudence, and more modesty [considering that Montesquieu, Voltaire, Rousseau and Buffon were regarded as the four jewels of the French prose during the Enlightenment, this is very high praise coming from a Frenchman]. In a country where eloquence is scarcely deemed admissible in scientific subjects, his passes for simplicity.

Hutton's theory may be unknown to the generality of foreign readers, and the most probable of any existing explanation of the awful revolutions our world has evidently undergone, must be of sufficient interest to its inhabitants, to encourage an expectation that the short and simple account I shall give of it to the unlearned may be found acceptable.

The solid crust of our globe is formed of rocks in great irregular masses, as granite, and of hard substances, arranged in beds parallel to each other. The latter are composed of fragments of other rocks more or less broken and attenuated, united by a common cement, as sandstone, for instance,—of hardened earths, or of calcareous substances, containing often shells and bones, as well as remains of known and unknown plants and fruits. Fossile [sic] coals, with indications of vegetable substances in their formation, form some of these beds; and, finally, sea salt is found also in alternate and parallel strata. These, and many other appearances, afford irresistible evidences of the mode of formation of stratified rocks. Rains and torrents, frost and moisture, are constantly wasting the surface of the continents to an extent, proved, by incontrovertible facts, to be wonderfully great. Mr Playfair very justly compares such insulated mountains as shewn, by the undisturbed horizontally of their strata, that they retain their original level, to the pillars of earth which workmen leave behind them, to afford a measure of the whole quantity of earth which they have removed. The apparent slowness of the progress affords no presumption against its reality, and only marks the comparative evanescence of our own duration. The materials of the surface of continents, and of their shores, are thus incessantly washed away into the ocean, and form in the undisturbed repose of its unfathomed depths, strata similar to those of a prior formation which we see around us. The various theories of the earth agree nearly in this respect. The strata, although parallel to each other, are scarcely ever found to preserve the level of their original formation, but are more or less inclined to the horizon,—are broken, and sometimes bent,—every appearance indicating the action of an irresistible force acting from below, and capable of disturbing and changing the whole level of the surface of the globe. These changes may have been slow or local. There are numerous instances of land invaded by the sea, and of sea receding from the land, which suppose either an elevation or depression of the local level. They may have been sudden and general, or extend to great portions of the globe; and the universal traditions of deluge seem to refer to catastrophes of this sort.

Thus far Hutton and his disciples, called Plutonists, do not differ altogether from the disciples of Werner or neptunists; for the latter suppose, in some limited degree, the breaking and sinking of the external crust of our globe into certain internal caverns, explaining by that means the derangements, and inequalities of the strata. But the Huttonians, pursuing the investigation of terrestrial appearances, say, that although the agency of water might be sufficient for the formation of the strata, yet it could not have indurated them into rocks, and still less bent, broken, and overturned them; and that it is inadequate to explain many other phenomena. Some other cause must have joined in the operation, and it can be no other than fire, at least an internal heat, disengaged by causes far from inexplicable, at certain periods, generally or locally, which heat they suppose capable of producing a fusion of the whole or of certain parts of the internal substance of the globe at an unknown depth, into which the continents, or elevated parts of the outward crust, pressed down by their own weight, would sink, while the lower and thinner parts, forming the bottom of the seas, would be buoyed up, and, discharging the weight of the superincumbent waters into their new bed, would be suddenly elevated to the original height of the old continents. They suppose that the crust thus elevated, breaking and opening in some parts, let the fused or softened matter force its passage upward, which, hardening as it cooled, formed the highest ridges of granitic and other mountains, called primary, although in some respects secondary. The following figure will render the relative situation of both stratified and erupted rocks more intelligible [Fig. 27 herein]:

(A) Represents rocks in parallel strata, forming the plain or level country; broken and turned up at the base of its highest mountains[124]. (B) Granitic erupted mass, forming the highest mountains, as well as the base of the stratified rocks. (D) Fragments of the stratified rocks found occasionally on the highest granitic summits, as if carried up, at the time of their eruption through the strata. (E) Inferior mountains over which the strata bend, without eruption of the granitic matter. (F) Clefts through the strata, produced by their violent bending; and it is extremely remarkable, that such of these clefts as have their openings downwards, are filled by a continuation of the granitic matter, as injected in its liquid state, or with metallic substances, forming the veins of mines, which are generally found in such clefts, diminishing upwards, inclined to the horizon, across, and never in the direction of the strata; sometimes disposed in steps, (G) and, where there has been a sliding of the strata one against the other, the sections on each side of the vein do not correspond, as in (H; [not indicated on the figure, evidently as an oversight]). The miners often find in these clefts small fragments, or even large blocks, of a nature totally different from the strata into which they appear to have penetrated from below.

This figure is not in the work of Mr Playfair; I have introduced it in order to facilitate the explanation, but without pretending that the relative order of the different substances is generally found so exactly defined in nature.

Dr Hutton guessed at a new principle, established since by experiments. The effect of compression on substances exposed to the action of heat, answering beforehand the objections he anticipated. He conjectured, that the strata of calcareous substances, shells and madrepores for instance, lying under a great depth of sea or land, and exposed to the action of heat, instead of losing their carbonic gas, and being calcined into lime, would fuse, and, in cooling, would crystallize into marble and other calcareous rocks, or form nodules and veins of spar insulated into other rocks, preserving not only the impression of plants and animals, mixed with the strata, but frequently the substance itself, which could not be volatilized. He conjectured that sea-salt and other substances penetrated and dissolved by heat, under the same circumstances of great superincumbent weight, would likewise undergo, in their respective prisons, a local decomposition and new combination, without loss of any of their elements.

Dr Hutton was asked, what sort of thing this internal fire could be?—how it was kindled,—and how supported?—whether it burned always, or was lighted and went out, and burned again, as he happened to want it, to bake his remoulded worlds ? He might have answered, and probably did answer, that the action of fire is quite as distinguishable as that of water on the face of our globe;—that if the existence of a subterranean fire is difficult to understand, that of the ocean on the tops of high mountains, where it has deposited entire strata of shells and marine plants, 15,000 feet above its present level [at this point, Simond introduces the following footnote: "Mount Rosa in the north of Italy, and in the Andes"], is no less so;—that, uncertain as the respective modes of action of the two agents certainly are, that action is not the less evident; and that we are not to reject what is known, merely because more is not known.

Some conception of an internal fire under large sections of the crust of our globe, or under the whole of it, may be formed, if we consider that a very analogous phenomenon is continually going on under our eyes. The volcanoes of Etna and Vesuvius have simultaneous eruptions, and consequently internal communications ; moreover, this communication seems to extend to Hecla. The tremendous eruption of 1783, which shook the south of Italy to its foundations,—whole towns and villages being swallowed up in yawning gulfs, suddenly opening in the earth,—was accompanied with similar convulsions in Iceland. Its volcanoes vomited lava and ashes in unprecedented quantities; islands arose from the sea, forty miles from the coasts; and for three successive years the inhabitants did not see the sun [at this point Simond introduces the following footnote, referring to the famous Laki eruption without naming it: "This

Figure 27. Simond's (1815) rendering of the uplift theory by Hutton. This is the first pictorial representation of Hutton's theory of mountain uplift that I am aware of: "They [Hutton and his supporters] suppose that the crust thus elevated, breaking and opening in some parts, let the fused or softened matter force its passage upward, which, hardening as it cooled, formed the highest ridges of granitic and other mountains, called primary, although in some respects secondary. The following figure will render the relative situation of both stratified and erupted rocks more intelligible."

great eruption of 1783, was attended with a haziness of the atmosphere, very perceivable over the greatest part of Europe, of which the writer retains a distinct recollection"]. At the time of the memorable earthquake of Lisbon in 1755, the lakes of Scotland, Loch-Ness in particular [Simond's footnote here: "Pennant and Gilpin. The latter mentions the circumstance of a boat in Loch Tay being thrown up twenty fathoms above the usual level of the water"[125]] were strangely agitated; pouring their waters alternately from one extremity to the other, as if the earth under them had changed its level. Many more instances of simultaneous eruptions and earthquakes, in places very distant from each other, might be adduced to prove internal communications; and when it is considered that there are at present, according to Werner, [Simond's footnote here: "Professor Jameson's Geognosy"] 193 volcanoes known to be in activity on the surface of the earth, besides a much greater number extinguished, or dormant, it seems as if there was not such a distance between this and a universal conflagration, that the one might not enable us to conceive the possibility of the other [Simond here introduces another footnote and informs his readers of the famous 1811 and 1812 New Madrid earthquakes in the United States: 16 December 1811, 23 January 1812, and 7 February 1812; their moment magnitudes may have been anywhere between 7.0 and 8.6, although the lower magnitudes are now preferred: "Since this was written all that part of North America, through which the Mississippi flows, has been convulsed by earthquakes, felt simultaneously in the West Indies; a tract of near 3000 miles. These earthquakes recurred with little intermission during several months, till a volcanic eruption in one of the West India islands relieved the internal dilatation."]

Notwithstanding what I have said of volcanoes, it is but fair to state, that the Huttonians, in the plenitude of their faith in a central fire, seek no assistance from such skin-deep means as these ; and they are even anxious to distinguish between the productions of common volcanoes,—lava,—and that of their volcano, *par excellence*, viz. granite, and all the unstratified rocks. Of all these, the rocks called in Scotland whin, are most like lava. They are a sort of basalt in great masses, often flattened at top, with perpendicular sides, in the terrace shape, and frequent indications of prismatic pillars. The environs of Edinburgh, as I have observed before, abound with abrupt protuberances of this sort of rock. That of Calton Hill [footnote by Simond: "Analysed by Sir James Hall."] presents, in its composition, a striking resemblance to the lava of Etna, with a single point of difference still more remarkable;—in the whin or basalt, the calcareous fragments have retained their carbonic gas, and are crystallized into spar; while in the lava they have lost it, and are calcined into lime. This fact affords undoubtedly a most curious confirmation of the Huttonian theory. The great depth, and mass, and consequent pressure, having, in one case, prevented the escape of the carbonic gas, and *vice versa*,

Whin-rock	Lava
Contains Silex - - 50 parts	Silex - - 51 parts
Argil - - 18.50	Argil - - 19
Oxide of iron, 16.75	Oxide of iron 14.50
Spar - - 3	Lime - - 9.50
Water – 5	Soda - - 4
Muriatic acid 1	Muriatic acid 1
98.25	99

The fracture of both these substances is of a bluish or greenish black, strewed with light specks. Both are decomposed by long continued contact with the atmosphere.

Unfriendly critics chose to understand the result of the system to be, a constant increase of the dimensions of the globe, puffed up by the repeated action of heat, without ever subsiding; and they expressed a fear that the earth, thus distended, must at last come in contact with the moon, and derange the system of the universe![126] It is to the Edinburgh Review of October 1802, that the Huttonians are indebted for this gratuitous supposition. Mr Playfair does not say anywhere, I think, that the parts raised never subside again. His book did not suggest the idea to me. I did not understand that the general level of the surface of the globe was permanently elevated, or its dimensions at all increased by the Huttonian process, continued ever so long. The greatest depth of the sea, according to La Place, is eleven miles, and the greatest height of mountains three or four miles; therefore the extreme points of the inequalities of the solid surface of our globe may be estimated at fourteen or fifteen miles [footnote by Simond: "There is a very extraordinary instance mentioned by Pallas, of the depth of stratified formation being traced to the incredible depth of 61 perpendicular miles!"] a range of hills, on the south east side of the Tauride, which is cut down perpendicularly towards the sea, and offers a complete section of parallel beds, inclined at an angle of 45° to the horizon, and 80 miles in length, as regular as the leaves of a book. Allowing for the slope, this shews as much of the formation of the strata as if a shaft of 61 miles was sunk perpendicularly into the earth. Mr Playfair seems, however, to suppose that there had been some considerable shifting or sliding of the strata, one against the other, unperceived by Pallas, which would alter the calculation entirely[127]; and, upon the whole, the fact is too improbable to be admitted. What wearing of continents could there have been, capable of supplying the materials of such a depth of strata formed at the bottom of the sea, and what prodigious depth does it not suppose to the sea? The difficulty, however, is not peculiar to the Huttonian theory, but applies equally to all those which suppose the agency of water in the formation of the strata. The separation of the different materials in distinct beds, instead of being confusedly mixed together, or arranged entirely according to their respective weight, seems to be a very great difficulty, also common to all these water theories. Its diameter being 9000 miles, these greatest inequalities are 1-600th of that diameter, which is probably less than the wrinkles on the thin skin of the smoothest orange. Now, let us suppose that orange subject to an internal heat, capable of liquefying its pulp, and very considerably softening and distending the skin, and communicating to the whole heated body considerable agitation, and some sort of boiling up motion. Suppose, after a while, the cause of the internal heat, whatever it might be, subsiding, and the pulp rapidly cooling and hardening, will not the shrunk skin be found to have contracted new wrinkles and inequalities, greater perhaps, than those it had before, although not the same;—parts that were high having sunk in, and others that were low having been raised. It would, no doubt, be very difficult to understand how the orange came to be so hot all at once, to melt and boil up of itself in that manner ; but, suppose it to have been an orange so strangely constituted, that it had been in the habit of emitting at times, from 193 little openings on its surface, streams of fire and brimstone, then the great fire would undoubtedly occasion much less surprise; very little more, indeed, than its diminutive prototypes[128].

It must be acknowledged that the Huttonian theory, ingenious as it is, rested in a great degree on assumptions and conjectures. It was a fine building on slight foundations; but a patient investigator undertook to strengthen it *sous autre*, and has added several stout props and abutments. The theory owes as much to the very important series of experiments of Sir James Hall as to the inventor himself; and it will be a consequence of this, as of all other systems, that, whether they stand or fall, they promote experiments and researches, and leave after them an invaluable treasure of facts. Sir James Hall found that fused rocks, instead of passing to the state of glass by rapid cooling, become rocks again, and precisely what they were before, when cooled slowly; very large masses must be a long time in cooling, therefore such fused masses as were thrown up from the interior of the earth became rocks again, and not glass. It is indeed true that he has only thus reproduced whin, a more homogeneous substance than granite. Should he succeed in making granite, incredulity must yield to the miracle [Footnote by Simond: "Sir James Hall has approached this miracle. Felspar [sic] and quartz, reduced to a powder and mixed, were fused together, the one serving as a flux to the other; but the two substances crystallized distinctly in cooling, and the crystals were closely joined, and set one within another, as in granite."] Granite, however, is far from forming exclusively the substance of the highest mountains. The Cordelieras [sic] of South America, which are the highest of all mountains, and some other very high mountains in the Sandwich Islands [i.e., the Hawaiian islands[129]], are mostly, as I understood Professor Jameson to say, of the clink-stone, which is nearly similar to the Scotch whin or basalt.

Sir James Hall obtained a still more important result, by melting calcareous substances. What neither the fire of volcanoes, nor the burning lens could do, has been effected by a very moderate degree of heat under great compression. This chemist made marble with shells, and actually did what Hutton had said could be done!

The Huttonians, not contented with the degree of probability which belongs of right to their system, see everywhere dikes and junctions;—that is to say, the places where their central lava, tearing up the solid pavement of the earth, has insinuated itself among the strata, and turned up their broken edges in the manner I have endeavoured to explain in the preceding figure, leaving indications of heat and calcination in those parts of the strata nearest to the fiery injections [Simond's footnote: "Mr Allan of Edinburgh, a learned amateur of mineralogy, and who has formed a very valuable collection of specimens, had the goodness to shew me, among other instances of this kind, nodules of flint decomposed into red earth,—where they happened to touch or come near whin dikes, or veins of erupted matter in fusion, while, a few inches farther, the flint had undergone no such decomposition. Beds of coal thus traversed by veins of whin exhibit, in the adjacent parts, the appearance and properties of coke or charcoal."] Some of these phenomena are satisfactory, but the greatest part require the eye of faith to discover in them what they are supposed to indicate. The great rocky mass of Salisbury Craig, close to Edinburgh, presents some appearances certainly less inexplicable by the Huttonian theory than by any other. The general form of this mass is that of a stupendous terrace, the top of which dips towards the north-east, presenting to the west a perpendicular face about 800 feet high, one-half of which is masked by an accumulation of earth and stony fragments, on a very steep declivity. This [Fig. 28 herein] is the appearance a part of this face of rock presents[130].

(A) Perpendicular whin or basalt of a greenish black, said to be a porphyritic aggregate of hornblend and feldspar. (B) Thin parallel strata of indurated, and apparently baked clay, interrupted in (C) by the basaltic mass which has penetrated between and round the broken and disturbed edges of the strata, which it seems to have split, and penetrated as a wedge. (D) Mass of basalt *en boule*, which has made its way through the strata of indurated clay. (E) Parallel strata of the same indurated clay and sand-stone, about twenty feet high, diminishing towards the top, and in (F) leaving the surface of the basalt covered only by the mould produced by its decomposition. There is, on the perpendicular face of Arthur's Seat, a very large fragment, fifteen or sixteen feet in length, of the same sand-stone, forming the stratum at the base, which seems to have been carried up the ascending mass of basalt.

Dr Hutton, as we have seen, supposes the action of two opposite principles, in his theory of the earth; one of destruction, and the other of regeneration, like the good and the bad principles of the Persians. But the entire destruction and regeneration are the extreme terms of his theory, and do not form a necessary condition of it. The internal fusion may take place before the old continents are entirely worn down;—raising or sinking old and new formations indifferently;—crossing, and mixing them, and throwing on the details of the mineral world an appearance of disorder, confusion, and want of design, quite opposite to the general character of all the other works of nature. This theory, however, far from suffering from these apparent irregularities, is confirmed by them; they are the signs of the very revolutions it supposes,—world after world succeeding each other, and a circulation of ruins. The imagination stands appalled on the brink of this abyss of time, where human reason dares to lead us!

The remains of plants and animals, so profusely scattered over regions quite foreign to their species, seem to point out, still more forcibly than the appearances of the mineral world do, some sudden, general, and tremendous revolution of our globe. Plants and fruits of India found in France and all over Europe; skeletons of crocodiles in England,—of elephants in Siberia; whole islands composed of the ivory of their teeth under the pole; the entire carcass of a rhinoceros, in regions of perpetual frost [Footnote by Simond: "The head of this rhinoceros was carried by Pallas to the museum of St Petersburg, where it is preserved. The accidental thawing of the

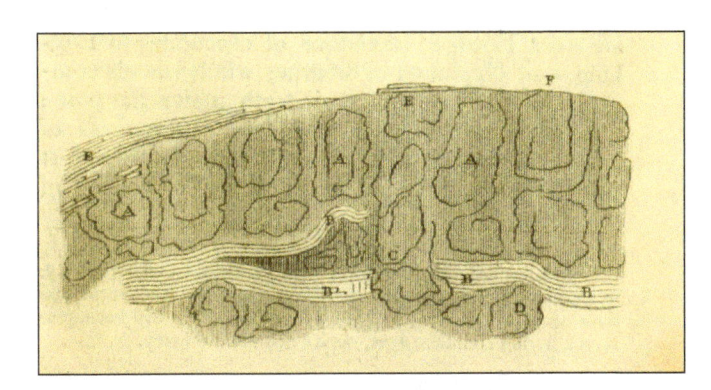

Figure 28. A sketch of an igneous intrusion disrupting strata at the Salisbury Crags near Edinburgh by Simond. Compare with John Clerk of Eldin's figure of the same locality drawn upon Hutton's instruction in Craig et al. (1978, figure 4).

frozen earth, in which the carcass had been preserved during countless ages, occasioned the discovery."] still covered with the greatest part of its hide; and many other examples of this sort speak a language which it is impossible to misunderstand. These animals never could have lived where their remains are now lying; the plants still less, if possible. These wonderfully curious and interesting facts are accounted for in no other way, with half so much probability, as by one of the overwhelming floods of the Huttonian theory, for which a very inconsiderable change of level would suffice. The ocean, pouring from its heaving bed over the sinking land,—tearing up mountains by the roots,—furrowing the strata into profound vallies [sic], and scattering the ponderous materials of the earth, like chaff in the wind!—plants and animals, yielding without resistance, would be swept off at the first onset, with the first terrible wave, to the extremities of the earth; from the torrid zone to the plains of Siberia, or of North America,—and, whirling in heaps, would fill holes and caverns [Simond's footnote: "A very remarkable collection of bones is found in the caves of Bayreuth, in Franconia, of vast size, and mostly of carnivorous animals, and having very little affinity to any now known. Incredible quantities of bones, the broken and confused relics of various animals, concreted with fragments of marble, are found in various parts of the coasts of the Mediterranean; leaving it doubtful, says Mr Playfair himself (page 459), whether they are the work of successive ages, or of some sudden catastrophe, that has assembled in one place, and overwhelmed with immediate destruction, a vast multitude of the inhabitants of the earth."[131]] or remain scattered among the earthy sediment subsiding in strata.

The commentator of Hutton has not made use, I think, of this obvious means of accounting for the strange situation in which the wrecks of the organized creation are found; he seems even to have adopted the unreasonable, and, I must say, extravagant idea, that elephants and rhinoceroses, lions, tigers, and crocodiles, might formerly have lived under the pole, where, no doubt, their native groves of palms and mangoes flourished likewise! I take some merit to myself for having been able to point out to the Huttonians one of their own overlooked resources, diligent as they have obviously been in mustering all their strength. Thus unexpectedly associated to the fame of the system, I shall, of course, quite believe in it myself; and really it appears to me already infinitely superior to the water-crystallization system of the German geologists. Dr Hutton admitted the agency of water in all the extent warranted by facts and experiments; but thought heat, acting under particular circumstances of pressure and time, indispensable to account for the whole of geological appearances. He thought its effects plainly discernible to the eye in many instances, and its presence in the interior of the globe explicable by existing phenomena. This system has also the great merit of accounting satisfactorily for the oblate figure of the earth. That this hard and unyielding mass should have precisely such a form as a liquid under similar circumstances would assume, was a fact so very remarkable in itself, that Buffon and Werner supposed it must have been in a state of fluidity, the one by fire, the other by water. Hutton, in this, as in other instances, retains the advantages of both their systems, without the same difficulties.

Werner's nomenclature is founded on the mere external appearances of the substances, without any regard to their composition. For instance, it places sapphire in the flint genus, although it contains 98/100 of alumina,—and opal in the clay genus, although it contains 98/100 silica, or matter of flints [Simond gives here a reference to Jameson's *Mineralogy*].

Mineralogy, in its present state, is really a very barren and uninviting science. We have names of substances; but as to their relative situation, and other facts leading to the true theory of their formation, contrary assertions are brought forward and denied with equal positiveness on either side:

"And all a rhetorician's rules
Teach nothing but to name his tools."

Werner and Hutton differ on a most material point of fact. The former insists that the wide extremity, or openings, of mineral veins, are all turned upwards, and the latter all downwards. They agree that these veins are accidental breaks or rents through the strata, and nearly perpendicular to them,—clefts open at one end, and closed at the other; but it suits one of the systems to have the opening below, in order to administer with convenience their hot subterranean injection while the other system requires the mouths to be turned up, ready to take in a certain solution of minerals in water;—and the rocks seem willing to accommodate both parties. The situation of Werner is, however, much the most critical; for a single vein filled from below overturns his theory irrevocably; while every vein filled from above, save one, answers every purpose of Hutton. Therefore the Wernerians spare no pains to maintain their ground; and being better stored with mineralogical details, they come forward in great strength, and furnish abundance of facts. They tell you, for instance, of an enormous cleft in a schistous mountain in Germany, opened upward of course, and filled with wakke; in this wakke, at the depth of 150 fathoms, trees are found, with their bark, their branches, and even their leaves, in a state of half-petrifaction. Wakke is a sort of argillaceous basalt, and precisely one of those substances the Huttonians are in the habit of injecting from below. The latter may say, there has been some eruption of our wakke in a state of fusion, and the overflowing stream meeting your cleft, fell into it. But the trees—would they not have been charred instead of petrified? O! the trees might be petrified already, before the introduction of the melted wakke. Well, but the leaves—how could the leaves, petrified or not, resist the shock of such a tremendous cataract of melted mineral? From this there is but one escape, which is, to question the fact, or hint at exaggerations. But the Wernerians are inexhaustible; and if this does not do, they are ready to produce hundreds of other facts, still stronger, and which you cannot reject absolutely, without a previous examination of the mines of Germany, from which they were obtained. This puts me in mind of a story-teller, who used to relate very strange anecdotes. At the first intimation of doubt, or incredulous look, he never failed to add some new circumstance still more wonderful, in order to make the first appear less improbable; and used to introduce this corroboration by—*et même*, &c,—and his hearers could always make him say "*et même*" at pleasure. (Simond, 1815, p. 1–18)

In 1819 the two-volume textbook by the mining engineer and geologist Jean-François d'Aubuisson de Voisins (1769–1841; Fig. 29) came out as one of the earliest textbooks of the subject in France, "one of the best general treatises on Geognosy extant" (Scrope, 1827, p. 38). D'Aubuisson started life as a law student in

Figure 29. Jean-François d'Aubuisson de Voisins (portrait in the Musée Paul Dupuy de Toulouse, document 57.44.1295; copied from Jean Gaudant, 2013).

the military school of Sorèze, but then wished to become an artillery officer. The Revolution intervened and the aristocrat d'Aubuisson found himself in exile as a member of the small counter-revolutionary army founded by the Prince Louis V Joseph de Bourbon-Condé (1736–1818). The army existed between 1791 and 1801, but d'Aubuisson left the army four years before its dissolution to study under Werner in Freiberg between the years 1797 and 1802. He returned to France in 1805 and occupied himself mainly with applied geological problems. Although d'Aubuisson never had a regular teaching position he wrote his book to teach what he had learnt from Werner in Freiberg and whatever had happened since he left Werner's school. What happened later was his complete conversion from Werner's neptunist teaching to those of the vulcanists as a consequence of a tour he took through Auvergne, Velay and Vivarais:

> ...a tour the result of which is well known to have been his speedy conversion from a thorough belief in the Neptunian origin of Basalt and the other members of the Flœtz Trap formation, to an entire conviction of their having flowed in a state of igneous liquefaction, and their complete identity with volcanic lavas; a conclusion to which indeed no one could fail to arrive who would pursue the same straight road to the truth as M. D'Aubuisson. (Scrope, 1827, p. 38)

D'Aubuisson wished to describe geognosy in such a way that even the followers of other systems, "such as Hutton's or Herschell's[132] [sic]" (p. xviij) would need to alter only a few paragraphs to adapt what he wrote to their books. In his very short review of the history of geology, he wrote:

> While among us geognosy has been making such progress, one saw, with some surprise, that it stagnated in England, where there are so many men of distinguished merit who has contributed in recent times to the advancement of physical sciences. But then it has made such rapid progress to bring it to the same level with Germany and France, which now promises the science the most positive success—at the end of the last century, Hutton in Edinburgh published a geological system entirely different from those hitherto appeared. Supported and commented upon by Mr. Playfair, distinguished as mathematician and author, it is generally accepted in Scotland and even in England. The beautiful experiments of Hall on the effects of heat on objects under great pressure seem to give it a new degree of probability and were greatly in its favour. (d'Aubuisson de Voisins, 1819, p. xxii–xxiij)

In the main body of the text in volume I, on p. 421–422, d'Aubuisson discusses what he calls Hutton's theory of geogeny and gives a competent summary of it, adding that its popularity is due to Playfair's eloquence and that it is largely accepted in England.[133] At the end he says that Hutton's repeated cycles of destruction and construction will continue until it pleases God to put an end to it. Had it not pleased God that such an order of things ought to have had a beginning? He doubts it. In the second volume, p. 30 and 40, d'Aubuisson discusses Hutton's observation on granitic dykes intruding into overlying schists. He then cites Jameson's and Jean-François Berger's (1779–1833) objections. Jameson, he says, simply denies their existence and Berger, in his paper published in the *Transactions of the Geological Society* (Berger, 1811), on the basis of the observations in Cornwall cited by others to be in favor of Hutton's theory, thinks the dykes are simple "protuberances" of the granite (d'Aubuisson de Voisins, 1819; see Berger, 1811, p. 145–156). D'Aubuisson repeats the same things, except the discussion on granitic dykes, verbatim in the first volume of the second edition (d'Aubuisson de Voisins, 1828, p. xxviij–xxix and 435–436). The second and the third volumes of the second edition of d'Aubuisson's book (1834 and 1835 respectively) were written by the French mining engineer and geologist Amédée Burat (1809–1883) and appeared under his name, because d'Aubuisson's interests had changed from geology to engineering matters before he could finish the second edition. In these volumes there is no mention of Hutton any longer, but it is clear that Burat had adopted his opinions on the igneous rocks and dykes and sills.

The Huguenot geologist Ami (Amédée) Boué (1794–1881; Fig. 30) had come to Edinburgh to study medicine after having completed his primary and secondary education in Hamburg, where he was born, and in Geneva. While in the University of Edinburgh, he heard Robert Jameson's (1774–1854) lectures and decided to become a geologist instead of a physician. As a result of Jameson's teaching, he became a convinced neptunist and it is with neptunist eyes that he toured much of Scotland and the Hebrides and ended up writing a book entitled *Essai Géologique sur l'Écosse* (Geological Essay on Scotland; Boué, 1820). The book was dedicated to Boué's teacher Jameson and is full of references to his work and to the Scottish physician and geologist John MacCulloch's western islands book (MacCulloch, 1819)[134]. However,

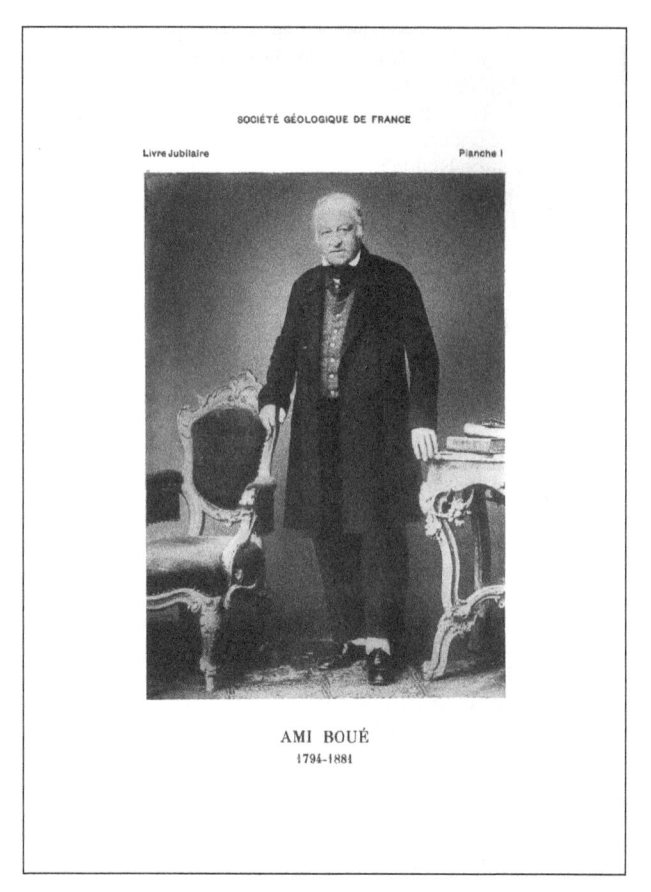

Figure 30. Ami Boué. (From *Livre Jubilaire 1830–1930: Centenaire de la Société Géologique de France*, v. 1: Paris, Société Géologique de France, Mémoire hors-série 10, plate 1, 662 p.)

In his *Mémoires Géologiques et Paléontologiques*, Boué has a review of the two volumes of Lyell's *Principles*, in which he identifies Hutton as one of his forerunners (Boué, 1832, p. 317).

In 1827, Pierre Louis Antoine Cordier (1777–1861; Fig. 31) published his famous memoir on the temperature of the interior of the earth arguing in favor of a central heat within the planet. In the beginning of his memoir, he mentioned the then current theories of geology. He says that in the latter half of the eighteenth century, neptunism, supported by Peter Simon Pallas (1741–1811), Horace-Bénédict de Saussure (1740–1799) and Abraham Gottlob Werner (1749–1817) prevailed, but continuing observations on the geology of various portions of the globe, the falsification of the theory of aqueous crystallization of the crystals of such rocks as granite and basalt and the observation that everywhere the temperature increases as one descends toward the interior of the earth finally overthrew neptunism:

> These remarkable facts, considered separately and in connexion have brought back all who have examined the subject, to the hypothesis of central heat. The general conclusion is, that the interior of the earth possesses a natural temperature incomparably higher than the variable temperature found at the surface, and even as some suppose, that beyond a certain depth, the whole mass exists, as at the beginning of things, in a state of fusion.
>
> La Grange and Dolomieu were the first who revived the hypothesis of a central fire. Hutton should also be mentioned, and his ingenious commentator Playfair, notwithstanding the obscurities with which they veiled their opinions, and the errors in physics into which they have fallen by attempting to apply their results to geology. (Cordier, 1827, p. 560; in the English translation, Cordier, 1828, p. 6)

Boué did read also Hutton and was unable to stay a neptunist. In the third and final part of his book there is a long theoretical discussion, in which Boué opens the subject with the following words:

> Everybody knows that there are only two dominant geological theories: one, the *neptunian* theory, the other *volcanic* and that this latter comprises a special geological system due to the genius of Dr. Hutton. (Boué, [1820], p. 394; italics Boué's)

Boué refers to Hutton's supporters as "huttonians" (Boué, [1820], p. 395). He says the neptunists claim that all rocks were precipitated out of water and that the vulcanists claim that all rocks solidified out of magma. He thinks these two extremes cannot be right and gives a long explanation of the origin of the volcanic and the sedimentary rocks, conceding the volcanic origin of basalts, trachytes, etc. He has a long discussion arguing that the red sandstones of Scotland contain products of volcanoes and that when the red sandstones were being deposited *as sedimentary rocks*, there must have been *active volcanoes* around. On p. 94, there is a reference to a locality of Hutton's using Playfair's 1802 book, p. 328.

Even J.-S. Bonnaire-Mansuy, a provincial naturalist and a corresponding member, among others, of the *Société Royale des Sciences et Belles-Lettres de Nancy*, also known as the Stanislas Academy in Nancy, who published his *Cosmogonie, ou de la Formation de la Terre et de l'origine des Pétrifications* in 1824 with a view to establishing the perfect agreement between the *Book of Genesis* and geology despite the high antiquity of the earth, which he acknowledged, was aware of Hutton and Playfair. He wrote:

> Doctor *Hutton* and after him his friend Mr. *Playfair*, established a system which was successfully combatted by Mr. *Deluc* using the observations of Mr. *de Saussure*; but this judicious and profound critique did not establish a decisive system after having destroyed a large number of errors. And geology, always a widow of principles, is still only a *collection of facts* whose causes are *unknown*. (Bonnaire-Mansuy, 1824, p. 5, italics Bonnaire-Mansuy's)

According to our learned author, the cause(s) were, obviously, the Creator!

Figure 31. Pierre-Louis-Antoine Cordier (engraved by Ambroise Tardieu from life in 1805). Courtesy of the Smithsonian Digital Libraries; Creative Commons CC0 1.0.

Cordier's paper brings us to the year 1827, in which Sir Charles Lyell first mentioned the "Huttonian theory" and made his own uniformitarian views, and his admonition of not mixing science with religion, public (Lyell, 1827, p. 473ff.).

After Lyell and before Geikie

The separate edition of Cuvier's *Discours* mentioned Hutton in its numerous editions from 1830 onwards (the ones I have been able to see are the following: 1830, 1840, 1850, 1854, 1861, 1863, 1864, 1867, 1881, 1889) always as it did in its initial edition as part of the *Ossemens Fossiles* in 1812 until its last edition in the late nineteenth century.

In the same year as the publication of the first volume of Lyell's *Principles of Geology*, the French army engineer and geographer Claude Antoine Rozet (1798–1858) brought out his *Cours Élémentaire de Géognosie* (1830) as a textbook for his lectures to the officers in the *Dépôt général de la Guerre de Terre et de Mer et de la Géographie* (General archive of land and sea war and of geography; this was really the geographical survey and archive of the French military). In his preface, Rozet says that in many places he followed verbatim the texts of d'Aubuisson, Cordier, Alexandre Brongniart (1770–1847), Alexander Baron von Humboldt (1769–1859), François Sulpice Beudant (1787–1850), Auguste-Henri de Bonnard (1781–1857), Voltz and some others. Philippe-Louis Voltz (1785–1840) was particularly helpful and even read his proofs. The paleontological part was assisted by the great conchologist and paleontologist Gérard Paul Deshayes (1795–1875), Lyell's counselor in Cainozoic biostratigraphy. In his *Discours Préliminaire*, Rozet gives a short history of geology with some small excursions into general geography (he even mentions the German geographer Bernhard Varenius {1622–1650}, the great popularizer of the concept of *geographia generalis*, created earlier by the protestant theologian Bartholomäus Keckermann {?1572–1608}, which I have not encountered in any of the other geological treatises of this time) and therein on p. 7 and 8 he mentions Hutton and Playfair, closely following d'Aubuisson:

> Whilst everywhere the exact study of the interior of the earth was being studied, in England hypotheses were still being made. Playfair supported with much energy and talent Hutton's system and the celebrated Hall rendered to its principal consequences a high degree of probability by his experiments on the influence of heat and pressure.

In 1837, Rozet's much enlarged version of his *Cours* came out in two volumes under the title *Traité Élémentaire de Géologie*. Its *Discours Préliminaire* is much the same as that of the earlier book, with some small but significant changes. The sentences about Hutton are as follows:

> Whilst everywhere the exact study of the interior of the earth was being studied, in England, which had remained behind, hypotheses were still being made. Hutton was dead, leaving behind him an ingenious theory of the formation of the earth, based only on some facts. Playfair supported it with much energy and talent and the celebrated Hall rendered to the principal consequences of the Huttonian system a high degree of probability by his experiments on the influence of heat and pressure. (Rozet, 1837, p. x)

Rozet then mentions Jameson and the foundation of the Wernerian Society in Edinburgh "before the very eyes of Playfair and Hall" and says that the shock this created was "violent, but the light came out; one abandoned hypotheses for exact observations" (Rozet, 1837, p. x). He lists a few of the British observers and then says that finally that famous book by Cuvier and Brongniart came out and illuminated the importance of the fossils in the rocks. The second volume of the *Traité* has no reference to Hutton.

The next significant French textbook of geology came out in two volumes in 1837 and 1839 by one of the former collaborators of the *Encyclopédie Méthodique*, Jean-Jacques-Nicolas Huot. It is called *Cours Élémentaire de Géologie*. In its second volume, Huot discusses Hutton's ideas in some detail (p. 693–694). At the end he quotes Conybeare and Phillips' evaluation of Hutton's ideas as reflecting his own assessment:

> "He has the merit" they say "being the first to direct the attention of geologists to the important phenomena of metallic

veins that emanate from granitic rocks to cut those above them and to have emphasized via new and clear views the igneous origin of the trap rocks. Nevertheless, the weakness of some of his theories, however, did not fail to lessen the usefulness of the new facts that he collected through observation. He, who would see in geological phenomena only the results of what is still happening before our eyes; the continuation of the same phenomena during an infinite accumulation of centuries without it being possible to assign the beginning or to foresee the end, could only be considered as having examined them through the prism of a hypothesis conceived in advance and of which he would remain preoccupied." (Huot, 1839, p. 694)[135]

This is pure Conybeare. He could never reconcile himself to uniformitarianism as he published in 1832 in his British Association report (Conybeare, 1835) and also later told Lyell (Wilson, 1980) again.

After this presentation, Huot considers the progress made since Hutton and Werner. He praises de Saussure's studies in the Alps and points out that he assigned to internal heat or to some sort of "elastic substance" the elevation of mountains. He finds the hypotheses of De Luc very peculiar and similarly has difficulty following Delamétherie's crystallization theory. He ends with a review of his former teacher Faujas' work, which he thinks excellent, although overtaken in part by the progress that has occurred since his death. His own views on the origin and development of the earth are based on the assumption of an originally fluid state of rocks caused by high temperature. In the sequel he points out that at his time there are still those who try to reconcile the Mosaic account with geology. He finds that the succession of life on earth is not in violent contradiction with the succession of events in the Bible and if by "days" "ages" is understood, then one may be allowed to admire the ingenious author for having foreseen some of the results of modern geology.

In 1844, Felix de Boucheporn (1811–1857), superb field geologist, mining engineer and natural philosopher, published his remarkable *Études sur l'Histoire de la Terre et sur les Causes des Révolutions de sa Surface* (Studies on the history of the earth and the causes of the revolutions of its surface) in which he developed his theory of repeated bolide impacts creating equatorial mountains and defining the period boundaries by changing the flora and fauna of the planet. This idea was not entirely new and had affinities to similar suggestions by William Whiston (1696), Gottfried Wilhelm Leibniz (1749a, 1749b), Edmund Halley (1724–1725), Pierre Simon Marquis de Laplace (*An IV* [1796]), and especially to that published anonymously in the little book by Éléonor-Jacques-François de Sales Guyon de Diziers, Count Montlivault (1821), although de Boucheporn's treatment of the subject was much more sophisticated than those of his predecessors. De Boucheporn was particularly interested in making mountains by lateral thrust and it is in that context he referred to Hutton. Since de Boucheporn's book is not widely known, I quote below what he says about Hutton, after having listed him among those who "attached the glory of their works" to the "beautiful ideas" developed during the eighteenth century (p. 60):

Hutton, ... observed in Scotland also the twisting of the strata, but related them to the apparent intrusion of the traps, basalts and even granites, thus conceived the earliest idea of the influence on the structure of the sedimentary terrains of the thrust of these igneous rocks. Prompted by the nature and the usual frame of his observations, Hutton considered physical phenomena from a point of view narrower than that of de Saussure, and was thus led in this respect to give greater importance to local causes. He attributed the twisting of the layers to an action of this essentially limited kind. If he had studied nature in a country where she could show herself to him in all the majesty of her deformations, perhaps the grandeur of the phenomena would have inspired him to other ideas. In any case, however, we owe it to him to have drawn our attention to the necessity of the lateral forces for the uplifting of the strata, although the spirit of his system made him restrict this action within the limits of too limited a compass. (de Boucheporn, 1844, p. 64)

Here, de Boucheporn adds the following footnote (p. 64; italics de Boucheporn's):

Here is indeed what his commentator Playfair wrote (*Explication of the theory of the earth by Hutton*): "Although the primary direction of the force which thus uplifted the strata must have been from below upwards, it was combined with the gravity and the resistance of the masses in such a manner as to give a *lateral* and oblique shock and to produce all the contortions which, on a grand scale, are amongst the most curious and most instructive phenomena in geology."[136]

Then de Boucheporn describes the "beautiful experimental study" by Sir James Hall "who gave the greatest support to his [i.e., Hutton's] grand ideas" (p. 64–65) showing how lateral shortening leads to folding. He says that one should take these facts of such a special nature into account in the study of the causes that produced the structure of the mountains. It is odd that in his chapter on faults and dykes and sills containing a sharp criticism of Werner's ideas, there is no reference to Hutton. Possibly this is so, because, by that time, Hutton's interpretation of dykes and sills had become commonplace and de Boucheporn here wished to emphasize their directional characteristics and relations to one another, which Hutton had not emphasized, as Sainte-Claire Deville later pointed out (1878, p. 165).

One of the dominant figures in geology during the middle quarters of the nineteenth century was the great French geologist Jean-Baptiste Armand Louis Léonce Élie de Beaumont (1798–1874), almost an exact contemporary of Sir Charles Lyell (1797–1875) and frequently his adversary in a number of fundamental issues in geology. He was appointed professor of natural sciences in the Collège de France in 1832 (succeeding Cuvier; from 1837 on he became the professor only of inorganic natural substances, but his sole subject was geology) and remained there until his death. A part of his lectures were published in two volumes, the first in 1845 and the second in 1869 (although the date given on the cover is 1849). In these lectures there is not much historical material and Hutton is mentioned among other luminaries of the natural sciences in volume 1, p. 19. However, Élie de Beaumont gives a very short summary of the work of de Saussure and Pallas as traveling geologists and Werner and Haüy as what he calls "sedentary geologists." His two volumes deal exclusively with external phenomena.

In 1860, the future successor of Cordier both in the chair of geology in the Museum National d'Histoire Naturelle and in the Academy of Sciences, the distinguished French geologist Gabriel-August Daubrée (1814–1896; Fig. 32), one of the founders of experimental geology and meteoritics, published his much admired *Études et Expériences Synthétiques sur la Métamorphisme et sur la Formation des Roches Cristalliens*, in which he not only discussed Hutton's ideas and his theory and his successors in a number of places (e.g., p. 4, 8–17, 25–26, 28, 30, 32, 45) but identified him explicitly as the founder of modern geology. In fact, in the first part of his book, the first chapter (p. 4–8) is entitled "State of geology up to the appearance of the system of Hutton," covering the period from Descartes to Werner. The second chapter (p. 8–13) reviews Hutton's system at some length:

> The importance of the work of Hutton and Playfair, in which the fundamentals of modern geology are established and developed for the first time, ... requires that I here summarise its principal propositions.... (Daubrée, 1860, p. 9)

At the end of his five-page presentation of Hutton's geological ideas, Daubrée summed up:

> In summary, Hutton explains the history of the globe with both simplicity and grandeur. The atmosphere is the region where rocks are decomposed; then their debris goes to accumulate at the bottom of the sea. It is in this grand laboratory that the mobile materials are finally mineralised and transformed, under the double action of the pressure of the ocean and the heat, into crystalline rocks having the appearance of ancient rocks, which are later uplifted by this same internal heat and are demolished in their turn. Thus, the degradation of one part of the globe serves the ongoing construction of other parts and the continuous absorption of the lower deposits relentlessly produces new molten rocks, which can be injected into the sediments. This is a system of destruction and renovation of which one can discern neither the beginning nor the end. As in the planetary movements in which perturbations correct themselves, one sees continual change, but contained within certain limits of such a kind that the globe shows no sign of infancy nor of senescence.
>
> So Hutton is the founder of the fecund principle of the transformation of sedimentary rocks under the action of heat.
>
> Nevertheless, we shall see below that there are many reservations to be voiced on such absolute conclusions. Like most of the men of genius who opened new paths, Hutton indeed exaggerated the scope of the ideas that he had conceived. One

Figure 32. Gabriel-August Daubrée. Unknown author. Public domain (PD-US-expired).

> cannot, however, consider without admiration with what profound penetration and what rigor of induction this farsighted man imagined, at a time when accurate observations were still very few, a system which, for the first time, called for the simultaneous competition of water and heat and embracing the entire history of the globe. He put forward principles that are today universally accepted, at least in fundamental issues. (Daubrée, 1860, p. 12–13)

The third chapter (p. 13–23) is devoted to Hutton's successors, in which the works of William Thomson (1760–1806)[137], Sir James Hall, John MacCulloch, are recounted, plus Giovanni Arduino's (1714–1795) pre-Hutton recognition of contact metamorphism in the Southern Alps (in 1779). After 1815, when relations between the UK and continental Europe were normalized following Napoleon's defeat at the Battle of Waterloo, says Daubrée, the studies of Hutton and his disciples began to be more widely known in Europe. The publications of Ami Boué, Louis Albert Necker-de Saussure (1786–1861), Sir Charles Lyell, Baron Christian Leopold von Buch (1774–1853), André-Jean-François-Marie Brochant de Villiers (1772–1840), Bernhard Studer (1794–1887), Peter Merian (1795–1883), Jean-Baptiste Armand Louis Léonce Élie de Beaumont (1798–1874), and a number of others in Italy, in the Pyrenees, in Scandinavia, and in Nova Scotia are mentioned by Daubrée as supporting Hutton's theory of metamorphism. In fact, the whole of Daubrée's book is soaked with a sentiment of admiration for Hutton and his theory of the earth.

In 1867, Daubrée published a summary report on experimental geology as a part of a general, multi-volume report on the progress of letters and sciences in France (Daubrée, 1867). The third part of that report is devoted to a review of metamorphism and there Daubrée repeats, almost verbatim, what he had written in the 1860 book. But in the second part, dealing with the "application of the experimental method to physical and mechanical phenomena" Daubrée starts discussing folding with Sir James Hall's experiments and reminds his readers Descartes' interpretation of crustal deformation as shortening resulting from radial diminution of the planet.

> But these ideas seem to have fallen into oblivion, when, more than a century later, Hutton announced that heat not only consolidates and mineralises the sediments of the bottom of the sea, but, in addition, it uplifts and tilts the beds that were originally horizontal. Saussure recently observed the tilting of the celebrated puddingstones of Valorsine, without, however, offering an opinion as to the cause.
>
> Sir James Hall corroborated this fundamental hypothesis experimentally, well-known today, the simplicity of which in no way diminishes its merit. (Daubrée, 1867, p. 33–34)

Daubrée put together the results of his 30 years' work on experimental geology in one of the great classics of this branch of geology in 1879 (Daubrée, 1879). As in his earlier books, there is repeated mention of Hutton in this newer book. Already in the introduction, Daubrée points out, after having said that Werner started teaching in Freyberg toward 1780 acquiring great fame and his fervent students spread his teaching to their own countries, that in 1785 Hutton published in Edinburgh "the first edition of his book" (p. 2) on the theory of the earth and "his persistent explorations in the mountains of Scotland and his genius in induction led him to judicious and profound views" (p. 2). On p. 7 there is a reference to Hutton's actualism:

> When it comes to understanding the mode of formation of rocks of all kinds, stratified or eruptive, one looks at the present-day phenomena that are nothing but the continuation of ancient phenomena. This is what Hutton showed so nicely. The actions happening now on the surface of the globe are indeed comparable to the events that continue before our very eyes.

The next mention of Hutton in Daubrée's massive book is in the section on metamorphism:

> The initial ideas on metamorphism are not much older than this century. They appeared at a time when the doctrines established by Werner were reigning: they are characteristic of the Scottish school founded by the genius of Hutton.
>
> Hutton had barely announced an ensemble of ideas on the transformation of rocks that came to be epoch-making, that his most distinguished pupil, James Hall, tried to test the new doctrine experimentally.
>
> The presence of crystalline limestones in the field, which the chief of the Scottish school considered as sedimentary rocks transformed by the heat, had given rise to the objection that in similar conditions, limestone would have lost its carbonic acid and changed into burnt lime. Now, as a result of a series of experiments, suggested to him, as he says, by Hutton since 1790, Hall reached the unexpected conclusion that chalk, when heated in an hermetically closed vessel, can soften without being decomposed. This experiment, already a classic, was the first example demonstrating how much pressure must be taken into account in geological phenomena. (Daubrée, 1879, p. 132)

On p. 353, Daubrée wrote on Hutton's role in the recognition that the earth's crust is highly fractured and faulted:

> Without being very apparent at a first glance, fractures of diverse classes, or lithoclases, cutting the constitutive rocks of the crust of the earth everywhere, are not less important for the explanation of the morphology of the ground.
>
> Already Descartes, after having conceived the great idea that the earth is an extinct sun with a surface covered by a crust, concluded from that with an intuition characteristic of genius that the voussoirs of the earth's crust, playing against each other, had to have produced asperities.
>
> Positive observations have taught since the end of the last century, following Hutton, de Saussure, Leopold von Buch, Élie de Beaumont and other geologists that the crust of our globe displays numerous lines of fracture to which the morphology adapts.

Thus, in this last book by Daubrée on experimental geology, summarizing his work on this topic, Hutton is referred to not only as an historically important figure, but as one whose ideas and interpretations are still valid forming the basis of the common understanding of geologists.

Between 1862 and 1865, the stratigrapher and professor of geology and mineralogy in the University of Besançon, Alexandre Vézian (1821–1903), published his modestly entitled *Prodrome de Géologie* in three stately volumes, the largest of its kind in France since Huot's treatise of a quarter of a century earlier. Referring to the geological ideas of de Buffon, who "perceived ruins on the surface of the earth, which, far from resulting from a recent catastrophe, showed traces of regular phenomena, which, in his mind, presided over the formation of the vestments of our globe" (Vézian, 1862, p. 38). He then says that these ideas went to inspire Hutton and Lyell (p. 39). About Hutton himself and his ideas, Vézian wrote the following:

> I recall under which circumstances the *Theory of the Earth* of Hutton was published.
>
> A careful examination of this work shows us that of all the geologists of his time, it was him who had the clearest ideas on geological phenomena and the structure of the earth's crust. Through his doctrine concerning the general system of Nature,

he was the successor of Buffon and the precursor of Lyell, C. Prévost and the supporters of the theory of actual causes.

The earth, he says, has been the theatre of the greatest revolutions and nothing on its surface has been exempt from their effects; but, for him, the word revolutions expresses the sense of the astronomer: he thought that these transformations happened slowly and regularly. Hutton's predilection for slow causes made him find everywhere the action of fire and the interior agents. "Every valley" he says "is the work of the stream that drains it. When a river flows through the narrow defile of a mountain, it is easy to reckon that this mountain had once been continuous across the narrow space where this river now flows; and, if one ventures to hazard a guess on the cause of such a prodigious event, one is prompted to attribute it to some great convulsion of Nature, which has broken this mountain into pieces to give a passage to the waters. Philosophical reflection alone, which has pondered profoundly on the possible effects of a continuous action and on the simplicity of the means employed by nature in all its processes, alone, I say, sees nothing in it but the gradual work of the stream that once flowed as high as the edges it now cuts so deeply and made its way through the rock, in the same way and with the same instruments that the lapidary uses to cut a block of marble or granite."

Here is another passage indicating the essential and distinctive trait of Hutton's system: "Mineral substances are subjected to a series of metamorphoses; they are alternately destroyed and renewed… It is not for us to determine how often these vicissitudes of destruction and renewal have been repeated.: they form a series in which we can see neither the commencement nor the end, which is in perfect accord with what we know of other parts of the universe. The author of Nature did not give the world laws resembling those of human institutions that carry in them the germs of their own destruction. His works show no character of infancy, nor of decline, nor any sign that might enable us to estimate the duration of time past or time to come. As he has given a beginning to the present system, without doubt he will also put an end to it at a determined period. But we can conclude with certainty that that grand catastrophe will not take place by means of any of the natural laws now operating."

We have just seen that the idea of permanence, I would almost say immutability of the natural phenomena, are embedded in the theory of Hutton. This idea, developed in its consequences, led Sir Lyell [sic] to formulate clearly the doctrine of the actual causes already vaguely mentioned by C. Prévost. (Vézian, 1862, p. 40–42)

When it comes to express his own position, Vézian says openly that Hutton's ideas "lack neither grandeur, nor originality" (p. 50) and in the beginning of his own studies he was seduced by them. He agrees with Hutton that the earth probably had a beginning and will have an end, but we cannot find traces of these events. He presents a wonderful discussion on a hierarchy of geological phenomena saying that the largest change neither their nature nor their energy. At the other end of the hierarchy are the small events that always change their nature and energy. Geological events generally happen very slowly, with exceptions, but they never happen instantaneously. What Vézian says, acknowledging Hutton's influence on his views, is what I had been taught as a geology undergraduate in the seventies of the twentieth century.

Vézian was one of the early protagonists of the anatectic origin of granites and believed that at least some of them formed in metamorphic terrains as a consequence of metamorphism. It is in this connection that he returned to a discussion Hutton's ideas later in his book:

Theory of Hutton on the origin of granite.—In the last century, the central heat was far from being the reigning theory in science. One should not be surprised that the Freiberg school had no notion of igneous rocks and eruptive processes. Without any qualms, they regarded the porphyries and even the basalts as masses that originated entirely in water. Here the error soon became obvious, and it was on the subject of this rock that the controversy between the neptunists and the vulcanists erupted. This war became livelier when Hutton declared that the granite itself was, like basalt, originally in a state of paste, which is incontrovertible, and that that state of paste could only be created by igneous melting, which is true only in a very limited degree. Hutton added that granite has been formed in numerous epochs and far from constituting the oldest of rocks, it is commonly among the younger ones. This way of viewing is justified up to a certain point. (Vézian, 1862, p. 166).

Vézian then proceeds to describe his own theory of granite genesis which supposedly occurred only during one period in the history of the earth, which he calls the plutonic period. Since then, Vézian thinks that conditions to generate granitic magma had not recurred. The original granitic magma, however, cooled so slowly that it formed for the longest time an immense subterranean reservoir that from time to time fed individual plutonic intrusions throughout the geological time. He says that what he calls the Huttonian theory explains very well the "plasticity of the granite and its transition to eruptive rocks; it also accounts for the metamorphism that it inflicts on the rocks with which it comes into contact" (p. 167). Vézian's main objection to Hutton's theory is its perceived inability to explain the passage of granites gradually to layered rocks such as gneisses and schists.

My purpose here is not to discuss Vézian's theory, nor to point out that Hutton's theory of granite formation was actually better than Vézian's. My reason for citing Vézian is to emphasize that in the beginning of the second half of the nineteenth century, he regarded Hutton as the architect of the views current in his time and that he wished to amend Hutton's theory of granite genesis. In his subsequent volumes, Vézian does not refer to Hutton again, not even when he discusses the origin and significance of dykes. He simply says that Werner's theory of the filling of the dykes from above was shown to be wrong and that in fact the opposite is true.

Shortly after Vézian's Prodrome was published, the famous yet unfortunate French geologist Viscount d'Archiac (Étienne Jules Adolphe Desmier de Saint-Simon, Vicomte d'Archiac:

1802–1868; a life-long bachelor, he committed suicide by drowning himself in the Seine on the Christmas eve following an episode of severe depression) brought out his *Géologie et Paléontologie* in two parts in one volume in 1866: the first part was devoted to a history of these two sciences and the second to their modern state. In the first part, the Viscount summarizes Hutton's theory between the pages 236 and 243. He opens his account with the following words:

> It is especially to the observations and theoretical ideas of Hutton, then to the experimental researches of Sir J. Hall that the Scottish school owes its originality and its greatest brilliance. (d'Archiac, 1866, p. 236)

After giving a competent summary of Hutton's ideas, the Viscount turns to Playfair's explication of that theory and presents Playfair's abstract of it followed by his own assessment of the place of Hutton's work in the history of geology, which, according to d'Archiac, represents the beginning of modern geology.

> As to the simplest and most explicit summary of Hutton's ideas, his pupil [i.e., Playfair] expresses it thus: "Hutton" he says "attributes an order to geological phenomena similar to those existing in the natural operations most familiar to us. It produces the seas and the continents not by mere coincidence, but by the action of regular and uniform causes. It serves to build one by means of destroying another. It thus gives a stability to all, not by perpetuating individuals, but by reproducing them in succession." This is indeed the most comprehensive and the simplest manner that conforms to the general facts known and the subsequent discoveries went to justify it.
>
> The incandescence of the interior of the globe and its gradual cooling, although excessively slow today, leads, by the expansion of gases and fluid materials, to displacements, uplifts and the tilting of the stratified rocks, complement a system perfectly ordered in its various parts, because the elements and the forces work simultaneously in their peculiar qualities, owing to their real properties, within their potential. This system is thus preferable to that of Werner, because it is less exclusive, it accounts better for the common operation of competing agents in producing a common result. Thus it must be closer to the truth. (d'Archiac, 1866, p. 242–243)

D'Archiac then outlines Sir James Hall's contribution that he made by means of his experiments on melting and recrystallization of basalt, the origin of marble and the cause of the folding of strata. The Viscount thinks they remain fundamental. He mentions Sir James' point that in physics there is hardly a truth better established than Hutton's theory and that the publication of the *Theory of the Earth* must mark a very important epoch in the history of this branch of natural history (p. 243). He allocated three full pages to Sir James' contributions. By contrast, for example, he recounts Robert Jameson's work in half a page and mainly to point out his mistakes under Werner's influence (p. 246).

In the Viscount's earlier, eight-volume *Histoire de la Progrès de la Géologie de 1834 à 1845* there is no mention of Hutton because of the temporal limits imposed onto the work.

In 1867, the French miner and journalist, prolific author of books on diverse topics ranging from geology through mining, travel to politics, later instructor of geology in the *École Spéciale d'Architecture*, Louis Laurent Simonin (1830–1886), a pupil of the *École des Mines de Saint-Etienne*, published his little book entitled *Histoire de la Terre—Origines et Métamorphoses du Globe*, dedicated to Élie de Beaumont, in which he considered Hutton one of the founders of geology. He emphasized in particular, most likely following Daubrée, that Hutton was the "creator of the theory of metamorphism" (p. 189). Simonin's book proved a great success and had at least nine editions before the death of its author.

The French Abbot Casimir Chevalier (1825–1893), educator, historian and writer, published a very noteworthy defense of actualism in his *Géologie Contemporaine—Histoire des Phénomènes Actuels du Globe* of 1868. An enlightened clergyman, he saw no problem in accepting the modern geology of his time *in toto* and pointing out that god could have created the world in whatever way he wanted and taking the six Biblical days of creation literally was not a sensible interpretation of the Holy Writ. He said that the world could have been created exactly as geological reasoning demanded and yet remained a glorious establishment "as the abode of man as its masterpiece and king" (p. 366). In the first chapter, he reviewed Hutton's ideas and wrote that the discussions Hutton's ideas gave rise to formed the basis of the geology of the writer's time:

> What was lacking in the cosmogonic theories of Leibniz and Buffon was observation. The mineralogists and the factory directors took it onto themselves to supply this want. Werner, named professor of mineralogy at the mining school in Freyberg in Saxony in 1775, studied with care the surrounding country and thought that all rocks without exception, even basalt, trapp, granite and porphyry were exclusively chemical precipitates of material found dissolved in water. According to Werner, water was the sole cause of all the deposits of the crust of the globe, the unique agent of geological modifications. His school (it contained fervent disciples) took the name of *neptunists*. The arguments of the learned professor of Freyberg were excellent for demonstrating the origin of certain aqueous beds, those that we call today *sedimentary*; but it did nothing for a certain number of rocks of clearly crystalline origin. The *plutonists*, who attributed an igneous origin to these rocks originating from the activity of the central fire, tried to demonstrate this and a lively discussion arose between the two systems. The English geologist Hutton entered the fray in 1788 and by numerous examples taken from Nature demonstrated that some rocks are of igneous origin, others of aqueous origin and that some rocks of sedimentary origin had been altered profoundly, modified and metamorphosed at their contacts with the molten igneous rocks.
>
> These discussions established in an irrefutable manner many important points that form the basis of current geology. (Chevalier, 1868, p. 13–14)

Only four years after the publication of Chevalier's book, the German-born French physician, lexicographer of science, and historian of science Jean-Chrétien-Ferdinand Hœfer (originally Johann Christian Ferdinand Höfer, 1811–1878) published his small book entitled *Histoire de la Botanique de la Minéralogie et de la Géologie Depuis les Temps les Plus Reculés Jusqu'à Nos Jours* (=History of Botany, Mineralogy and Geology from the Earliest Times until our own Day) in the series *Histoire Universelle* edited by the historian and one-time French minister of education Victor Duruy (1811–1894). In his book, Hœfer devoted 285 pages to the history of botany and only 119 pages to the history of geology. In this restricted space, Hutton's contribution is reviewed after Hœfer presents the geological theories of antiquity to Renaissance and then those of Burnet, Whiston, Woodward, Steno, Leibniz, and de Buffon. It is important to note that Hutton's ideas are no longer presented as a "theory of the earth," but instead they appear under the section heading "volcanism." Hœfer rightly points out that Hutton, starting from the inference that crystalline rocks were once molten "lava," argued that the causes which now modify the terrestrial crust would suffice to explain its past revolutions. Hœfer goes on to provide some details of Hutton's interpretations and proceeds to show how Playfair supported them. On p. 380–381 he summarizes Werner's theory under the heading "neptunism" and immediately proceeds to another section entitled "adversaries of the Wernerian school." That section opens with the sentence:

> It was not the vulcanists who dealt the most severe blows to the system of Werner; it was the pupils of the Freiberg professor who brought him down. (Hœfer, 1872, p. 381)

Hœfer then cites a letter that von Buch wrote to his friend Alexander von Humboldt from Pergine Valsugana (*Pargine en Italie* in Hœfer, p. 382; some 10 km east of Trento and immediately north of the double lakes of Caldonazzo and Levico) around 1798 saying

> I found the layers of porphyry overlying Secondary limestone and the micaceous schists overlying the porphyry[138]. Do not all of these threaten to reverse the beautiful systems by which one pretends to explain the epochs of formations? (Hœfer, 1872, p. 382)

After this quotation, Hœfer writes that von Buch's visit to Italy made him realize that tranquil deposition in water was not the whole of geology as the Freiberg school had imagined. Below we shall see some details of von Buch's gradual conversion to Hutton's views as he himself said explicitly. I should note, in passing, however, what Gian Battista Vai reminded me (written communication, 24 August 2019): Hoefer omits to say that Arduino as early as 1782 had already described contact metamorphism and in 1769 and 1774 ancient stratiform volcanic rocks alternating with marine fossiliferous strata. The same is true for Fortis having first in 1778 reported on marine fossil shells cemented by basaltic flows in the Venetian Tertiary hills. Regrettably, scientific intercourse does not follow the rules of a perfectly efficient market!

After the death of Élie de Beaumont in 1874, his creole student and subsequent adjunct in the Collège de France from the Caribbean island of St. Thomas, Charles Sainte-Claire Deville (1814–1876; Fig. 33), was appointed to his teacher's chair of natural history of inorganic substances in the Collège after having given the geology courses there for the past 21 years as Élie de Beaumont's assistant. Sainte-Claire Deville was able to fill that position only for 18 months. His last course of lectures was devoted to a history of geology and particularly to the work of his teacher Élie de Beaumont. Regrettably, Sainte-Claire Deville was unable to finish his work on the book that resulted from his lectures. He was able to read the proofs of only the first three chapters before a sudden death took him away and it was not possible to insert any figures, which makes understanding of his summary of Élie de Beaumont's famous theory of the *réseau pentagonal* by an uninitiated impossible, despite the clarity of his text. His manuscript was edited by his student and eventual successor Ferdinand Fouqué (1828–1904) and published in 1878 under the title *Coup-d'Œil Historique sur la Géologie et sur les Travaux d'Élie de Beaumont* (A historical overview on geology and on the work of Élie de Beaumont). In this book Hutton plays

Figure 33. Charles Sainte-Claire Deville. Public domain: Principales Découvertes et Inventions Dans Les Sciences, Les Arts Et L'Industrie Par Adolphe Bitard Avec Photo Gravures Dans Le Texte, Mégard et Cie, Libraires Éditeurs, 1881, ch, VIII, p. 148.

a very important rôle. Sainte-Claire Deville devoted a chapter-and-a-half to him (i.e., as many lectures out of his nine annual lectures at the Collège de France) and quotes his contributions also in the subsequent chapters.

Sainte-Claire Deville opens his lecture on Hutton with the following words:

> As I said in finishing the last lesson, Hutton also presented a complete theory of the earth from this viewpoint, which played an important rôle in the aetiological school of geology. But he clearly distinguished himself from all his predecessors and, one can add, all those who followed him in this order of considerations, in that the facts that he cites and knows, almost all as eye-witness, are admirably observed and the deductions made from them generally betray an excellent logic. (Sainte-Claire Deville, 1878, p. 145)

Sainte-Claire Deville then says that Hutton's ideas led to a heated war in Europe with the adherents of Werner's views. With time, "the learned" forgot these conceptions, extremely remarkable for their time, under the pressure of other preoccupations. In his day, Sainte-Claire Deville says, away from the passionate controversies of those days and the disdain exhibited by those who represented the factions, one has discovered two very real and very diverse merits in Hutton's work: (1) Hutton generalized and made more explicit the more or less vague views of his predecessors and contemporaries; and (2) based on his own conclusions he ventured to anticipate facts and experiments, realized only after him and some of which were undertaken expressly to test his ideas. Despite much talk about Hutton's theory, Sainte-Claire Deville thinks that it is really little known. His adversaries ascribe to him ideas he never held, whereas his supporters have abandoned those parts of his system, which Sainte-Claire Deville finds to be the best. He thus proposes to examine Hutton's ideas in the order presented by Playfair. He first reviews Hutton's definition of a stratum. He shows that Hutton documented that most were deposited in a submarine milieu, but his familiarity with coal also allowed him to posit deposition on continents. He then presents Hutton's theory of consolidation of beds and shows how Hutton relied on heat to bring it about, because he was unconvinced that beds could be indurated in an aqueous milieu. Here he says that we must consider the time in which Hutton was writing and finds his reasoning perfectly acceptable. He then proceeds to Hutton's claim of the original horizontality of strata and discusses the few exceptions Hutton pointed out. This led him to the question as to why most strata were now found not only upturned, but even convoluted in diverse ways. Why were marine strata today found at elevations of thousands of meters above sea-level? Hutton thought it implausible that the recession of the sea could possibly be the cause. In fact, the high mountains all showed convoluted strata, not flat-lying beds. Hutton insisted on the fact that the deformation was "angular." Older strata were not only uplifted, but upturned also. Newer strata were laid down on them "transversally"; Sainte-Claire Deville says that we now call that relationship "discordant." What could have pushed them up and deformed them? Heat, Hutton thought, is the only agent capable of bringing about that kind of deformation. Sainte-Claire Deville then reviews Hutton's interpretation of the whinstone and granites and finds them little different from the modern views of the nineteenth century. He says Hutton was the forerunner of Leopold von Buch's primary vertical uplift theory of mountain-building. He concludes with the following assessment of Hutton's theory:

> Of this account it clearly emerges, I think, that the founder of the Scottish school of geology was at once an excellent observer, a great logician and that in a large number of questions he proved to have an admirable perspicacity that allowed him to foresee and announce many of the discoveries made after him. (Sainte-Claire Deville, 1878, p. 174–175)

Sainte-Claire Deville's next chapter (and his next lecture at the Collège de France) is headed "End of the [account of the] Theory of Hutton; Werner." He introduces this chapter by pointing out that Hutton had concluded that the earth had its own central heat on the basis of his observations and deductions. He says that to reach that conclusion Hutton, "does not disguise the fact that he leaves the field of observation and that of logical deductions to appeal to hypothesis." (Sainte-Claire Deville, 1878, p. 176). Hutton, he says, was aware that this central heat had not been demonstrated. That is why he searched for observations such as hot springs, volcanic phenomena and earthquakes to support that hypothesis. He added, Sainte-Claire Deville quotes:

> The cause of the earthquakes is certainly one that resides very deep below the surface; without such a cause the extent of the commotion would not have been such as is observed in many of the instances. (Hutton quoted by Sainte-Claire Deville, 1878, p. 176)

Sainte-Claire Deville is surprised that Hutton was unaware of Athanasius Kircher's work of 1644[139] and the observations by the mineralogist and mining engineer Antoine de Gensanne (deceased in 1780) made in the lead, copper and silver mines of Giromagny (with a little gold) to the south of the Hercynian massif of the Vosges, in Burgogne-France-Comté (for the geology of this area and its mines, see Karpinski, 1931). De Buffon reports de Gesanne's observations in his *Époques de la Nature* (de Buffon, 1778, p. 495–496) and Sainte-Claire Deville chides Hutton and Playfair for not referring to a source so strongly in support of Hutton's views and that also happens to be familiar to most. But he hastens to point out that Hutton was very much superior to de Buffon in that he pointed out that his internal heat should not be imagined to be "fire" or "flame," because these require air to burn. Hutton then pointed out, says Sainte-Claire Deville, that fissures in the earth's crust brought to the surface sometimes melted rocks, sometimes hot vapors. Such, according to Hutton, are the changes experienced by the mineral substances in the depths of the earth. They are different for stratified and unstratified substances. The latter fuse completely, whereas the former simply become soft and cement the pieces that were detached and free earlier. But they all obey the same principle and are explained by the same cause. Hutton then considers, Sainte-Claire Deville tells us, what happens once the rocks reach the surface. He considers the influence of the air on them. He exaggerates, like de Buffon, the importance of the destructive effects of water. He describes how flowing water patiently excavates valleys during a long interval of time. He likens the mountains to pillars of land that workers leave behind as a measure of the amount they have removed. Here the plutonist becomes a neptunist and uses, with some exaggeration Sainte-Claire Deville says, the everyday events to explain the results he attributes to them. But destruction is not the only thing that happens at the surface. There is also works of "reparation and reconstruction" by means of accumulation of detritus, and of corals by an enormous secretion of calcareous material. All these events constitute in the eyes of Hutton a sort of peculiar life of the surface. All of these phenomena tell us of the flow of time. Hutton warns us, Sainte-Claire Deville emphasizes, that the time he is talking about is not one to be measured by the rotation of the Sun or the Moon. There is no synchronism between the most recent epochs of the mineral kingdom and the most ancient of our ordinary chronology.

Sainte-Claire Deville concludes his account of Hutton with the following words:

> By the nature of his spirit, by the way he approached the problems, and especially from the aetiological viewpoint that was always dominant with him, Hutton was a descendent of the ancient schools. But, by the soundness of his observations, by the rigour of his deductions and by that gift of synthesis, of which he gave the most remarkable example and which allowed him to announce in advance the practical consequences of a cause once determined, he belongs to the modern school, which had already produced a Guettard and a Desmarest and, during his time, Werner was the highest manifestation. (Sainte-Claire Deville, 1878, p. 182)

With these words, he proceeds to recount Werner's work for the rest of his lecture, except the end, where he wraps up again a comparison between Hutton and Werner and their immediate successors. In the following, the eighth, lecture, Sainte-Claire Deville treats the topic of "cause" in geology and there he goes again back to Hutton and to his adversaries.

> We have devoted the two last sessions to placing before us, so to say, the two imposing figures of Hutton and Werner.
>
> If I emphasized especially the first of these two great minds, who, at the end of the last century, shared the glory of placing geology on a solid base with de Saussure, it is because to me his system seems not well understood and not appreciated, albeit much talked about. (Sainte-Claire Deville, 1878, p. 205)

Sainte-Claire Deville pointed out that all of Werner's publications were written from the double viewpoint of cryptoristic and troponomic researches[140]. In them, he says, there are hardly a few pages devoted to the explanation of phenomena. Werner never wrote down his theory of the earth and left it to his students "charmed, in a way, seduced" by him (p. 206). Sainte-Claire Deville then proceeds to relate the attacks of Kirwan and De Luc on Hutton. He points out that they accused Hutton of having erected an atheistic system and berated him for irreligiosity thereby forgetting

> the sage maxims, which, a long time earlier Bacon expressed in *Novum Organum*: "This vanity must be all the more condemned and repressed, because not only the whimsical philosophy, but also the heretical religion are born of the unhealthy mixture of divine and human things. That is why it is very salutary for a sober spirit to present an article of faith only as an article of faith."
>
> These systems, opposed to that of Hutton with such animosity, contain, by the way, nothing new or original. Like Werner the two of them explain the universality of the formations by means of aqueous masses and Kirwan's ideas are hardly superior to those of Whiston and Burnet. The theory of Deluc is definitely well below that of Buffon. (Sainte-Claire Deville, 1878, p. 207)[141]

Thus ends Sainte-Claire Deville his discussion of Hutton, although in a number of places in his later lectures he quotes geologists who cited him approvingly. He adds that De Luc's separation of times in which other laws governed geology than the ones now operating from those governed by laws now observable "a scientific heresy" (p. 208).

In the eighties of the nineteenth century, the latter age of grand treatises in geology began with Albert Auguste Cochon de Lapparent's (1839–1908; Fig. 34) much acclaimed *Traité de Géologie* that appeared in 1883 and, with ever increasing number of volumes, went through five editions until 1906. In its first edition there is an overview of the history of geology, in which de Lapparent wrote:

> A new era opened up for geology when the progress of public security, together with the improvement of the means of transport, made it possible for the curious about Nature to undertake these fecund explorations in which among others, a Pallas, a Dolomieu, a Desmarest, a Saussure distinguished themselves. The limited horizon of the first geologists was remarkably enlarged, and one learned to suspect that the theories most opposed to each other may well contain partial truths. It was in this way that the most convinced neptunists were able to appreciate, by means of a simple trip to Scotland, all that had been found in the doctrines, so decidedly

opposed to those of Freyberg, that Hutton and his followers were developing with so much success in England, while the almost entire continent blindly remained under the influence of Werner. (de Lapparent, 1883, p. 12)

Even in French high school textbooks on geology, Hutton and his friends find a place. In the first edition of his textbook *Cours Élémentaire de Géologie* (Elementary Course of Geology), the French mineralogist François-Sulpice Beudant (1787–1850) mentions not only Sir James Hall's experiments on folding of strata (p. 108), but also Hutton himself, together with Playfair and MacCulloch, on metamorphism (p. 137; Beudant, 1841). This small textbook was so successful that it underwent 17 editions until 1886, i.e., 36 years after Beudant's death, from time to time updated and improved by his colleagues. In the later editions the names and the observations made by the persons (e.g., the Glen Tilt observations by Hutton) to whom the names attach continue to be mentioned (see, for example, the fifteenth edition, published in 1877, p. 171).

The popular book writers, however, were not always so conscientious or scrupulous as Beudant. For example, in his extremely popular *Lettres sur les Révolutions du Globe* (Letters on the Revolutions of the Globe; it had ten editions between 1824 and 1879, although the author had died 48 years previous to the last edition!), the French physician Alexandre Jacques François Bertrand (1795–1831) mentioned Hutton only because he quoted, at the end of his historical review, Cuvier's assessment of various systems, Hutton's being one; but P. von Maack, the German translator/adapter of Bertrand's fifth posthumous edition (Bertrand, 1839), thought this inadequate and added (roughly corresponding to the place between paragraphs 2 and 3 on p. 27 in Bertrand, 1839) four paragraphs to Bertrand's summary reviewing Deluc's, de Saussure's, Pallas', Werner's and Hutton's contributions. This is what he wrote about Werner and Hutton:

In an historical summary of geology the name of an Abraham Gottlob Werner cannot be overlooked. By introducing mineralogy and geognosy into the circle of sciences he gave geology a secure basis and made an end to all empty hypotheses by considering the various bodies in the crust of the earth in their spatial relations and internal composition. According to him all rocks and the entire crust of the earth had been built in water by sedimentation. This view he abstracted from his researches in the mountains of Saxony, the only ones he saw himself. But, because he lacked any precise knowledge of volcanic activity and their rocks, he necessarily entered a wrong path. What he saw himself, he observed well, but he ignored the observations of others that contradicted his own. Although the neptunist system erected by him can no longer be maintained in the advanced state of the science, no one can dispute his position as the scientific founder of geognosy and mineralogy, so that through him the small Freiberg became a widely known European name.

What Leibnitz [sic] suspected and expressed prophetically a hundred years earlier, the molten initial condition of the earth,

Figure 34. Albert Auguste Cochon de Lapparent (*Leipziger Illustrierte Zeitung*, 19th century).

Hutton proved with irrefutable facts. Although the volcanic origin of certain rock formations (traps, basalts) had been shown before him, it is to his credit alone to have recognized and proved the same of granite, this basal foundation of the crust of the earth. He also strictly excluded from geology or the evolutionary history of the earth all geogenetic hypotheses (i.e., those about the origin of the earth). His "Theory of the Earth" appeared in 1788.

Thus the basis, on which a scientific geology could be constructed, was laid down by Werner and Hutton towards the end of the eighteenth century. (von Maack, 1844, p. 19–20)

With de Lapparent's treatise and its two enlarged editions in the nineteenth century (1885 and 1893), in which he repeated exactly what he had written in 1883 about Hutton, we almost have reached the temporal limit of this paper.

THE RUSSIAN EMPIRE[142]

Before Lyell

The Russian academicians in St. Petersburg, of diverse ethnic origins, have become aware of Hutton's theory of the earth already in the eighteenth century. Princess Dashkova's early contacts with the Edinburgh scientists may have had a rôle in this early awareness.

On 10 January 1798, the chemist Alexander Nicolaus Scherer (1772–1824; Fig. 35) wrote a long letter to the mathematician Johann Albrecht Euler (1734–1800), Leonhard Euler's eldest child, concerning diverse scientific topics. After having mentioned Hutton's chemical work on alkalis in zeolites, Scherer writes

In Scotland, through the Theory of the Earth of the late Hutton a dominant awareness for volcanism has become disseminated, and even Black and Hope belong to its most enthusiastic supporters. Naturally, opponents have not been lacking; so, for example, B. Thomson in Edinburgh, a few years ago, read a paper against it. But partly the reputation of the person is important in this matter, and partly a great lack of a thorough knowledge of geognosy seem the reason that people do not want to acknowledge how impressive Hutton's theory is. (Scherer, 1798, in Kopelevich et al., 1987, p. 207)

It is remarkable that Scherer, who had a reputation for not mincing his words, ascribes the motive of Hutton's opponents to their fear for their own reputations and their ignorance of "geognosy"!

Dimitrii Ivanovich Sokolov (1788–1852) seems to be the first Russian geologist who mentioned Hutton. After graduating from the Mining Cadet Corps in St. Petersburg (1805), this remarkable man taught there until 1841. Between 1822 and 1844 he was also a professor at St. Petersburg University. Sokolov worked in the Urals. He was a member of the Russian Academy of Sciences but he belonged to the linguistic branch with publications also on this subject.

In 1825 Sokolov published a paper under the title *Uspekhi geognozii* (Advances in Geognosy) in the most famous Russian geological and mining periodical, the *Gornyy Zhurnal* (i.e., Mining Journal). At the time Sokolov was still under the influence of neptunism, imported from Germany. The following is the section in which Hutton is mentioned:

Figure 35. Alexander Nicolaus Scherer about a quarter of a century before his death. No known copyright. Inscription below portrait: "F.W. Nettling sc. 1800. D. Alexander Nicolaus Scherer."

England, having recovered from a deep slumber, into which it had fallen at a time when the French and the Germans were diligently working on the development of geognosy, astonished the world by its quick success in geognosy. And perhaps no country has so much success in geognosy compared with England. It was an unexpected incident that produced such an important triumph.

At the end of the last century, Hutton unveiled in Edinburgh a geological system that was completely different from all previously known and did not agree with the system of Werner, the oracle of our times. Hutton argued that the interior of the Earth consists of substances melted by the strongest fire. He held that the crust originated partly from the gradual cooling of that molten material, partly from the destruction of the cooled layers by the action of the atmosphere and water, and partly from the secondary welding of the annealed parts. These processes were accomplished and to this day still occur on the ocean floor, where the elastic force of the underground heat acts in conjunction with the pressure of water. The disintegrated parts, which are carried to the sea by terrestrial waters, being once the highest, most remote from the inner fire and now occurring in direct contact with water, are subjected to softening and then compaction under the great weight of seawater. These parts form rocks, more or less than mechanical composition—our sandstone, clay, and limestone. The earth masses, which are beneath and which are closer to the fire while being protected by upper layers from the influence of water, melt completely, then crystallize during cooling, using the same pressure as an agent, and produce bulks of crystalline composition. So our granites, gneisses, porphyry were formed. The elasticity of the underground heat raised the molten substance along the cracks in the earth's crust, producing veins and veinlets in it. This force lifted up various layers from the initial horizontal to tilted position.

Hutton's system, backed up by Hall's experiments on the effects of hot bodies exposed to great pressure, and assisted by Playfair's eloquence, greatly shook the minds of the British, flattered their vanity, promising them a crowd of followers, and was adopted unanimously not only in Scotland, but also in England. Suddenly, brave Jameson enters the field. A zealous follower and student of Werner he is eager to protect the glory of his teacher from eclipse, and he establishes the Society of Wernerians in Edinburgh (in 1808) under the eyes of the very opponents such as Hall and Playfair. The war between volcanists and neptunists flared up, and the truth derived its benefit from it: disputes always render triumph. Werner's impartial advocates were not blind in reasoning to the shortcomings of Werner's teaching; and, to the contrary, in the merits of his teaching they found a strong weapon to defeat their opponents. The latter strained all their strength for defense. The geological practice was a mediator and reconciliation of opponents. Werner's teachings remained victorious and returned from the battle field decorated with new perfections, the fruits of his exploits. The geological practice gained new fame and the Geological Society was established in London, whose single goal is to encourage explorations of the structure of mountain ranges around the globe. The provinces began to imitate the example of the capital, and a new society of geologists was formed in Cornwall, striving for the same goal. (Sokolov, 1825, p. 23–27)

Professor Natal'in provided the following commentary on this text:

> I read this sentence several times. Werner is considered the winner! There is no mistake in the translation, but its sense contradicts the second paragraph, in which Hutton's theory is highly prized too. Povarennykh (1953) claims that in 1825 Sokolov was still Werner's follower, but later changed his mind. There were good reasons for this change. Sokolov was associated with the *Gornyy Zhurnal*. In this journal, I found a paper written by an ordinary mining engineer, N.N. Kovrigin (1835–1836). In the Kitoy region, in the Sayan Mountains in Siberia, he describes a system of fractures that can be confounded with bedding (I think that is cleavage which was also described by Sedgwick in the same year; Sedgwick is usually given priority for its discovery) and occurrence of limestones structurally below granites. Obviously, Sokolov was aware of these descriptions before writing his later books published in 1839 and 1842. (Boris A. Natal'in, written communication, 22 July, 2019)

In 1830, a certain A. Kemmerer published a paper in the *Gornyy Zhurnal* comparing the theories of Werner and Hutton. On Werner's theory he wrote:

> However, despite the fact that Werner's, or the so-called Neptunic theory, was accepted by many geologists and has been jealously protected for a long time, it did not deserve universal approval. Moreover, upon a more accurate examination of this theory, paradoxes appear and they cannot be completely resolved. (Kemmerer, 1830, p. 8)

Kemmerer then lists the important objections to Werner's theory mentioning Alexander von Humboldt as one who gathered much data against it. He says that basaltic volcanoes the world over show the same composition and repetitive activity. He finally says that

> …many geologists of our time began to follow a different theory of the Earth's formation, although even before Werner's time some ideas about it had been around. Hutton was the first who established it by many observations, shortly after the appearance of Werner's theory (Kemmerer, 1830, p. 12)

Professor Natal'in added the following comments and further citations from Kemmerer:

> I list the components of Hutton's theory as Kemmerer presents them (his p. 13):
>
> "…The earth was originally a hot mass and at the same time it was surrounded by water."
>
> Rivers erode and destroyed the "continent" moving its parts to the sea. There, because of various reasons but mainly because "the heat of the internal part of the Earth these parts are welded in a single mass" (p. 13). Extending crust (because of its hot state) rises above the sea level creating tilting of layers. Rocks of upper layers were subjected to softening, while deeper layers experience melting. Upper layers were broken into plates then uplifted, tilted, and bent but lower ones became foliated (Natal'in: a specific Russian word is here used which means both bedding and foliation! I interpret it here as foliation).
>
> "Moreover, the flow of heat pushed the molten masses into the layers above. When in this way, through the elevation of mountains, new continents were uplifted from the sea, the former were covered with water" (p. 13–14).
>
> "Finally, Hutton assumes that the present continents, resulting from the destruction of the former continents, themselves are destroyed by rivers and streams, and will produce new continents" (p. 14).
>
> "All geologists, following Hutton's theory, accept that since the creation of the earth to our times the underground heat has continuously decreased, thus the current volcanoes must be considered as the latest works of this heat" (p. 15).
>
> In the following text Kemmerer describes the opinion of plutonists on the succession of formation of plutonic rocks in the earth's history. Granites and syenites were the first to be formed, then porphyry, then basalt, etc. although we know that Hutton did not suggest such a succession. Kemmerer notes that Hutton's theory does not explain the compositions of the magmatic rocks and timing of their formation. I suppose that here he means this to be a difference from Wernerian stratigraphy. (Boris A. Natal'in, written communication, 26 July 2019)

Kammerer's paper is remarkable, both because of its early date in Russia, comparing Werner's and Hutton's theories and clearly favoring Hutton's and its statement that "many geologists of our time began to follow a different theory of the Earth's formation" thereby explicitly meaning Hutton's.

After Lyell and before Geikie

In 1839, Sokolov published the first Russian textbook on geology, aimed not only at students, but also the general public, with the title *Kurs geognozii, Sostavlennyy Korpusa Gornykh Inzhenerov Polkovnikom, Sanktpeterburgskogo Universiteta professorom D. Sokolovym* (A course of geognosy, compiled by the Corps of Mining Engineers Colonel, Saint Petersburg University, Professor D. Sokolov), in three volumes. In 1842 a second edition appeared under a somewhat modified title *Rukovodstvo k Geognozii* (Guide to Geognosy), in which the author seems much influenced by the work of Lyell. In the second volume, we read:

> Werner's theory rejected any participation of underground heat in the formation of the earth. Volcanoes were considered in it as only local phenomena related only to the earth's surface.

Meanwhile, as this theory prevailed in Europe, the Scotsman Hutton, a contemporary of Werner, tried to establish a completely different theory, which asserts the fiery formation of granite and trap rocks, as well as the fact that these rocks emplaced in the solid crust of the earth at different periods. Hutton described in great detail granitic veins and the changes that occurred in the layers that were cut by them. Hutton also suggested the current theory of metamorphism. He was the first to assert that the so-called primary or crystalline rocks had not settled out of water, but had formed under the influence of heat on the sedimentary rocks of a mechanical origin. Playfair was a zealous advocate and commentator for Hutton's theory, nicknamed as the volcanic theory. Finally, after some debate, Hutton's theory was accepted by all the geognosists…. (Sokolov, 1842, p. 5–6)

The great Russian geologist Ivan Vassilievich Mushketov (1850–1902; Fig. 36) published the first volume of his textbook *Fizicheskaya Geologiya* in 1899 to be followed by one more posthumous edition in 1926. A precursor book to these was published in 1880 under the title *Kurs Geologii, Chitannyy v Gornom Institute S. Petersburg* (The course of geology delivered in the Mining Institute St. Petersburg). In that book, on p. 50, we read:

Figure 36. Ivan Vasilievich Mushketov. Public domain (PD-US-expired).

In 1788, after numerous studies in England and Scotland, Hutton published a "Theory of the Earth." This was the first book, in which geology was recognized as a science that had no connexion with the question of the origin of the earth. In it, for the first time, we find an attempt to dispense with contrived hypotheses and to explain the primitive changes of the earth by relying only on natural causes. He divided the rocks into igneous and sedimentary. He was the first to consider granite, along with basalt, an igneous rock, based on studies in the Grampian Mountains, where he found granitic veins in metamorphosed limestone. Werner had considered granite to be a part of the original core of the planet, so it is not difficult to imagine how big the indignation was, when Hutton decided to oppose the teachings of the Freiberg school. His defender was his expositor Playfair. Hutton allowed periodic revolutions and in this respect he lagged behind Hooke, Moro and Raspe. Moreover, Hutton and Playfair explained all the valleys and sediments by the action of rivers. They did not allow the formation of any topographic irregularities on the ground by elevation and depression of the earth's surface.

I shall not go into the mistakes made by Mushketov in his presentation of Hutton, because my point here is to show Mushketov's perception of Hutton as a more modern geologist with respect to his predecessors and contemporaries.

On p. 61, Mushketov mentioned the neptunist/plutonist dispute and the foundation of the Geological Society of London, almost as a reaction, by those devoted entirely to observation not to get mired in controversies about hypotheses.

In 1885, the famous, but unfortunate Russian geologist, Alexandr Alexandrovich Inostransev (1843–1919; he took his own life as a consequence of the Russian Revolution), friend and correspondent of Eduard Suess, published his lectures under the title *Geologiya—Obshchiy Kurs Lektsiy* (Geology—General Course of Lectures). In that book Inostransev mentions Hutton "among the greatest scientists." In volume 1 of that book, he wrote:

Hutton was a vigorous opponent of Werner's hypothesis. In contrast to Werner, this scientist taught that most rocks such as granites and traps are volcanic rocks that poured out onto the earth's surface in a molten state at different times and in different parts of the earth's surface, and that gneisses and various crystalline schists, which are called primary rocks, do not represent chemical, or mechanical sediments of the primary ocean, but in fact are ordinary mechanical deposits, modified by the internal heat of the earth. Here, therefore, on the one hand, Hutton considered volcanism in a broad sense, on the other hand, he was the first to recognize metamorphic rocks such as gneisses and crystalline schists, that is, that is the rock, which underwent a certain kind of modification under the influence of later phenomena. One of the main reasons for objection to Werner's hypothesis was the combined presence in granites and gneisses of minerals that have so different solubilities as feldspar, mica, and quartz. According to Hutton, the joint formation of all these minerals from aqueous solutions is not conceivable, because their degree of dissolution in water is different, and therefore we would have to meet not a mixture of all these minerals together, but separate layers composed only of individual minerals, which is not found in reality. This objection gave rise to a dispute between Werner and Hutton supporters, known as the Neptunists-Plutonist dispute, which presented such a diversity of views on the origin of ancient rocks. (Inostransev, 1885, p. 383–384)

In the 1899 *Fizicheskaya Geologiya* of Mushketov, there is a short, but excellent historical introduction to the first volume: there, the great Russian master wrote:

But Werner's extreme neptunic view caused a fierce struggle with extreme vulcanists that were led by Hutton. The fierce and abundant quarrels between neptunists and vulcanists once more undermined the esteem for geology, especially since religious issues interfered, as it had happened before. Werner's neptunism is a negative side of his activity, which however in no case can destroy his positive merit, as the founder of scientific geology, especially since he left behind two great students. (Mushketov, 1899, p. 6)

1899 is two years after my upper temporal limit of this paper. However, what Mushketov says about Hutton in his own book does not resemble what Sir Archibald had written in his publications between 1882 and 1897, but is in very much in line with the previous Russian views as they had developed in Russia. Mushketov presents Hutton as the one who helped to destroy Werner's neptunism, which he says was Werner's great mistake, and his mention of two Hutton converts, Leopold von Buch and Alexander von Humboldt, as Werner's students for their merit as the two great pioneers of geology corroborates what I deduce from his words. Thus, it seems clear that Mushketov's assessment of Hutton was independent of Sir Archibald's evaluation, as also shown by his 1880 statement.

GERMANY

Before Lyell

Werner dominated the German geological thinking in the last two decades of the eighteenth and the first two decades of the nineteenth centuries and, because first the French Revolutionary and then the Napoleonic wars cut off Great Britain from the rest of Europe, we see very little in the German geological literature concerning Hutton during these 40 years. A prominent example of this state of affairs is von Leonhard's oration on the occasion of the festivities for the name day of the Bavarian king (Maximilian I Joseph, 1756–1825; reigned as king: 1806–1825) before the Academy of Sciences in Munich on 12 October 1816 entitled *Bedeutung und Stand der Mineralogie* (Significance and State of Mineralogy) in which von Leonhard reviewed both mineralogy and paleontology plus geology (von Leonhard, 1816) and did not find it necessary to mention Hutton, although, already in the last decade of the eighteenth century, the famous German physicist and satirist, the first holder of a chair of experimental physics in Germany, Georg Christoph Lichtenberg (1742–1799) not only had read Hutton, but also mentioned him in his lectures on physical geography and included his *Theory of the Earth* paper of 1788 in the list of books he used for his lectures (Gamauf, 2008, p. 938–953, Hutton's 1788 paper on p. 942, item 39; see also von Engelhardt, 1986/1987, p. 34). His student, the theologian and naturalist Gottlieb Gamauf (1772–1841) remembered him saying:

> Many, among whom Doctor Hutton, believe that the sea-floor was raised by fire. (Gamauf, 1818, p. 405; 2008, p. 953)

However, Lichtenberg did not approve of this and thought that it was more likely that parts of the earth's crust must have subsided and drawn the oceanic waters away from its former floor so as to expose it to our access. He says for such a phenomenon we have actualistic examples: one reported by Lulof in his *Einleitung zu der Mathematischen und Physikalischen Kenntnis der Erdkugel* (translated from Dutch in 1755 and published in Göttingen): in 1421 in Holland an entire area containing 27 villages was supposed to have "gone down." Also a landslide in Winzigerode near Duderstadt, in the vicinity of Göttingen, in 1798, he thought could be an example of such subsidences of the crust and asked his student Benzenberg to investigate it, publishing it later in the *Hannöverisches Magazin*: von Engelhardt, 1986/1987, p. 27). He also mentions the creation of the small Bay of Dollart[143] in the Waddenzee in Holland in 1277 as a consequence of a catastrophic storm surge, during which some 43 settlements had perished and ~80,000 people lost their lives. Lichtenberg was also aware of volcanic uplifts and cited the origin of Monte Nuovo volcano in Pozzuoli near Naples in 1538 and the origin of a new island near Terceira in the Azores in 1720[144]. But he thinks they are not to be equated with what he calls terrestrial revolutions (Gamauf, 2008, p. 953–954).

But this was not the only thing Lichtenberg mentioned about Hutton's theory of the earth. In his posthumously published notes that he had prepared for his lectures he classified Hutton's theory among those that took an empirical route to geological investigations:

> To observe carefully the form of the crust and to deduce from it the immediate conclusions, and from these the next immediate and from those the next … so to proceed backwards. In this regressive evaluation there are, as one can imagine, multifarious limitations and also, unfortunately, differences in the precision of the observations. To this group belong [the theories of] Dr. Hutton, von Saussüre, La Metherie, Dolomieu and DeLuc.[145]

Lichtenberg pointed out that in 1789 in his *The Natural History of the Mineral Kingdom*, the mineral surveyor John Williams (1730–1797) had mentioned Hutton's theory (see above). Lichtenberg presents Hutton's theory by pointing out that it could be distilled into four propositions to all of which he also cites William's objections: (1) All rocks and strata formed in the sea, to where their components, as ruins of a previously existing land, had been carried by floods on land; Williams admits that mountains and rocks are indeed weathered and carried to the sea by rivers, but he denies completely that strata form in the sea and in this he is supposed to be supported by Dolomieu and others. The author (i.e., Williams) proves that all that had been so carried is thrown back by floods etc. into bays and mouths of the streams and onto the shore in general. Williams thinks thus forms new land, especially in deltas. (2) Strata under the sea is indurated by subterranean fire. This is considered unnecessary, because we see strata harden every day in caves. Williams does not consider basalt as "molten stone"; he says whoever considers basalt volcanic and a lava, he must also consider the strata above and below it so also. (3) Mr. Hutton thinks that strata under the sea is thrown up by subterranean fire. Mr. Williams says that all effects of fire are violent and we see nothing of the sort in the strata. (4) Through the three propositions above one sees that there is a continuous cycle of erosion and deposition, consolidation and uplift. Williams objects and says that such erosion does not extend far as Dolomieu also excellently demonstrated. ...

Finally, Lichtenberg says that Williams' book contains much that is excellent, but his theory cannot stand, because he has little general knowledge, especially physical, to be able to handle scientifically a chaos of facts. Although Lichtenberg does not approve of Williams' objections to Hutton, he likes Deluc's theory more (despite the fact that Lichtenberg himself was an atheist), as propounded in the first two volumes of his *Lettres Physiques et Morales sur l'Histoire de la Terre et de l'Homme adressées à la Reine de la Grande Bretagne* (all published until that time), mainly because subsidence of the crust, as proposed by Benjamin Franklin (whom Lichtenberg very greatly admired) in a letter to the French naturalist Abbott Jean-Louis Giraud-Soulavie (1751–1813) seems more congenial to him than uplift by igneous agencies (see Lichtenberg's relation of Franklin's theory of the earth in Gamauf, 1818, p. 379–403).

Lichtenberg's notes are disorderly, in places self-contradictory, not very rarely also mocking; his German is careless, abounding in grammatical and orthographical (even for his time) mistakes, simply because they were never intended for publication. They were his *aides-mémoires* for his lectures and as such they are admirably rich in information. His coverage of the literature was truly amazing. It is hardly surprising that he was aware of Hutton's theory and liked many aspects of it.

In 1817, Carl Ritter (1779-1859), one of the two founders of modern scientific geography together with Alexander von Humboldt (1769-1859) published the first volume of his famous *Erdkunde im Verhältnis zur Natur und zur Geschichte des Menschen oder allgemeine, vergleichende Geographie, als sichere Grundlage des Studiums und Unterricts in physicalischen und historischen Wissenschaften* (Geography in relation to Nature and human history, or general, comparative geography as a secure basis for the study and instruction in physical and historical sciences). In the Introduction, Ritter reviews the development of geology as part of physical geography. After having related the accomplishments of Leopold von Buch in overcoming his teacher Werner's neptunism, he continues:

> From the north, the Scottish naturalists reached him the hand, unbeknownst to him, through ideas inspired by him. The famous hypothesis of Hutton, based on the granite dykes of Cornwall, their explication by Playfair, based on more methodology and a better overview of the facts, and the chemical experiments by J. Hall on the products of fire that must have formed under pressure, enriched the physical geography with ideas and hypotheses, which not only in Iceland through Olafsen, now again by Mackenzie, but in almost all the Atlantic islands have been proven right. (p. 46–47).

This is a somewhat bungled account of the accomplishments of Hutton and his friends, as Ritter seems to think that they had been inspired by the work of Leopold von Buch. Here is the German original of the first sentence in the above quote: "Vom Norden her boten die schottischen Naturforscher, ihm unbewußt, seinen durch ihn aufgeregten Ideen die Hand." Is it possible that here *ihn* may refer to Hutton in a carelessly constructed sentence? Although Ritter's prose style is well-nigh impenetrable and such errors in long, run-on sentences are not rare, I must point out that in the second edition, the same paragraph is repeated verbatim (Ritter, 1822, p. 46-47). Ritter also erroneously thinks that it was the granitic dykes in Cornwall that formed the basis of Hutton's theory. The reason that I quote Ritter here is that although he clearly knew of Hutton only second hand, yet realized his importance in toppling the Wernerian geology and thus enriching physical geography.

Only five years after Leonhard's oration, a pamphlet appeared anonymously in Bonn entitled *Kritik der Geologischen Theorie besonders der von Breislak und Jeder Ähnlichen* (Criticism of the geological theory especially of Breislak and other similar ones: Anonymous, 1821). The author of the anonymous pamphlet was actually Karl Wilhelm Nose (1753–1835; Fig. 37), a German physician and mineralogist, correspondent of Goethe and Alexander von Humboldt and a convinced neptunist, who first described the mineral sanidine (a potassium feldspar: (K, Na) (Si, Al) $4O_8$)) in 1789. The purpose of the pamphlet is to criticize both the neptunists and the vulcanists and to argue that all rocks on

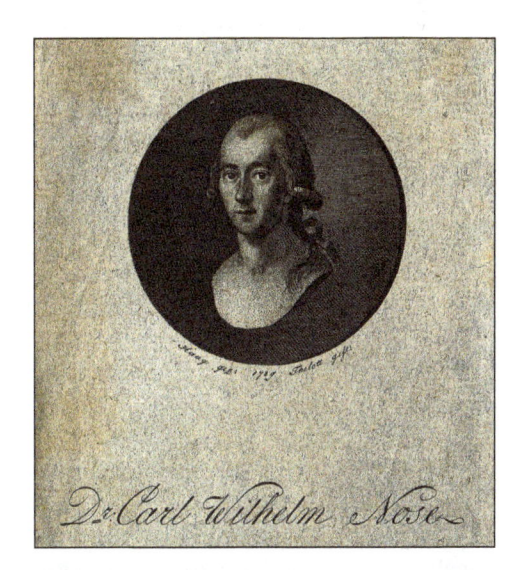

Figure 37. Karl Wilhelm Nose. Stipple engraving by E.K.G. Thelott after Haug, 1789. Credit: Wellcome Collection. Attribution 4.0 International (CC BY 4.0).

earth are of two types: a primitive group created by an almighty cause not further specified (not necessarily supernatural) and a secondary group that forms by the transformation ("Umformung": Anonymous, 1821, p. 35) of the primary material. Nose allows both neptunian and volcanic processes in transforming the primitive rocks. In this pamphlet there is much talk of neptunists and vulcanists as Nose's main target is the Italian geologist Scipione Breislak, the ultravulcanist, but as yet no mention of Hutton specifically. In 1822 Nose published a continuation of his earlier pamphlet, again anonymously (Anonymous, 1822) where one of the earliest mentions of Hutton's name in the German geological literature appears. On p. 4, Nose quotes from the introduction of Friedrich Karl von Strombeck (1771–1848), the German jurist and amateur geologist who translated into German an improved version of Breislak's *Introduzione alla Geologia* (1811; see below). In that introduction, von Strombeck (1819, p. VI) mentions Hutton as an immortal geological celebrity along with de Buffon and Delamétherie. Then, Nose (1822) adds on p. 5 that "von Strombeck is an enthusiastic vulcanist." Then Hutton is further mentioned in the body of his text (p. 12, 13, 21, 25, although he does not appear in the author index!). In fact, on p. 21, there is a statement concerning a comparison between "neptunism and Huttonism." I do not recall seeing neptunism contrasted with "Huttonism," instead of "vulcanism" or "plutonism" anywhere else.

Already in 1822 Christian Leopold von Buch, Baron of Gelmersdorf and Schöneberg (1774–1853; Fig. 38), one of Werner's greatest students and one of the most influential architects of the magmatic uplift theory in the nineteenth century, was converted to Hutton's views on the volcanicity of basalts and openly said so years before Lyell published anything on the subject. Von Buch followed the vulcanist-neptunist debate with interest, and indeed with involvement, initially on the side of the neptunists (von Buch, 1800[1867]), then with disappointment, yet still with some stubbornness (von Buch, 1809[1867], p. 518):

> So we stand dismayed and crushed over the results, to which the appearance of Montdor has led us. If the porphyry of the Puy de Dôme, of Sarcou, of Puy de Nugère originated from granite, so could also the beds of the Montdor owe their origin to the metamorphosis of the granite (not melting) and the basalt could be a fluid product of these rocks. But even the most enthusiastic vulcanists would not dare to view this result as a general one and to apply it to the German basalts.

In 1805, however, another Werner student (and later fierce opponent in the basalt dispute), Johann Karl Wilhelm Voigt (1752–1821) reported already, that von Buch had told him that the lavas of the 1760 and the 1794 eruptions in Vesuvius could not be distinguished in any way from basalts (Voigt, 1805, p. 41). Von Buch finally gave in completely to Hutton's ideas (among many other publications of his, for example, von Buch, 1818–1819[1877], p. 17, where he explicitly and "gladly agree[d] with Dr. Hutton")[146].

Figure 38. Christian Leopold von Buch, Baron of Gelmersdorf and Schöneberg. Used with permission, Bibliothèque nationale de France.

The last publication mentioning Hutton in Germany before the publication of Lyell's *Principles* is actually a part of an encyclopaedic textbook of forestry. The eighth part, volume four of the multi-volume *Die Forst- und Jagdwissenschaft nach allen ihren Theilen für Angehende und Ausübende Forstmänner und Jäger* (Forestry and Science of Hunting in all their parts for prospective and practicing Foresters and Hunters) is devoted to geology and is entitled *Lehrbuch der Gebirgs- und Bodenkunde in Beziehung auf das Forstwesen* (Text-book of Rock- and Soil Science in Relation to Forestry: 1826) by Stephan Behlen (1784-1847), chief forester and professor in the Forestry School in Aschaffenburg (*Forstliche Hochschule Aschaffenburg*) in northwesternmost Bavaria. Behlen made some contribution to the geology of Spessart, but he was clearly not in the forefront of the science of geology when he wrote his book. He was stuck in various teachings of the Wernerian school (although he himself had studied in the University of Mainz), but, in his textbook, he nevertheless wrote:

> Hutton, with a view to explaining the present state of the earth, observes Nature and sees a state of decay everywhere and in this apparent destruction finds the real source of its renewal. The continents were destroyed by the atmosphere and water, its mass was decomposed and collapsed and the ruins were carried into the sea and were spread out on its floor. Not without influence is the subterranean heat, which penetrates and softens the ruins, indurates the upper ones forming the layered deposits and melts the lower ones building the granites, the greenstones etc.
>
> This expansive force is responsible for uplifting those layers and masses above the level of the waters and thus new continents come into being.
>
> Hutton assumes an uninterrupted sequence of such destruction and reformation until the creator of the world will be pleased to set an end to all. (Behlen, 1826, p. 218)

Behlen cleary noticed in his book that the Wernerian theory of the earth was deficient, because one needed a mechanism for uplifting mountains and he evidently found it in Hutton's ideas. What is of the greatest interest here is Behlen's view of the role of the subterranean heat. He seems to assume a thermal gradient that melts rocks at depth and just indurates them higher up, indicating the way, perhaps, to a more clement interpretation than usual of Hutton's statements on the heat induration of strata. This is not too different from what the Count of Collegno wrote 21 years later (see below), clearly with no knowledge of Behlen's book.

After Lyell and before Geikie

One of the most influential geology textbooks published in Germany after the publication of Lyell's *Principles* was Karl Cäsar von Leonhard's (1779–1862; Fig. 39) *Geognosie und Geologie* that appeared as a part of a multi-volume series with the title *Naturgeschichte der drei Reiche. Zur allgemeinen Belehrung* [Natural History of Three Kingdoms. For general education ...] *bearbeitet* [edited] by G.W. Bischoff, J.R. Blum, H.G. Bronn, K.C. von Leonhard and F.S. Leuckart. Leonhard's first edition came out in 1835, in which he referenced Hutton's two volume book, Playfair's explanatory volume and its translation into French (see von Leonhard, 1835, p. 25) in his 75-item bibliography. In this book, he considers Werner as a founder of a considerable part of the science of geognosy. In this edition there is no history of geology, neither a special comment about Hutton elsewhere. However, the book contains a sort of prologue published in 1832 by the famous zoologist Friedrich Sigismund Leuckart (1794–1843). The prologue has two parts: a foreword (p. 1–6) and a "Speech" (p. 7–129). It is in the "Speech" that Hutton is mentioned as an opponent of neptunism:

> Others, by contrast, for example Hutton and his supporters, think that the main metamorphoses and changes of the earth up to the present day were caused by the activity of fire. Those, who accepted this theory are called vulcanists. Buffon, Lagrange, Dolomieu, Laplace, Cordier and others even think that the earth has an inner, central fire. (Leuckart, p. 38–39)

What von Leonhard says in the second edition of his successful textbook is of much greater interest:

> Rightly, Werner is considered the founder of a considerable part of geognostical science. His works and their influence remain among the most beautiful victories of the German mind. But, James Hutton must be regarded as the founder of modern geology, without belittling the services of Alexander von Humboldt, Leopold von Buch, Cuvier, Élie de Beaumont, Murchison and a few other luminaries of our times. (von Leonhard, 1849, p. 8)

Note that this statement was published nine years before Sir Archibald Geikie wrote anything about the Scottish school of geology or Hutton himself. Von Leonhard then adds that the

Figure 39. Karl Cäsar von Leonhard. Photo of an original lithograph by Rudolf Hoffmann, 1857; Austrian National Library, Vienna.

following parts in his book will bring forward sufficient proof of that statement. In their praise of Hutton, Leuckart, and von Leonhard were not alone in Germany, however.

Friedrich Hoffmann (1797–1836) one of Germany's most promising geologists in the first decades of the nineteenth century, who regrettably very early succumbed to a chronic disease, most likely some sort of cancer that destroyed his vital organs as revealed by an autopsy, wrote one of the earliest histories of geology. He presented it in his lectures in Berlin in 1834, but his text had to be published posthumously in 1838. It is a superb summary of the development of geology from earliest antiquity to his day: in it, this is what Hoffmann had to say about Hutton:

> As it happens, a very remarkable view of the construction of the earth, inspired purely by vulcanistic ideas, was developed in Scotland. Had it not coincided in time with the heyday of Werner's school, it would have had a much greater influence. The representative of this view was James Hutton (*Theory of the Earth*, 1795, two volumes). John Playfair (*Explications*, 1802) added very attractive explications confirming the principles enunciated by Hutton and they contain a large amount of very important and appealing facts for physical geography. This whole work was translated into French under the title *Explication de Playfair sur la Théorie de la Terre par Hutton*, traduit de l'Anglais par C. A. Basset, 1815.

> J. Hutton was motivated by the observation of numerous dykes in Scotland and the neighbouring islands, consisting so commonly of all the older volcanic rock types such as granite, greenstone and basalt. He very quickly noticed that they were formed by tearing of the original continuity and the alteration and surficial melting of the surrounding rocks. He was therefore the first not only to grasp enthusiastically this phenomenon, which later was employed to support views on the abnormal rock types, but also correctly interpreted it in that he thus proved that these older rocks had risen along

fissures once in a molten state and spread over the surface. Naturally, the nature of his homeland greatly facilitated his researches, as in deeply cut and nakedly exposed coastal profiles sections are seen that show the structure of the earth's crust in a much better manner than ever possible in our own country. While studying these volcanic phenomena Hutton did not at all neglect the neptunian products in his fatherland. By contrast, it was expressly a fundamental tenet of his system that the mostly younger, conspicuously-stratified rock types must have formed as deposits on the sea-floor. He believed that the induration of these rocks, the consolidation of the sandstones and the conglomerates, must have happened as a result of heating, even approaching melting, after their formation. He was also very attentive to the bending and steepening of strata. At a time when Werner, as we saw, claimed the opposite amidst enthusiastic support, he asserted that all mountainous masses had risen from below to the surface of the sea. Originally horizontally layered rock masses at the bottom of the sea were later dislocated from their original positions, broken and bent through later volcanic phenomena. He tried moreover to prove that lands so constructed were later disintegrated by volcanic shaking and floodings of the sea and that from the accumulation of such debris and products of destruction newer continents would form and, as far as we know from geography, such transformation phenomena had already happened at least twice before.

These views make Hutton without a doubt the creator of the ideas now generally accepted in our science. We therefore can argue that Scotland is the cradle of the new geology. (Hoffmann, 1838, p. 208–210)

Figure 40. Siegmund August Wolfgang Baron von Herder. Unknown author.

An interesting point made by Hoffmann is that Werner's dominance in Germany had hindered the spread of Hutton's ideas in that country. This is similar to the deleterious effect of the great German physicist Wilhelm Eduard Weber's (1804–1891; one of the "Göttingen Seven") idea on the instant and direct action at a distance had against the acceptance of Maxwell's electromagnetic interpretation of the nature of light in Germany (Planck, 1949, p. 272). When Hoffmann pronounced these views in his lectures in Berlin in 1834, Sir Archibald Geikie had not yet been born and when they were finally published in 1838, he was only three years old!

In 1837, the German geologist and director of mines in the Kingdom of Saxony Siegmund August Wolfgang Baron von Herder (1776–1838; Fig. 40), the second son of the influential German philosopher Johann Gottfried von Herder, published a proposal for a deep shaft with a view to secure the future of the Freiberg mining district under the title *Der Tiefe Meissner Erbstollen—Der einzige, den Bergbau der Freyberger Refjer für die fernste Zukunft sichernde Betriebsplan*. Of interest to his undertaking was the knowledge concerning the depth ore-bearing veins could reach. To that end he reviewed the then common theories of vein origin (von Herder, 1838, p. 27–32). There were four of them: the congeneration theory that assumed that the veins originated together with the rocks in which they are now found. The second theory assumed that the vein contents originated by lateral secretion from the country rock. The most popular in Freiberg for a long time was Werner's "descension theory" that assumed that the veins represented open fissures and that their contents were fills that settled as sediment from above. Finally the ascention or plutonic theory argued that veins represented conduits for magmatic rocks supplied by plutons at depth. He came to the conclusion only the last of the theories he listed, the ascension or plutonic theory, "which makes veins get filled and formed by means of watery-muddy, or fiery-fluid, molten or sublimated masses or through mineral waters:—which theory is adhered to by the most accomplished geologists of our days, namely von Humbold [sic Alexander von Humboldt, 1769–1859], von Buch [Leopold von Buch, 1774–1853], von Dechen [Heinrich von Dechen, 1800–1889], Hutton [James Hutton, 1726–1797], Parrot [Friedrich Parrot, 1792–1841], Brongniart [Alexandre Brongniart, 1770–1847], Fournet [Joseph Fournet, 1801–1869], Long-Champ [de Longchamp], Beaumont [Léonce Élie de Beaumont, 1798–1874], Hausmann [Friedrich Hausmann, 1782–1859], Nöggerath [Johann Jacob Nöggerath, 1788–1877], Naumann [Carl Friedrich Naumann, 1797–1873], von Leonhard [Karl Cäsar von Leonhard, 1779–1862], Bischoff [sic Karl Gustav Bischof, 1792–1870], Weiss [Christian Samuel Weiss, 1780–1856] and many others. This theory is to be preferred to the other three mentioned before it" (von Herder, p. 29–30). I note that here Baron von Herder names Hutton among "the most accomplished geologists" of his time. When the Baron published his list, Sir Archibald Geikie was barely two years old.

The last mining director of Saxony and a former student of the Bergakademie in Freiberg, the geologist and jurist Friedrich Constantin Baron von Beust (1806–1891; Fig. 41) wrote a critique of Werner's theory of veins in 1840 and pointed out that from the viewpoint of neptunism, Werner's interpretation made perfect sense, but, since little had remained of Werner's overall theory of geology, it had become difficult to defend his ideas on the origin of veins. Already on the third page of his book, von Beust wrote that in order to uphold Werner's geology, one had to ignore what had since been done by "Alexander von Humboldt, Leopold von Buch, Heim [Johann Ludwig Heim, 1741–1819], Studer [Bernhard Studer, 1794–1887], Weiss, Hausmann, Hoffmann [Friedrich Hoffmann, 1797–1836], G. Rose [Gustav Rose, 1798–1873], von Dechen, von Oeynhausen [Karl Baron von Oeynhausen, 1795–1865], Noeggerath, von Leonhard, Walchner [Friedrich August Walchner, 1799–1865], Naumann, Cotta [Bernhard von Cotta, 1808–1879], Élie de Beaumont, d'Aubuisson [Jean-François d'Aubuisson de Voisins, 1769–1841], Fournet, Brogniart [sic], Boué [Ami Boué, 1794–1881], Beudant [François Sulpice Beudant, 1787–1850], Dufrénoy [Ours-Pierre-Armand Petit-Dufrénoy, 1792–1857], Prévost [Constant Prévost, 1787–1856], Thurmann [Jules Thurmann, 1804–1855], Graf Montlosier [François Dominique de Reynaud, Comte de Montlosier, 1755–1838], Lyell [Sir Charles Lyell, 1797–1875], Macculloch [sic John MacCulloch, 1773–1835], Greenough [George Bellas Greenough, 1778–1855], Buckland [The Reverend Dr. William Buckland, 1784–1856], Conybeare [The Very Reverend William Daniel Conybeare, 1787–1857], Hutton, Sedgwick [The Reverend Adam Sedgwick, 1785–1873], Murchison [Sir Roderick Impey Murchison, 1792–1871], Scrope [George Julius Poulett Scrope, 1797–1876], De la Bèche [sic Sir Henry Thomas De la Beche, 1796–1855] and many others". This list is clearly influenced by von Herder's list. It is a much embellished republication of the same and yet, here, Hutton's position in the eyes of a German mining geologist emerges even more forcefully as one who prepared the modern geology of those days. Baron von Beust emphasizes that if one wanted to follow Werner's theory in geology, one had to ignore all the work done by the names he cites, which amount to almost all the founders of modern geology (with the exception of paleontologists and biostratigraphers).

What Friedrich Hoffmann presented in his Berlin lectures were popularized by Hermann Hauff (1800–1865; Fig. 42), the learned translator of Alexander von Humboldt's *Relation Historique,* in the same year as Baron von Beust's small book. In the second volume of his *Skizzen aus dem Leben und der Natur. Vermischte Schriften* (=Sketches from Life and Nature. Miscellaneous Writings: Hauff, 1840) Hauff has an entire section entitled "Geologische Briefe" (=Geological Letters) in which the following are the chapter headings: "Das Verhältnis der Geologie zu unserer Zeit" (=The relationship of geology to our time), "Orientirung" (=Orientation), "Aeltere Ansichten" (=Older views), "Werner," "James Hutton. Leopold von Buch," "Élie de Beaumont," "Neueste Beobachtungen" (=Newest observations). Hauff introduces this

Figure 41. Friedrich Constantin Baron von Beust. Source: Walter Fischer: Aus der Geschichte des sächsischen Berg- und Hüttenwesens, Hamburg 1965, S. 14. Public domain (PD-US-expired).

section by pointing out the great progress geology had made during the last decades of the eighteenth and the first decades of the nineteenth centuries and ascribes it largely to the theories of Werner, Hutton, Cuvier, von Buch and von Humboldt. Interestingly, he does not here mention Lyell nor does he mention Spallanzani, Brocchi, William Buckland, William Smith, Adam Sedgwick, Roderick Murchison, Élie de Beaumont (although he has a special chapter for him later on), or Constant Prévost. It is clear that Hauff was interested listing in his introduction only those whom he believed to be the *originators* of the ideas that had led to the progress he wished to emphasize. The review he presents is based very largely on Hoffmann's history, referenced above.

Hauff then goes on to compare the developments in astronomy with those in geology. Whereas astronomy took entire stars and planets as its subject matter with a view to establishing their present positions and movements with respect to one another, investigating the history of our planet using a similar holistic approach had led to failure and it had become clear that one had to start observing the relationships of the rocks at the surface with each other to erect a history. Copernicus, says Hauff, created the modern astronomy. What he then says concerning geology is of the greatest interest from the viewpoint of this paper:

Figure 42. Hermann Hauff. Portrait of Hermann Hauff, before 1865, unknown. Source: Karl Fricker: Wilhelm Raabe's Stuttgart years reflected in his poetry. Stuttgart 1939, p. 12.

> The history of the earth in time is now the task of the present and the future and James Hutton is the Copernicus, Leopold von Buch is the Galileo of the new episode of research. (p. 410)

He compares Werner's "exclusively neptunist system" with the Ptolemaic theory of the heavens (p. 410–411). Hauff points out that during the first decades of the nineteenth century, geology was the most popular natural science. Its discoveries were reported in newspapers of a rapidly growing journalistic profession. However, popular science, he says, always lags behind the current state of research and he observes that, despite the new developments, the old Wernerian neptunism and the idea of a last great catastrophe that allegedly created the present face of the earth still circulate in the popular press and even among the better educated public. His purpose in his "Geological Letters" is to emphasize the difference between the old, exclusively neptunistic theories and the modern views.

In the letter entitled "Orientirung" Hauff gives an admirable synopsis of the science of geology; its part relating to external phenomena and paleontology can even today be given to a first-year geology student to be read with benefit. The internal processes are based on the igneous intrusion-motored uplift theories then prevalent, where Hauff again emphasizes the inadequacy of Werner's older neptunistic theory. He deals with "older views" in his next letter where he contrasts the Egyptian neptunistic theories because of the influence of the Nile with the Greek vulcanist theories inspired by the live volcanoes and strong earthquakes in the places where Greeks lived. In Hauff's history of geology, neptunist and vulcanist theories are presented as eternal opponents,

oft reflecting the personalities of their defenders. He points out the profound influence of the Great Geographical Discoveries on geological research and emphasizes Steno's rôle in being the harbinger of both the neptunistic and vulcanistic theories well into the nineteenth century. With regret he notes that almost all the successors of Steno until the middle of the eighteenth century were his intellectual inferiors and they tried to erect theories that agreed with the letter of the Holy Script. He quotes the great physicist and famous wit Georg Christoph Lichtenberg (1742–1799) who had said that all such bibliolatrous theories of the earth were not contributions to the history of the earth, but contributions to the history of the aberrations of the human mind[147]. After a brief review of the ideas that regarded fossils as "plays of Nature" he points out that from the middle of the eighteenth century onward, people realized that holistic theories based on the biblical creation myth had led geology into an erroneous path and began looking at the rocks themselves to erect an earth history. Here Hauff sees Werner's service as decisive in giving geologists a common language. He says Werner is the real founder of scientific geognosy. He admires his mineral system, but then points out that Werner betrayed Steno's principles in his extreme neptunistic earth theory that was unable to explain the geological processes, very especially those related to the internal dynamics of the planet. It is here that he introduces Hutton:

> James Hutton directed his attention to a phenomenon that repeated itself in a conspicuous manner in his homeland Scotland; to the fact that the so-called abnormal, unstratified rocks, which Werner explained, as we saw, as having formed in water—although they had been regarded as old volcanic products before him—namely granites, greenstones, porphyries, basalts filling in multifarious ways dykes and crevasses in stratified rocks. Hutton noticed that thereby the intruded rocks always appeared dislocated and torn. He recognised the original incandescent fluidity of the intruded materials in that the country rocks were often chemically and mechanically altered as if their surfaces had been fused. Hutton thus was the first to pronounce that such abnormal rocks were raised through the fissures of the earth as melts and often spread out onto the surface, where they were immediately covered by deposits of water so that they often appeared as intercalations between beds. This situation was one of the most embarrassing puzzles in Werner's theory, because thus granite, supposedly the oldest rocks, appeared concordantly interlayered with the deposits of much later times. (p. 469)

Hauff then says that Hutton published these ideas in 1795, at a time when Werner's ideas had begun to conquer the opinions of geologists and that is why initially Hutton's views had little impact, but, "Hutton, is however, the father of modern geology" (p. 470).

Here is an extremely well-informed and well-connected German author acknowledging the paramount importance of Hutton's position in the history of geology to the point of calling him the "father of modern geology" (when Sir Archibald Geikie was only five years of age!). In this, Hauff was no doubt greatly influenced by Leopold von Buch's conversion to Hutton's views as I quoted above and by Hoffmann's history.

In the same year as Hauff's book came out the influential little book by the autodidact Christian Keferstein (1784–1866) entitled *Geschichte und Litteratur der Geognosie, ein Versuch* (=History and Literature of Geognosy, an Essay), in which there is a lengthy and laudatory section on Hutton. That excellent book remained the standard work on the subject for decades after its publication. Before the publication of his "History," Keferstein had published a two-volume *Naturgeschichte des Erdkörpers in Ihren ersten Grundzügen* (=Natural History of the Earth in its First Principles) in 1834, in which he mentioned Hutton's contributions with emphatic approval:

> Since the keen observations by the ingenious Hutton, one is more and more convinced: that granites and similar crystalline rocks are genetically very similar to basalts and we have come so far as to be able to conclude that at a certain epoch a series of granitic rocks and entire mountain ranges were raised, because it turned out that rock types previously regarded as primordial rocks are of very young age. (Keferstein, 1834, p. 72)

In a later passage, Keferstein referred to Hutton's 1788 paper again, writing:

> James Hutton, in the *Theory of the Earth*, 1788, was also convinced that the Earth is a large organism the internal life of which expresses itself in constant generation and decay, in that layers of the earth disintegrate and then form anew. This happens mostly at the bottom of the sea, where calcareous remains of organisms are mixed with sands etc. and are melted by the internal heat of the earth. For such underground operations, the volcanoes are the breathing holes. The basalts, porphyries, granites also form like the lavas, in that they acquire their crystalline fabric by melting under high pressure. That granite is not a primordial rock and does not form by aqueous precipitation the author proves partly on the basis of its fabric and partly its occurrence in dykes intruding into schists. This theory led to a large number of experiments on melting, whereby it was really proved that the product of melting becomes sometimes glassy, sometimes crystalline and sometimes massive, depending on the variation in the pressure and the kind of cooling. (See in Hall in Experiments of Whinstone 1806; G. Watt in numerous articles in 1805; Fleurian de Bellevue in Journal de Physique 1805. All three in numerous articles in 1808. Guiton-Morveau in the Annal. de Chimie, part 73. Bose d'Anlic in the Journal de Mines no. 177.)
>
> Hutton's theory was further developed by Playfair and Thomas Beddoes confirmed it in 1791 by means of nice observations. (Keferstein, 1834, p. 129–130)

Keferstein thus ascribed to Hutton the discovery of the dynamic behavior of the earth and what we today call, after a suggestion by Sergei Ivanovich Tomkeieff (1946), the geostrophic cycle. Keferstein gave a more detailed account of Hutton's discoveries in his *History*, whereby he ended up emphasizing that Hutton had introduced into geology entirely new ideas, echoing Desmarest's judgement almost half a century earlier:

> In England[148] geognosy did not have the same general participation and development as in Germany, which appeared first in the beginning of the subsequent century. But exactly at the same time as Werner there appeared a geologist here, who got into geology deeply and generated ideas that were in sharpest contrast to those of Werner. The Doctor James Hutton (born 1726, died 1797), a private gentleman in Edinburgh, observed with great industry and clear eye the conditions in England and Scotland. On their basis he erected a distinctive history of the formation of the earth in his brilliant Theorie [sic] of the Earth (in the Transact. of the royal Soc. Of Edinburgh 1, 1788) and as an independent work in 1795, of which a German extract was published in Voigt's Magazin der Physik in 1789[149]). By studying closely the whinstone and the toadstone, i.e., the trap rocks, that occur commonly near Edinburgh and in England, Hutton convinced himself that these, just like lavas, had risen from below upwards by cutting through and dislocating strata and thus forming the well-known and commonly basaltic dykes. "The strata" he says "appear to have been broken and the two correspondent parts of those strata are separated to admit the flowing mass of Whinstone." He showed further first that the trap rocks do not cut through the Flœtz rocks only perpendicularly, but also the originally fluid masses intrude between horizontal strata, so that they now appear like Flœtz layers between strata. "In the case" he says "the strata are not broken in order to have the Whinstone introduced, they are separated and the Whinstone is interjected in the form of strata, having various degrees of regularity and being of different thickness." He realized that whinstone, trap, almond-stone and basalt are only slight modifications of the same mass; that porphyry and granite had a complete analogy with them and thus probably the same origin. But to prove this latter claim by means of the structural relations was the great task he set to himself. He thus went to the Grampian Mountains, studied carefully the contact between the granite and the country rock and made here, at Glen Tilt, the important discovery that distinct dykes and debris come out of the main stock of the red granite cutting the country rock, the micaschist, limestone etc. and that near the granite contact the country rock appear considerably altered. This discovery, which clearly proved his conjectures, filled him with the greatest pleasure and after having found the same corroborated many times elsewhere, he ventured to pronounce that granite, as porphyry, almond-stone, basalt, etc. that are related to it, *are of newer origin*, younger than the strata they cut and *are pyric* [Keferstein wrote "pyrotypisch"], *lava-like rocks*. He then made the important statement for the first time that granite, porphyry, basalt etc. are of the same kind with respect to their massive character and according to their essence and have acquired their different forms owing to different conditions in which they occur. The real lava ("erupted lava") appears scoria-like because it flowed in the open air, the atmosphere affected it and it cooled quickly. However, those older rocks such as trap, granite etc. were subterranean molten lavas ("subterraneous lavas") that had been fluid under higher pressure. They did not reach the surface via volcanoes and thus have a different character. Owing to high pressure, even limestones can become fluid without

changing chemically and the crystalline marble—as seen from its geognostic conditions—one such product of melting under the ground that created by the granite and also by lavas. It is not a product of subterranean volcanic fire, but of the peculiar inner heat of the earth that is a part of its live activity expressed in eternal formation and decay. The demolished rock masses that now appear as sand on the seabed, can one day appear as granite; because everything is subjected to an eternal cycle. Hutton regarded basalt, porphyry, granite etc. as pyric rocks, but not as products of burning volcanoes; the supporters of this theory were later called plutonists to distinguish them from the vulcanists.

Whereas Werner infinitely reduced the activity of vulcanism and the pyric rocks, Hutton accorded them so much a larger field of activity. He was the first to prove definitely the intimate connexion of all massive, crystalline rocks positing them as contrasts to the stratified rocks and completely undermined the idea of "primordial rocks," thus introducing into geology completely new ideas. (Keferstein, 1840, p. 80–83)

Keferstein then says that Werner's ideas originally had found a much wider circulation than those of Hutton "for the reasons discussed before." The reasons he had given before was that Werner was a good and inspiring teacher with a large school; his "element was teaching and systematising" (p. 67). Keferstein gave a scathing criticism of Werner's methods and observations and ascribed to his "school" the widespread employment of his ideas. Keferstein pointed out emphatically the weakness of Werner's position with respect to igneous rocks and tectonic phenomena and that in these subjects he actually set back the development of geology. He concluded his account of Hutton with the following words:

Only in the following century [i.e., the nineteenth] these questions were discussed in greater detail and Hutton's views were confirmed, but it took another 30 years before his ideas were accepted generally. (Keferstein, 1840, p. 83)

In 1841, one of the postgraduate theses in the University of Leipzig directed by the historian Friedrich Christian August Hasse (1773–1848) dealt with the history of geography. I here deal with the second part of the thesis, the first part of which had been already presented in 1837 and dealt exclusively with the history of geography. The author of the thesis is not recorded, but it was presented by Hasse under the title *Quantum Geographia Novissimis Periegesibus et Peregrinationibus Profecerit* (How much has geography progressed with the newest descriptions and discoveries). Its second part is devoted to geology, under the title *Quae ad rationes globi terrestris naturales, maxime ad Geognosiam et Geologiam, seu Geographiam subterraneam spectant* (Subjects examining the natural causes of the terrestrial globe, mostly on geognosy and geology i.e., subterranean geography). The candidate wrote:

In the same manner it is evident, as this suitable study and comparison by no means may be found elsewhere, geology itself was built and likewise destroyed by ingenious hypotheses and ideas, not to mention those of Thales, Anaxagoras and the teaching of others, workshop of Hephaestius, poems of Empedocles or Lucretius, but also from times of Burnet, Woodwards, Whiston, Count Buffon, with comments of many others. It is truly evident that the history of the origin of Earth's formation—of course unless men penetrate to the inner parts of Earth—even after geological theories of Werner, Hutton, Breislak and others, which examine thoroughly Earth's foundations and sources, are not concluded today and won't completely be arranged or completed yet. (Hasse, 1841, p. 5–6)

Here, the candidate mentions Hutton as one of the moderns who put forward geological theories, yet they cannot be considered conclusive, because we do not know what the Earth looks like in its deeper layers. The candidate then says that:

The English emulated the Germans. Right after 1785 Hutton among the Scots, the author of systematic geology, opposed Neptunism with what is called Plutonism. (p. 9)

This is remarkable. The candidate, a historian of the earth sciences in the last decade of the first half of the nineteenth century, not only regards Hutton as an opponent of neptunism, to which he opposed what is called his plutonism, but at the same time *as the author of a systematic geology*.

In 1837, the professor of chemistry and technology in the Königlichen Rheinischen Friedrich-Wilhelms-Universität zu Bonn, Gustav Bischof (1792–1870), one of the founders of neo-neptunism, published his study of the internal heat of the earth concluding that the interior of the planet was indeed very hot. Bischof has no historical comments in that book, but refers to Hutton and Sir James Hall for their work that showed that the fabric of igneous rocks depended on the ambient temperature and pressure of their formation (Bischof, 1837, p. 313). His book had an English edition in 1841 published in London, but of its planned two volumes, only v. 1 was published. Bischof's great reputation (and the cause of his Wollaston Medal from the Geological Society of London, which he received in 1863), however, is based mainly on his multi-volume *Lehrbuch der Chemischen und Physikalischen Geologie* (=Textbook of Chemical and Physical Geology) that had two editions and an English translation and which supported the neoneptunist position that had been revived in Germany in the middle of the nineteenth century.

In the foreword ("Vorrede") of the first edition he criticizes Hutton's and Sir James Hall's experiments on the origin of marble (Bischof, 1847, p. XI). In the first part of v. 2, p. 767–768, Bischof refers to the discovery, which he ascribes to Hutton and Sir James Hall, of the contact metamorphism of sills in the Salisbury Crags near Edinburgh and argues that this is an universal characteristic seen at the contacts of dykes (Bischof, 1851). Then, in the second part of v. 2, p. 962–963, he describes the marble experiment, again with reference to Sir James Hall and James Hutton, and finds the requirement that marbles in Bohemia could have been buried under a rock overburden of 1–1.5 km impossible.

Although Bischof was wrong in his marble criticism of Hutton and Sir James, what is of interest for the purpose of the present book is that he treats their work not as of historical interest, but as of current relevance in the middle of the nineteenth century. In the second edition, there is again reference to Hutton and Sir James in relation to the Salisbury Crags in the context of a critique of Leopold von Buch's theory of mountain uplift by magmatic intrusions (Bischof, 1866, p. 264). An English edition of Bischof's book came out in three volumes (1854, 1855, 1859) under the auspices of the Cavendish Society between the two German editions; it is not a mere translation, but essentially a new work as Bischof himself points out in his "Introductory Remarks" (Bischof, 1854, p. vii). In that somewhat shorter book I have not come across Hutton's name.

Only two years after Keferstein's *History*, Bernhard von Cotta[150] (1808–1879; Fig. 43), professor of geology in Freiberg, i.e., in Werner's own institution, published his textbook *Anleitung zum Studium der Geognosie und Geologie, besonders für Deutsche Forstwirthe, Landwirthe und Techniker* (=Guide for the Study of Geognosy and Geology for German foresters, farmers and engineers). There is a section in it on the history of geology, based mainly on Hoffmann's and Keferstein's books. Von Cotta restates what they wrote, says for a long time Hutton's ideas were not popular, but eventually they became the "presently dominant school with some deviations from it" (Cotta, 1842, p. 420). Cotta further mentioned a calculation of the average density of the earth, which a "Hutton" had found, together with Playfair, to be 4.7 (p. 12; this calculation is also referred to by Alexander von Humboldt in his *Kosmos*, v. 1, p. 424; von Humboldt reports the value found as 4.713: von Humboldt, 1845; see also von Humboldt, 1858, p. 32, where only Playfair is mentioned), but the "Hutton" who did that work was the English mathematician and geodesist Charles Hutton (1737–1823), not James Hutton and the work referred to was the famous Schiehallion Experiment of 1774 some 48 km northwest of Perth in Scotland (Maskelyne, 1775; Hutton, 1778). Cotta refers to James Hutton's idea of the origin of marble (p. 377–378). In the second edition of his book with the somewhat altered title *Grundriß der Geognosie und Geologie* (Outline of Geognosy and Geology), he repeats what was said in the first edition, except the origin of marble (Cotta, 1846, p. 52, 367). Charles Hutton's density calculation for the planet is also noticed by Gustav Leonhard (1863, p. 5–6), Karl Cäsar von Leonhard's son (his dates are 1816–1878).

Bernhard von Cotta published in 1866 a somewhat philosophical work under the title *Die Geologie der Gegenwart Dargestellt und Beleuchtet* (Geology of the Present Described and Illuminated) in which he praised Hutton for pointing out the igneous origin of the crystalline rocks and also the difference between volcanic and plutonic rocks (von Cotta, 1866,

Figure 43. Bernhard von Cotta. Photograph by Emil Rade (1832–1931). Plate 10, Photograph album of German and Austrian scientists, 1 January 1866 (darwinproject.ac.uk).

p. XIX–XX). This book had five editions till 1878 and in all von Cotta simply repeated what he had said in the first edition.

Also in 1846 came out the first volume of the famous zoologist, geologist, and politician, Darwin's great champion, August Christoph Carl Vogt's (1817–1895; Fig. 44) *Lehrbuch der Geologie und Petrefactenkunde. Zum Gebrauche bei Vorlesungen und Selbsunterrichte. Theilweise nach L. Élie de Beaumont's Vorlesungen an der Ecole des Mines* (Textbook of Geology and Palaeontology to be used for Lectures and Private Instruction. Partly after the Lectures by Élie de Beaumont in the École de Mines), the second volume of which was published in the subsequent year. In that second volume there is a long section on the history of geology and paleontology (p. 375–435) and there Vogt lauds Hutton's contributions. He points out that Hutton first noticed the plutonic origin of crystalline rocks and that their intrusion deformed the intruded sedimentary layers. Although Vogt mistakenly thinks that Hutton had ascribed a plutonic origin also to layered rocks, he nevertheless praises Hutton for having resuscitated Steno's idea of the original horizontality of strata and their subsequent deformation. Vogt praises Hutton's realization that crystalline rocks had been originally molten.

Figure 44. August Christoph Carl Vogt. (Manuscripts and Archives Division, The New York Public Library; New York Public Library Digital Collections, http://digitalcollections.nypl.org/items/510d47dd-faa6-a3d9-e040-e00a18064a99, accessed 15 June 2020.)

Vogt gives much credit to Sir James Hall (1761–1832), Gregory Watt (1777–1804) and John Playfair (1748–1819) for making Hutton's ideas acceptable to the geological community at large through their observations and experiments. He particularly praises Sir James Hall's experiments and regrets that at his own time experimental geology was no longer pursued as it had been by Sir James. Vogt then points out that when Leopold von Buch was in London (it was in the winter of 1814) Hutton's ideas had already been generally accepted there and that von Buch was "inevitably" greatly influenced by them. As we saw above, when von Buch finally convinced himself of the volcanic origin of basalt, he said he "gladly agreed with Dr. Hutton." Vogt regrets that Leopold von Buch had no Sir James Hall!

Vogt's textbook made three other editions in 1854, 1866–1871 and 1878. What he says about Hutton in these are reprints of what he wrote in the first edition.

In 1860, the prominent German zoologist and paleontologist and the first professor of paleontology in Europe (cf. Wellnhofer, 1980), Johann Andreas Wagner (1797–1861), the discoverer of the famous Pikermi Fauna in Greece and the opponent of the Huttonian geology that triumphed during his lifetime in the whole world and the theory of evolution (that is why he refused to recognize *Archaeopteryx* as a bird and described it as a reptile with feathers, naming it *Griphosaurus*, i.e., puzzle lizard), summarized his geological remembrances in 1860 from his long professional life for the Bavarian Academy of Sciences in an extract from his *Geschichte der Urwelt mit Besonderer Berücksichtigung der Menschenrassen und des Mosaischen Schöpfungsberichtes* (History of the primitive world with special regard to human races and the Mosaic account of the creation; Wagner, 1845, 1857, 1858). These memoirs are in reality a tirade against vulcanism and a defense of Werner's neptunism. Although Wagner's deviant geological ideas, influenced not only by his religious commitment but also by a former student of Werner, the great mineralogist and chemist Johann Nepomuk von Fuchs[151] (1774–1856, also devoutly religious; the mineral fuchsite was named after him) had absolutely no influence in the geological world of the mid-nineteenth century (see especially Fritscher, 1987). I quote below the beginning paragraphs of his text, because it gives a rare eyewitness account of the downfall of Werner's theory of neptunism and the triumph of Hutton's vulcanism in Germany by a latter-day Wernerian—I am sure our recent revisionists' pronouncements on Werner and Hutton would have delighted Wagner, who regretted the Huttonian triumph his entire life, because he found it contrary to his religious beliefs:

> At the time when I began my university study (in 1814) Werner's theory of origin of rocks stood at its apogee; at least in Germany it was the only ruling theory. According to it, the entire rocky edifice of the earth had been deposited out of water and the original aspect of its surface had largely remained as it now appears, with the exception of the effects on it of the atmospheric phenomena, floods, volcanoes and earthquakes.

In England and France, however, Werner's theory did not enjoy such a general acceptance as in Germany. In the former country, in 1795, James Hutton had published his "Theory of the Earth" entirely on the basis of vulcanism. In England it found an enthusiastic acceptance, so that next to it Werner's theory could not gain much ground. By contrast, Hutton's teaching did not have much luck in Germany, because here it came up against Werner's immense authority. It would not be uninteresting to take a look at Hutton's theory somewhat more closely.

First it was basalt, for which Hutton assumed a fiery origin, as before him other naturalists had done and as the close relationship with lavas offered a well-grounded justification for such an assumption. Hutton often saw in his fatherland Scotland that basalts and trap rocks intruded into neighbouring rocks in the form of dykes and branching veins. Hutton could not imagine any other way than basalt being a fiery liquid during its intrusion. But this assumption was met with the objection of the neptunists that, as a result of melting of rocks, only glassy substances, but no crystalline rock could be recovered upon re-cooling. Hutton was saved from this dilemma caused by the objection with the help of the chemist Hall, in that Hall proved by his melting experiments on trap rocks and lavas that during a slow cooling of the molten materials no glassy substance, but a stony, partly granular mass formed that was similar to the ones he had fused. Hereby, for Hutton, the fiery origin of basalt was proven.

The second rock, for which he assumed a fiery formation was granite. He had already considered this earlier, because lack of bedding in granites and its granular structure, as seen also in some lavas and basalts, had led to an expectation of a similar origin as those. However, he became sure of it when he observed that granite, like basalt, sent apophyses into the schists and limestones covering it. With this observation the equivalence of the origins of granite and basalt was fully documented. This time he did not allow himself to be misled by chemical experiments, which, trying to obtain a granite-like rock in melting-cooling attempts, had totally failed. For Hutton, the apophyses of the granite were perfectly sufficient to be unshakably convinced of its volcanic origin and the majority of his compatriots agreed with him.

In Germany, where the new discoveries were examined somewhat more critically, it was soon seen that the implications drawn from them stood on shaky ground and that they allowed also another interpretation. For example, one such interpretation was that by Bischof and Delesse, who deduced from the granite apophyses the impossibility of a fiery formation of granite. Therefore, not only granite, but also basalt remained as a member of the neptunian rocks in Werner's school as before. D'Aubuisson, one of Werner's best students, ably defended in 1803 his teacher's view on the formation of basalt. But when he journeyed through the Auvergne and studied there the occurrence of basalts in numerous lava-like flows, what he saw surprised him to such a degree that he very soon abandoned the neptunist view of basalt formation and went over to the volcanists. Although d'Aubuisson warned that extending such a view also to the formation of granites was entirely unjustified, his warning was no longer influential. Not only basalt and trachyte, but also granite were carried by German geologists in increasing numbers from the neptunist camp to that of the volcanists.

Leopold von Buch dealt the final blow leading to the complete downfall of Werner's theory. This was occasioned by his observations in Southern Tyrol, especially in the Fassa Valley, in 1822. He found the thick occurrences of dolomite, augite porphyry and quartz porphyry in such peculiar relationships that he thought himself justified in pronouncing the most daring views. It was here that Leopold von Buch framed the theory of the uplift of mountain chains by means of subterranean volcanic forces; it was here that he established with evident proof the theory of dolomitisation of limestones by means of infiltration of the rising exhalations of magnesium oxide from the molten augite; it was here that the black and red porphyries acquired their world-altering importance. Twenty years earlier, when von Buch had visited Southern Tyrol, his studies naturally had reached entirely different conclusions. Then, the quartz porphyry, exposed in large masses near Bozen [now Bolzano in Alta Adige, northern Italy], had been for him decidedly neptunian formations. After the passage of two decades he spoke of this porphyry in a completely opposite sense in that he declared that it not only could uplift individual mountain chains, but entire continents. Not that this rock had since then changed its character, but the opinion of the famous geologist had changed. In the meantime he had studied with great eagerness the volcanoes of Italy and the Canary Islands and the violent volcanic effects he saw there had impressed him to such a degree that he became inclined to recognise their earlier actions even there, where no traces of extinct and still active volcanoes could be seen. Von Buch communicated his new, till then unheard-of, ideas with such power of conviction and such obviously enthusiastic descriptions that they were received in all countries with fervent acceptance. But with that the feverish combat between vulcanism and neptunism was brought to an end; the defeat of the latter could not have been more complete. (Wagner, 1860, p. 378–379)

In the summary quoted above, Wagner did not repeat his sentence published earlier, both in the first and the second edition of his book (1845, p. 8; 1857, p. 23):

The Scotsman James Hutton has to be regarded as the father of modern geology, who, in 1795, brought his *Theory of the Earth* in circulation by having it printed.

When Wagner first published that sentence, Sir Archibald Geikie was only ten years old!

The Swabian naturalist, physicist and philologist Wilhelm Constantin Wittwer (1822–1908) published in 1860, as a supplement to the famous *Briefe über Alexander von Humboldt's Kosmos*, possibly the best scientific biography of von Humboldt. Its superlative qualities are due to the fact that it was written only a year after von Humboldt's death and thus many later vanished documents were still available together with many people who knew von Humboldt personally, some even having worked with him closely; moreover, Wittwer was himself a scientist and a philologist sharing the same scientific milieu and interests as his subject. He was thus able to describe von Humboldt's work from within his own time and place. During his narrative Wittwer described various stages in the development of von Humboldt's ideas pertaining to geology and while doing so he inevitably came across Hutton and Werner. Wittwer simply says that Hutton's theory opposed Werner's and that during the first half of the nineteenth century the former eventually stamped out the latter.

While Werner's theory assigned a very subordinate role to volcanoes in the construction of the earth, the claims of the vulcanists on behalf of the effects of fire became ever greater and soon not only basalt, but also trap, dolerite, porphyry, even granite; in short all rock types without natural bedding and without petrified remnants of former organisms were ascribed to subterranean fire, a theory that found its main representative in Hutton, whose main work (Theory of the Earth) appeared five years later than the publication of Humboldt's writing on basalt. (Wittwer, 1860, p. 19)

After this introduction Wittwer describes the outlines of Hutton's theory and points out that because of Werner's personal authority, neptunism remained dominant in Germany, whereas the French, almost as a body, had become vulcanists. Later, on p. 25, while discussing von Humboldt's work on the "irritability" of living tissues, Wittwer says that the theory of electricity in organisms initially had the same kind of resistance that Newton's theory of gravitation and Hutton's theory of the earth had experienced, implying that Newton's and Hutton's theories eventually triumphed. During the course of his narrative in his book, Wittwer repeatedly returns to Hutton's triumph as he relates the evolution of Humboldt's ideas on geology. For example, on p. 241, where he discusses the significance of unconformities, he wrote:

Werner regarded all the greater inclination of beds as local, just as he did in the case of volcanoes. Hutton and with him the vulcanists insisted on the idea that the beds must have been originally horizontal and explained the inclination because a push from below must have pushed one side of the bed more than the other. Werner took the earth as rigid, Hutton thought the interior fluid and this latter assumption made his theory more flexible in dealing with multifarious observations than the crystallising Neptunism was able to be. That the beds were originally horizontal is now completely demonstrated.

Thus, as Wittwer beheld it in the middle of nineteenth century, geology had become entirely "Huttonian."

In 1867 the German geologist Hermann Vogelsang (1838–1874; Fig. 45), one of the pioneers of microscopical petrography and an "européen avant la lettre" (see Touret's delightful biography: Touret, 2004) published his *Philosophie der Geologie und Mikroskopische Gesteinsstudien* (Philosophy of Geology and Microscopic Studies of Rocks) with a view to combatting universal hypotheses in geology and emphasizing the great importance of observations. To that end he divided his book into two parts: the first reviews the disciplines that are indispensable to geology such as physics, chemistry, biology, etc. Here Vogelsang makes

Figure 45. Hermann Vogelsang. Unknown author. Portrait vignette; possibly a print from 1930 of an original from ca. 1870–1910.

the extremely important statement right at the outset, at the head of his second paragraph:

> A natural science that renounces genetic studies stops being a science. (Vogelsang, 1867, p. 7)

Vogelsang condemned at length in his book Werner's method of doing science with the following words that reminds one of Karl Popper's characterization of what science is:

> When we support such a justification of Werner wholeheartedly, we should not forget that this is only a personal defense, a sort of moral one, but not a scientific one. It is against all natural scientific principles to work a few observations into a general theory and then to declare it with apodeictic strictness to be an inviolable law. Going around the world to find facts supporting a preconceived notion and throwing everything away that contradict it cannot be called natural research. Such a procedure would only put on the old philosophical bugaboo a new garb: instead of testing the theory or explaining it using real occurrences, Nature is interpreted in favour of a preconceived idea. And can one call an idea in geology a theory that has no physical and chemical basis for the continuing changes of the surface of the earth? When one is satisfied that under suitable conditions all possible substances may be dissolved and distributed in water and then can be deposited from it; that at any given time any part of the earth can be covered with water, then one can accept such a geogeny. But who gives the right for such assumptions? Who ever dissolved basalt in water or saw it being deposited somewhere, somehow? There is a theory of theory and Werner and his entire school had no concept of it. Steno and Agricola were far more advanced in this sense than Werner. He could have learnt much from his immediate predecessors and his opponents. [Georg Christian] Füchsel [1722–1773], [Johann Karl Wilhelm] Voigt [1752–1821], [Johann Friedrich Wilhelm] Charpentier [1738–1805] have denounced Werner's way, but the name of the master, his entrancing lecturing, his unquestionable services, procured his fallacious theory a

resilient existence. Wishing to form a universal theory from observation was in those days incomparably more nonsensical than it is today, when we have greatly augmented the natural principles of geogeny. Such principles were then unknown and they should have been searched for and dealt with.

Thus, Werner's theory was wrong for his own geology. It did not belong there like a tumor that does not fit into the organism and creates ever greater frailties the greater the power is that placed the tumor and tried to keep it there.

So an unfortunate contrast was created: on one side a rich and methodical study of geognostical occurrences were aimed and accomplished and on the other in interpreting them the most important natural laws were turned upside-down.

We thus got a scientific geognosy and an unscientific geogeny. (Vogelsang, 1867, p. 77–78)

These are hard words. Vogelsang had a reputation of not mincing his words in criticism; he had the audacity and honesty of youth (when he wrote these words he was only 26: see Touret, 2004, p. 99; he died of tuberculosis when he was only 36). He admits, and rightly so, Werner's various techniques in the laboratory and in the field greatly improved geological research and his terminology, although based on a fallacious theory, served a useful purpose, but he cannot forgive him for having abandoned the basis of sound science when it came to dealing with processes and earth history. As Touret (2004) also emphasized, Vogelsang's place in geological philosophy was on Hutton' side. This is what Vogelsang said about Hutton in comparison:

> Hutton had a considerably richer and clearer geognostic outlook than Werner. His theory is much better worked through and based on the explanation of certain occurrences. (Vogelsang, 1867, p. 79)

In 1869, the German mineralogist, petrographer, volcanologist and seismologist of Lothringian French descent Arnold Constantin Peter Franz von Lasaulx (1839–1886; Fig. 46) published a small pamphlet on the controversy about the origin of basalt. As part of this history, he wrote:

Figure 46. Arnold Constantin Peter Franz von Lasaulx. (Only known image of him; author unknown.)

As Werner found in Scheibenberg the basis of his theory, Hutton in Scotland constructed an opposite model on the basis of the basalt cliffs of Arthur's Seat rising over Edinburgh. His writings, accompanied and supported by the explications of Playfair and the clever experiments of Hall on the rocky congealment of a slowly cooling molten glass, were the foundation of a plutonic school in England.

...

In England, the supporters of Hutton kept increasing in number, while in Germany Werner's supporters progressively became fewer. Such men as Macculloch, Mackenzie [Sir George Steuart Mackenzie, 1780–1848], Henslow [Darwin's mentor John Stevens Henslow, 1796–1861], Murchison, Sedgwick, Allan [Thomas Allan, 1777–1833] and many others were enthusiastic defenders of the plutonic origin of basalt. They were joined by the classic works by Conybeare and Lyell and the writings of Poullet Scrope and Daubeny [Charles Giles Bridle Daubeny, 1795–1867], especially important because of the excellent knowledge on the active volcanoes.

In Germany a number of geologists followed Leop. von Buch and Alex. von Humboldt, detailed descriptions by whom of the basalt mountains studied in all parts of Germany and all other countries and of the material evidence for the true volcanic phenomena accumulated enormously. (von Lasaulx, 1869, p. 13–14)

Von Lasaulx also emphasized the uniformitarianism of the Huttonian camp:

The smallest causes acting in extraordinarily long time intervals can be compared with full reason with the individual processes of creation having large energies. (p. 30)

Von Lasaulx's entire pamphlet is a thorough defense of Huttonism and a demonstration of the untenability of the Wernerian geology.

Another petrologist/geochemist, Justus Ludwig Adolf Roth (1818–1892) wrote his important paper on metamorphism and the origin of crystalline schists only four years after Vogelsang's book and two years after von Lasaulx's pamphlet came out. It was entitled *Über die Lehre vom Metamorphismus und die Entstehung der kristallinischen Schiefer* (=On the theory of metamorphism and the origin of crystalline schists) in the *Abhandlungen der königlichen Akademie der Wissenschaften zu Berlin*, for the year 1871 (published in 1872, p. 151–232). In that paper Roth has a long discussion on Hutton's ideas as they pertain to metamorphism (p. 157–168). He introduces Hutton's theory as "the only theory of the earth based on positive geological observations" (p. 157). At the end of his discussion, Roth noted:

Hutton's theory is the last of the geogenic theories that was formulated before the upswing of the natural sciences and from which even to our own day ideas have persisted. Playfair used for him the designation *osservatore oculatissimo* [most careful observer], whose geological studies in Scotland, England, France and in the Netherlands reveal the outstanding observer. (Roth, 1872, p. 168)

The textbooks published later in Germany until the end of the nineteenth century, such as those by Ritter Karl Wilhelm von Gümbel (1888) and Friedrich Heinrich Emanuel Kayser (1893) do not contain sections on the history of geology and there are no references in them to Hutton and neither to anybody else from the last decades of the eighteenth and the first decades of the nineteenth centuries. Gümbel's book was written for the readers of his multi-volume *Geologie von Bayern* [Geology of Bavaria] lest they be unfamiliar with the basics of geology. Kayser's book was a textbook for university classes for physical and historical geology, which proved immensely useful and underwent six editions of ever increasing size. Only in two books have I found a discussion of Hutton's ideas explicitly referring to him in the last quarter of the nineteenth century in Germany before the publication of Geikie's book in 1897. First, in the popular book *Naturgeschichte des Mineralreichs für Schule und Haus* (=Natural History of the Mineral Kingdom for school and home) by Gustav Adolph Kenngott (1818–1897) and Friedrich Rolle (1827–1887), Hutton is mentioned. Rolle (who authored the geology part) wrote that Hutton and Voigt objected to Werner's neptunism and that the later research by Werner's students Alexander von Humboldt and Leopold von Buch, vindicated Hutton (Rolle, [1888], p. 2). The second was in the semi-popular book by the natural history writer Wilhelm Bölsche (1861–1939) entitled *Entwicklungsgeschichte der Natur* (=History of the evolution of Nature) published in two volumes in 1894. In the first volume, Bölsche presents a detailed and competent history of the ideas pertaining to the evolution of the universe and the earth. As part of it he reviews the work of Werner (p. 170–171) and then writes that Hutton's ideas proved a progress upon Werner's teaching (p. 171). He notes that not only the igneous rocks found their correct interpretation in Hutton's theory, but that mountain uplift was recognized and ascribed to "volcanic phenomena." He calls Alexander von Humboldt and Leopold von Buch "Werner-pupils converted to Hutton" (p. 173).

In the nineteenth century, long before Sir Archibald followed the common thinking in considering Hutton the founder of modern geology, even the small, popular encyclopedias in Germany, commonly between one and three volumes, could not do without mentioning Hutton. Here is a sample:

The modern G.[eology] begins with *Werner* (1750–1817), whose neptunistic system founded the stratigraphic part of G.[eology]. *Heim, Voigt* and esp.[ecially] *Hutton* (1788 and 1795) appeared as opponents, and as *L. von Buch, Alex. von Humboldt* and *Weiss* adopted Hutton's plutonist teaching, neptunism collapsed. (Anonymous, 1874, p. 685; italics in the original)

By the last decade of the century, Hutton's reputation had acquired an almost mythical aura and, in Germany, as elsewhere, even things he certainly did not do, nay opposed, had begun to be ascribed to him. For example, the German mathematician, geographer and science historian Siegmund Günther (1848–1923) wrote in his *Lehrbuch der Physikalischen Geographie* (=Textbook of physical geography; 1891):

> Since the geological experiment, following the example of Hutton, F. Pfaff and others, especially Daubrée … has become a valuable tool of the science of the earth, it began to imitate the above-mentioned process [i.e., metamorphism] at a small scale. (p. 86)

One would think that here he confuses Hall with Hutton. But, on p. 104 he says that Hutton and Hall undertook experiments on folding! Did he not know that Hutton actually *opposed* experiments in geology and that Sir James Hall could do his experiments only after Hutton's death? In footnote 2 on p. 453 he mentions the German geographer of Ice Ages fame, Albrecht Penck's (1858–1945), reference to Hutton, among others, for the origin of valleys.

With Bölsche's book we are less than three years away from the younger time limit of the present book.

But before we take leave of German authors, mention must be made of the great German poet and polymath Johann Wolfgang von Goethe's (1749–1832) views on the neptunist-vulcanist debate as expressed in his literary works. Goethe, apart from his geological writings, first in *Wilhelm Meisters Wanderjahre* (Wilhelm Meister's Years of Wandering; first full publication in 1829), but in greater detail and theatricality in the second part of his great dramatic poem *Faust, der Tragödie*, Part 2 (1831) described the views of the vulcanists and the neptunists with his clear, but emotional, preference for the latter. He did not like the triumph of the Huttonian geology:

> Wie man die Könige verletzt,
> Wird der Granit auch abgesetzt,
> Und Gneiss, der Sohn, ist nun Papa!
> Auch dessen Untergang ist nah:
> Denn Plutos Gabel drohet schon
> Dem Urgrund Revolution;
> Basalt, der schwarze Teufels-Mohr,
> Aus tiefster Hölle brich hervor,
> Zerspaltet Fels, Gestein und Erden,
> Omega muß zum Alpha werden.
> Und so wäre denn die liebe Welt
> Geognostisch auch auf den Kopf gestellet.
> (*Zahme Xenien VI*[152])
> (=As one hurts kings,
> Granite too will be dethroned,
> And gneiss, the son, is now Papa!
> Its demise is also soon:
> Pluto's fork is threatening revolution
> Against the primitive ground;
> Basalt, the black Devil's blackmoor,
> Erupts from the deepest hell,
> Splitting craig, rock and earths,
> Omega must become Alpha.
> And thus the beloved world is
> Geognostically turned upside-down.)

He expressed his great disappointment upon the fall of Wernerian geology after Werner's death in 1817, in his famous short poem published in the *Zahme Xenien VI*:

> Kaum wendet der edle Werner den Rücken,
> Zerstört man das Poseidaonische Reich;
> Wenn alle sich vor Hephästos bücken,
> Ich kann es nicht sogleich;
> Ich weiß nur in der Folge zu schätzen.
> Schon hab ich manches Credo verpaßt;
> Mir sind sie alle gleich verhaßt,
> Neue Götter und Götzen
> (Scarce had noble Werner turned his back
> Poseidon's Empire was destroyed
> When all bow before Hephaistos,
> I cannot do it so fast.
> Only later can I see
> That I have already missed certain beliefs
> I hate them equally,
> New gods and idols.)

AUSTRIAN EMPIRE (AUSTRO-HUNGARY AFTER 1867)

Before Lyell

Geology in the Austrian Empire was in a very underdeveloped state in the beginning of the nineteenth century and the main reason for that was the Wernerian system of teaching in the universities and vocational, especially mining, schools in the monarchy. No one has described this state of affairs better than one of the greatest geologists who ever lived, Eduard Suess (1831–1914), in his memoirs:

> Austria includes, in a wonderful diversity, the old massif of Bohemia, the edge of the Russian table-land, the much younger Alps and the Carpathians and the western extremity of the Aralo-Caspian lowlands. The knowledge of these great natural units remained entirely foreign to the university teaching. Its aim was supposed to be to order the minerals, according to their natural historical aspects, i.e., the externally knowable characteristics, in a system, as it was done in zoology and botany. It was, however, overlooked that here the basic principle of any classification of organic beings was lacking, namely the parents; and the minerals were given very learned-sounding names. Instead of gypsum, the student had to say prismatoid euclase-haloid and, instead of siderite, he had to pronounce brachytype paracrossbarite.

Geognosy was defined "as the science of the composition of the earth deduced from the individual minerals" but it was never asked how such a composition had come into being in the first place [153]. In this, the historical element was entirely neglected. (Suess, 1916, p. 113)

Under such circumstances, the textbooks of the early nineteenth century in the Austrian Empire reflected the rudimentary state of the teaching. One of the earliest of such textbooks was that by Franz Ambros Reuß (1761–1830), entitled *Lehrbuch der Mineralogie nach des Herrn O.B.R. Karsten Mineralogischen Tabellen* (=Textbook of Mineralogy after the Mining Director Karsten's Mineralogical Tables) In this multi-volume work (1801–1806), Reuß remains very loyal to his former teacher Werner's ideas and uses the work of another Werner student as a basis, Dietrich Ludwig Gustav Karsten (1768–1810), considering basalt a sedimentary rock associated with greywacke and thinking volcanoes as surface expressions of large subterranean lignite fires (see, for example, Reuß, 1805, p. 401). So far as I was able to see, there is no mention of Hutton in this book.

Another textbook, *Anleitung zur Geognosie, insbesondere zur Gebirgskunde—nach Werner für die k.k. Berg-Akademie* (=Guide to Geognosy and especially to Rock-Science—after Werner for the Imperial and Royal Mining Academy), written for the students of the Austrian Imperial and Royal Mining Academy in Schemnitz[154] by Franz Reichetzer (1770–1835) and published in 1812 also does not mention Hutton and is based, as its title implies, entirely on Werner's teaching. However, by the time the first edition was out-of-print and a second edition was called for, things were not going well for the Werner School and Reichetzer was extremely disappointed as he openly said in his foreword to the second edition (Reichetzer, 1821, p. XI–XIV); from a footnote on p. 69 of the second edition (Reichetzer, 1821, p. 69), we learn that Charles Hutton's news had by that time reached Austria and at least some professionals had read him, because Reichetzer cites him by name about the structure of the terrestrial globe. However, I doubt if he read James Hutton, because he does not list Scotland among the places where graphic granite occurs (p. 64) and there is no reference to Hutton about the formation of dykes and fissures (p. 247 ff.). He further discusses the possibility of a young age for granite and thereby cites the observations "in Scotland" without giving a source (p. 123); but it is clear that here the reference is to James Hutton, probably second hand, most likely from his reading of Jameson's book cited below. He is also aware of Robert Jameson's (1774–1854) work on the islands of Scotland (*Mineralogical Travels Through the Hebrides, Orkney and Shetland Islands, and Mainland of Scotland, with Dissertations upon Peat and Kelp*, 1813b, Hutton's views on granite are discussed in Jameson's following pages: v. I, p. 79, v. 2, 168–170, 284 ; Reichetzer cites Jameson neither by name nor by the title of his book, nor by year of publication, though) and cites Richard Kirwan (1733–1812) by name twice only for the terms "granilite" and "granitelle" (p. 65 and 66).

After Lyell and Before Geikie

The tendency that prevailed in Austrian textbooks before 1830 continued into the early years of the second half of the nineteenth century. This, in fact, prompted Suess and his friends Christian Ferdinand Friedrich Ritter von Hochstetter (1829–1884) and Baron Ferdinand von Richthofen (1833–1905) to decide to translate Lyell's *Principles of Geology* into German to show the people in the Austro-Hungarian Monarchy what geology was really about (although its two different editions had already been translated in Germany). Neither Carl Friedrich Christian Mohs' (1773–1839; of hardness scale fame) *Geognosie* (1842), nor Johann Grimm's *Grundzüge der Geognosie und Gebirgskunde für Praktische Bergmänner* (1852: Fundamentals of Geognosy and Rock-Science for Practicing Miners) and his *Grundzüge der Geognosie für Bergmänner zunächst für die des Österreichischen Kaiserstaates* (1856: Fundamentals of Geognosy for Miners, especially for those in the Austrian Empire; this is really a second, enlarged edition of his earlier textbook) have anything about Hutton (or anybody else; these simple textbooks written for practical workers have no references). However, in one textbook, that by Franz Xaver Maximilian Zippe (1791–1863), Hutton is mentioned, together with Werner's student and later adversary in the basalt controversy, Johann Karl Wilhelm von Voigt (1752–1821), as Werner's opponent and Zippe gives him the credit for discovering the "firey" origin of the volcanic and intrusive rocks (Zippe, 1846, p. 334–335). Despite this, Zippe is considered as one of those who were held responsible for the primitive state of geology in the Austrian Empire during the second quarter of the nineteenth century.

Things began improving in Austria with the founding of the Imperial and Royal Geological Survey in 1849. Franz von Hauer (1822–1899), the "second geologist" of that venerable institution, published what may be regarded as the first "modern" geology textbook in Austria (von Hauer, 1875), but it too has no reference to James Hutton. In it, on p. 9, Charles Hutton is mentioned in relation to the Schiehallion Experiment. In the second edition of his book (1878), von Hauer repeated the same also on p. 9.

In 1872, a famous trio, the meteorologist Julius Ferdinand Hann (1839–1921)[155], the geologist Christian Gottlieb Ferdinand Ritter von Hochstetter (1829–1884) and the botanist Alois Pokorny (1826–1886) published their *Allgemeine Erdkunde—Astronomische und Physische Geographie, Geologie und Biologie* (General Geography—Astronomical and Physical Geography, Geology and Biology) that became very popular and went through five editions (with slight changes in the subtitles) until 1896 (in fact, the Tempsky imprint of this last edition is dated 1896/97). The book grew in size in every successive edition and beginning with the third edition (1881), Ferdinand von Hochstetter, who wrote the geology section, mentioned Hutton among those including John Playfair and John MacCulloch who opposed Werner's inert earth and ascribed the origin of mountains to magmatically driven vertical uplift (Hann et al., 1881, p. 293).

One of the grandest geology textbooks ever was published in the Austro-Hungarian Empire and came out in 1887 written by one of great geniuses in the history of geology, one of the most prominent of the Viennese giants, Melchior Neumayr (1845–1890; Fig. 47). Into this splendid, two-volume, book, simply entitled *Erdgeschichte* (Earth History) Neumayr inserted a 17-page history of geology. In it he lauds Werner as the scientific founder of geology and geognosy and characterizes him first and foremost as an organizer comparing him with Linné. He praises his system of mineralogy as the first of its kind and pays tribute to his teaching of the methods of field geology. Next he mentions his stratigraphic system, pointing out that many aspects of which he had inherited from Johann Gottlob Lehmann (1719–1767) and Georg Christian Füchsel (as we shall see below, Fitton, 1839, p. 37 {441}, footnote*, also emphasized Füchsel's priority in bringing some order into stratigraphy by defining the concept of a formation). However, Neumayr then says, that Werner was less successful in theoretical geology and relates the failure of his system and its reasons that basically amount to his inability to find adequate causes, *verae causae*, for his postulated processes.

He next comes to Hutton and this is what he wrote of his work:

Hutton, who published his views towards the end of the last century (1788), started mainly with dykes that cut various layered and massive rocks as fissure fillings. In the origin of the fissures he saw a violent tearing and in their filling an eruptive process, an injection of a melted mass, which, in most cases, is indeed correct. First, he assumed this for basalt, the dykes of which alter the country rock in a way as to betray the influence of heat. He then extended this interpretation to all massive rocks [*Massengesteine* = igneous rocks] from granite to basalt arguing that all had solidified from a molten mass. The crystalline schists he regarded as ordinary layered rocks altered by the internal heat of the earth, i.e., reworked and recrystallised long after their deposition, and thus became the founder of the theory of metamorphism that has been and still is controversial. Indeed, he has gone so far as to claim that the marine deposits became limestone, sandstone and claystone only through the action of heat. Also the formation of mountains Hutton ascribed to uplift and erection brought about by volcanic action applied from below. As an especial service it must be emphasised that he was actually the first to have an adequate conception of the immensity of the geological time.

.....

Hutton's theory represents as against Werner's position a very great progress, that lies mainly in the recognition of the internal heat of the earth and its effects, in the interpretation of massive rocks as products of cooling and the realisation of the importance of mountain-building forces.

......

...it can be said that Hutton gave geology its direction in many decades to come and even in its present form the influence of his teaching is not inconsiderable. The next success was the general victory of fire over water that came about after Werner's death and after a long and bitter struggle between the neptunists and the plutonists, i.e., Wernerians and Huttonians. (Neumayr, 1887, p. 23–24)

But Neumayr did not confine his reference to Hutton to the historical summary in his book. On p. 601 Hutton's interpretation of massive rocks as crystallization products of a cooling melt is mentioned. On p. 623 there is a criticism of Hutton's and the mineralogist Gustav Rose' (1798–1873) theory for the origin of marble. On p. 626, where Neumayr discusses metamorphism of rocks, he wrote:

The first founder of the theory of metamorphism was the brilliant Scottish geologist Hutton and since his time many of the distinguished representatives of the science have turned their attention to it.

On p. 122 there is mention of Charles Hutton's estimate of the earth's density at the ridge of Schiehallion which should not be confused with James Hutton's contributions. Neumayr's book became extremely popular and was translated into Swedish by the geologist and Arctic explorer Alfred Gabriel Nathorst (1850–1921) under the title *Jordens Historia* (1894), into Italian by the anthropologist and botanist Lamberto Moschen (1853–1932) as *Storia della Terra* (1896), into Russian by Vladimir Vladimirovich Lamanskii (1874–1943) and Alexander Pavlovich Nechaev (1866–1921), with additions on Russian geology and literature, under the editorship of the great Russian geologist Aleksandr Aleksandrovich Inostransev (1843–1919) as *Istoriya Zemli* (1903), and into Polish by Józef Morozewicz (1865–1941)

Figure 47. Melchior Neumayr (relief in the Arcade of the main building of the University of Vienna on the Universitätsring). Photo by A.M.C. Şengör.

and Karol Franciszek Koziorowski (1864–1933) with complementary pieces by Mieczysław Limanowski (1876–1948) as *Dzieje Ziemi* (1912) in addition to second and third editions in German by Suess' successor in the chair of geology in the University of Vienna, Victor Uhlig (1895) and Suess' geologist son Franz Eduard Suess (1920; only the first volume) respectively. The Russian and Polish translations were made from Uhlig's second edition of Neumayr's classic. If nothing else, at least the Swedish and the Italian translations of Neumayr's book must have contributed to the spread of Hutton's fame in these countries before Geikie's book came out in 1897 (but, as we shall see below, Hutton had been famous in Italy, a portion of which was then a part of the Austro-Hungarian Monarchy, even well before Lyell's *Principles*).

References to Hutton's ideas crop up even in places where one would not necessarily expect them: in 1887, the Austrian geographer and educator Friedrich Umlauft (1844–1923), a student of the great physical geographer Friedrich Simony (1813–1896), published his *Die Alpen. Handbuch der Gesammten Alpenkunde* (The Alps. A Manual of the Entire Alpine Science). The fifth chapter of that book is devoted to the geology of the Alps where Hutton is mentioned as the author of the magmatic uplift theory of mountain-building:

> In contrast to Werner[156] and his supporters, the "Neptunists," the Scottish geologists Hutton, Hall, Playfair, Macculoch (1795–1821), as "Plutonists," ascribed the origin of mountains to the extravasation of old eruptive rocks, which uplifted, folded the strata and metamorphosed them along their contact zones. This view was followed also by the great German researchers Alexander von Humboldt and Leopold von Buch, as well as by the French geologist Élie de Beaumont and the English [sic] geologist Sir Roderick Murchison. (Umlauft, 1887, p. 75)

However, Umlauft immediately pointed out that in the Alps this view, "entertained against all the facts" (p. 75) had been abandoned and replaced by theories that ascribed mountain-building to horizontally acting forces (p. 76–77). As protagonists of this new interpretation, he names, James Dwight Dana (1813–1895), Alphonse Favre (1815–1890), Eduard Suess (1831–1914), Albert Heim (1849–1937) and Johann August Georg Edmund (Ödön in Hungarian) Mojsisovics Edler von Mojsvar (1839–1907).

Only a year after the publication of Neumayr's great classic, the *professor extraordinarius*[157] for geology in the University of Vienna, Eduard Alexander August Reyer (1849–1914; Fig. 48) produced his magnum opus, the *Theoretische Geologie* (Theoretical geology; Reyer, 1888). Reyer was the son of a wealthy physician and had a respectable personal library and a good knowledge of the literature of geology (for the life of this now little-remembered geologist, see Proßegger, 2017; Sacchi, 1980, emphasized his current oblivion; as if to stress this point, Sacchi himself got Reyer's first name wrong!). In this remarkable book, Hutton is mentioned in 13 different places and *not only as historical reference, but as part of the then current scientific*

Figure 48. Eduard Reyer. (From Österreichisches Volkshochschularchiv; used with permission.)

discussion. On p. 60, he is presented as one of the three most distinguished spokespersons of the magmatic uplift theory of volcanoes, namely together with von Buch and Élie de Beaumont. On p. 150, Reyer writes about granite apophyses and points out that Hutton was the first who provided the true interpretation of them for the first time. On p. 205, Hutton is counted among those who assumed a central heat and ascribed vulcanicity, hot springs and earthquakes to it. What we read on p. 278 is the following:

> Hutton pointed out the similarity in the genesis of the crystalline and non-crystalline sediments and expressed the opinion that the heat of the depths had brought about this transformation. The objection that then the lime in the crystalline rocks ought to have been burnt he countered with the explanation that the high pressure had prohibited the escape of the carbon dioxide. Hall proved the correctness of this hypothesis.

Reyer gives two references for these statements in footnote 2: Hutton's 1788 paper and the 1795 book and Roth's 1872 paper mentioned above. It is remarkable that Hutton's publications are cited as source material along with a paper some 84 years younger! On p. 300, Reyer repeats the reference to Hutton's idea that high pressure inhibits the escape of carbon dioxide from metamorphic rocks and again references Sir James Hall's corroborating experiment. On p. 310, Reyer points out that Hutton had shown that calcareous and clastic components of sedimentary rocks are seen in the entire geological column (as opposed to Werner's hypothesis that they characterize distinct times in earth history). On p. 359, Reyer emphasizes that the important distinction between lavas and granite was really made first by Hutton. Reyer stresses further Hutton's ascription to pressure the difference between the fine texture of the lavas and the coarser crystalline texture of granite. On p. 375, Reyer points out that Hutton was among those who destroyed the idea that granite was the oldest rock by documenting its intrusion into schists. On p. 555, there is reference to Hutton's idea of contact metamorphism, and on p. 556–557, Reyer refers to his idea of regional metamorphism under the influence of the central heat of the earth. Reyer finds these two types inadequate as explanation for

Figure 49. Bernhard Studer. Image from: Die Schweiz im neunzehnten Jahrhundert. Herausgegeben von schweizerischen Schriftstellern unter Leitung von Paul Seippel. British Library HMNTS 9305.h.2, v. 02, p. 238, Lausanne, 1899.

Figure 50. Arnold Escher von der Linth. Courtesy of ETH-Bibliothek of ETH Zurich. CC-BY-SA.

all the features of metamorphic terrains and introduces the idea of mechanical metamorphism, i.e., one dependent of deformation. Unfortunately for Reyer, later work proved Hutton right and him wrong, except for very limited rock volumes around faults and ductile shear zones. On p. 569, there is talk of schlieren and flattening structures in some igneous rocks. Reyer says that Hutton had not been able to find an explanation for these within the framework of his theory of the earth, which was later supplied by Breislak (see below). On p. 797, Hutton is again referenced (footnote 4), as the first among others, for understanding the terraces around Great Britain. Finally, on p. 823, Hutton is declared as the cofounder, together with the German theologist and naturalist Johann Esaias Silberschlag (1721–1791), of the magma-driven uplift theory for the mountains. Reyer's reference (his footnote 2) is here to Silberschlag's v. 1 of his *Geogenie* of 1780. I repeat here that Reyer treated Hutton in his book not only as an important predecessor, but also as the author who established many of the theories Reyer himself dealt with in his book and whose hypotheses still needed serious consideration. No other eighteenth century author was similarly treated by Reyer.

The last edition of the Hann et al. book brings us to the date of publication of the first edition of Geikie's *Founders*.

SWITZERLAND

This small Alpine country has been celebrated for its epoch-making contributions to geology from the Renaissance (recall the great polymath Conrad Gessner of Zurich {1516–1565}) up to our own day, quite out of proportion to its size and population. Although many illustrious names, such as Gessner, the Italian count Luigi Marsili, the two Scheuchzer brothers, the Russian count Georgi Rasumovsky, Horace-Bénédict de Saussure and Hans Conrad Escher von der Linth illuminate the annals of pre–nineteenth-century Swiss geology, the systematization and institutionalization of this science in Switzerland was really the job of two great geologists in the first half of the nineteenth century: Bernhard Studer of Berne (1794–1887; Fig. 49) and Arnold Escher von der Linth (1807–1872; Fig. 50) of Zurich, closely followed by Louis Agassiz (1807–1873; Fig. 51) and Arnold Henri Guyot of Neuchâtel (1807–1884; both later emigrated to the United States, partly because of the turmoil caused by the failed 1848 revolutions; Agassiz found employment in Harvard, Guyot in Princeton).

After Lyell and Before Geikie

Hutton's theory of the earth had reached Switzerland at the latest by 1840, when Louis Agassiz published his famous *Études sur les Glaciers* in both French (published in 1840) and in German (translated by his friend Carl Vogt and published in 1841). In it, the great Swiss geologist and zoologist wrote:

> Playfair also thought that it was the glaciers that pushed the erratic blocks. (Agassiz, 1840, p. 12 footnote **)

The German edition has only a slightly different wording:

I also know that Playfair claimed that glaciers had pushed the erratic blocks to their present places. (Agassiz, 1841, p. 13)

Jean de Charpentier (1786–1855), the predecessor of Agassiz in glacial theory in Switzerland, published his ideas only five months after the publication of the French original of Agassiz's book, in which, he too touches on Playfair's mention of the Alpine glacial age (but dates it to 1815, because he must have become aware of it via Basset's French translation (see de Charpentier, 1841, p. V).

It is interesting that neither Agassiz nor de Charpentier mention Hutton, although it seems that they had read Playfair's book, which is only an explication of Hutton's ideas and not an original contribution, as Davies (1968) pointed out. In that paper Davies also quoted Hutton's statement that

There would then have been immense valleys of ice sliding down in all directions towards the lower country, and carrying large blocks of granite to a great distance, where they would be variously deposited, and many of them remain an object of admiration after ages, conjecturing from whence, or how they came. Such are the great blocks of granite which now repose upon the hills of Saleve. (Hutton, 1795, v. 2, p. 218)

Davies further quotes another passage from the same book:

But, in the Alps of Switzerland and Savoy, there is another system of valleys, above that of the rivers, and connected with it. These are valleys of moving ice, instead of water. This icy valley is also found branching from a greater to a lesser, until at last it ends upon the summit of a mountain covered continually with snow. The motion of things in those icy valleys is commonly exceeding slow, the operation however of protruding bodies, as well as that of fracture and attrition, is extremely powerful. (Hutton, 1795, v. 2, p. 296)

Although both Agassiz and de Charpentier quoted Playfair, they did so in apparent ignorance of the fact that they were in fact referring to Hutton's idea of a former ice cover on the Alps.

In 1844 came out the first volume of Studer's *Lehrbuch der Physikalischen Geographie und Geologie* (=Textbook of Physical Geography and Geology). On p. 363–364 Studer mentions Hutton's idea, repeated by Playfair that valleys are results of fluvial erosion. The second volume of Studer's book came out in 1847, in which on p. 118 there is a detailed discussion of Hutton's ideas on metamorphism pointing out that they are now generally accepted. On p. 180–181 Studer then presents Hutton's theory of mountain-building by magmatic intrusions in some detail as he himself had become a supporter of it.

Figure 51. Louis Agassiz around 1865 (photographer unknown).Courtesy of Cornell University Faculty Biographical Files, #47-10-3394. Division of Rare and Manuscript Collections, Cornell University Library.

From Scotland, the almost forgotten volcanic, or, after the new appellation, plutonic theory of the Italians gained a new power. J. Hutton, *Theory of the Earth*, 1795 and J. Playfair, *Explications*, 1802, concluded, on the basis of the tilted and vertical strata, on the basis of their bending and crushing and on the basis of the intrusion of granite and trap-rocks into the overlying formations, the existence of a force working from below and the cause of that force they in turn saw in heat that had indurated the sediments on the sea-floor into compact rock types. Next to the facts collected in Scotland, de Saussure's observations in the Alps served as the basis of Hutton's theory. Many pieces of evidence supporting it were multiplied and made more generally known by the work of MacCulloch, *Descr. Of the Western Islands*, 1819, Necker, *Voyage en Ecosse*, 1821, Boué, *Essai sur l'Ecosse*, 1822, *Mém. Géol. Sur l'Allemagne*, 1822, *J. de Phys*. But the most certain support for it came from the correct and comprehensive concepts on volcanic activity, earthquakes and the significance of igneous rocks that had become common knowledge since the beginning of the century through von Humboldt and von Buch.

Studer further discusses also Hutton's theory of rain in detail and with approval on his p. 423–424 of his second volume.

Studer published in 1863 a very detailed history of the physical geography (including geology) in his homeland up to 1815. Hutton is mentioned on p. 406 in his general review of the development of geology outside Switzerland as an opponent of Werner, who "based his theory on facts opposed to it" [i.e., Werner's theory of geology]. Studer writes that Hutton's theory did not become well known on the continent until after the Napoleonic wars, i.e., after 1815 (which, as we have seen above, and will see also below, is not entirely true).

In 1883, in the same year as the publication of de Lapparent's *Traité* in neighboring France, Maurice de Tribolet (1852–1929), younger brother of the geologist Georges de Tribolet and professor in the Faculty of Science in Neuchâtel published his pamphlet entitled *La Géologie—Son Objet, Son Développement—Sa Méthode, Ses Applications* (Geology—its objective, its development—its method, its applications) in which, after having reviewed Werner's contributions and having said that Werner is the father of the geological nomenclature (p. 17) and that modern geology had commenced with him (p. 18), he continues:

> Endowed with a spirit of observation not inferior to his antagonist, Hutton arrived at opposite conclusions concerning the fundamental phenomena and the two schools became established at the same time.
>
> At this moment a long discussion between the neptunists, partisans of the exclusive dominion of the waters and the plutonists was born and often degenerated into a very lively polemic, which, it must be pointed out, continued for more than half a century. Hutton rebutted the hypotheses attributing to water the origin of all rocks and ventured to declare that the continents have existed all the time. His glory is to be the first to recognize that heat played an important role in the formation of the earth's crust. According to his theory, the ruins of an ancient world were visible in the structure of our planet and that our present continents were only the debris of the older continents deposited on the floor of the seas, consolidated by the volcanic heat and subsequently raised. (de Tribolet, 1883, p. 19–20)

De Tribolet describes the rise of geology in the first half of the nineteenth century and concludes:

> The school of Saxony had finished its time. For half a century, according to Cuvier's words, one had interrogated Nature in the name of Werner. In 1830 things had changed their aspect so much that an author could write: "Freiberg, this ancient centre of enlightenment, became for the moment a geological China in the middle of an enlightened Europe." (de Tribolet, 1883, p. 21)

De Tribolet's booklet came out only a year before the great French genius Marcel Bertrand (1847–1907) published his theory of nappes on the basis of the observations in the Canton of Glarus (Bertrand, 1884) using Albert Heim's (1849–1937) magnificent field descriptions (Heim, 1878) and Eduard Suess' prescient theoretical deductions (Suess, 1875). After that, nothing was quite the same in the geology of Switzerland, when this tiny country took the lead, together with Suess' Austria-Hungary, in theoretical geology.

ITALY

Before Lyell

Modern geology was born in Italy, in the hands of two of Hutton's most important predecessors, Steno and later, Moro. Both did their main work in Italy. With its active volcanoes, frequent earthquakes and magnificent mountains with rock layers displaying vivid colors and bearing abundant fossils, Italy has always been a fine place to do geology in a most pleasant environment (see Whewell, 1831, p. 190). It is therefore hardly surprising that in Italy Hutton's ideas were picked up and discussed before Lyell began advertising them in his *Principles*. Here I discuss only Breislak's famous *Introduction to Geology*.

Scipione Breislak (1748–1826; Fig. 52) was born in Rome to an immigrant Swabian father and originally studied theology and philosophy to be attracted to natural sciences later under the influence of the Venetian naturalist, cartographer and traveler Alberto Fortis. After having had various teaching assignments in mathematics, physics and mineralogy in various parts of Italy, Breislak went to Paris and came into close contact with the chemist, physician and politician Jean-Antoine Chaptal, Count of Chanteloup (1756–1832), zoologist and geologist Baron Georges Cuvier (1769–1832), chemist and physiologist Antoine-François Count of Fourcroy (1755–1809), mineralogist Abbot René-J. Haüy (1743–1822) and geologist Alexandre Brongniart (1770–1847) (http://www.treccani.it/enciclopedia/scipione-breislak_(Dizionario-Biografico)/).

Having written a number of geological memoirs on various parts of Italy, Breislak published, in 1811, his famous two-volume treatise on geology, *Introduzione alla Geologia*. Let us note that this book came out 13 years earlier than the publication of the very first scientific paper by Sir Charles Lyell in 1824 on the freshwater limestones and marls of Scottish lakes, showing their similarity to the older freshwater formations of the Tertiary sedimentary rocks of the Paris Basin (Wilson, 1998). In his book Breislak begins, already in the preface, pointing out that the Wernerian theory of terrestrial structure and evolution is entirely indefensible[158]:

> There is in Europe a school (the Wernerian) very worthy of this science as well as of mineralogy, and very respectable for the celebrity of its founder, for the number and the knowledge

Figure 52. Scipione Breislak. Public Domain (PD-US-expired)

of his pupils, whose dogmas are so widespread that it will seem audacious to suggest any doubt around them. I am far from directing to it those reproaches which are in part fair, but a little too sharp, which were made by one of the most learned chemists of our age (Mr. Chenevix V. Annals of Chemistry, volume LXV[159]): I will only remind my readers that the place once occupied by the authority of the chair has long been given over to reason, to observation, to experience. Some very vague and uncertain principles of that school, many vague ideas—more or less, a little or a lot—a mysterious nomenclature, devoid of any reasonable meaning, uncouth to pronounce, just as difficult to remember, many absolute assertions, supported only by authority and lacking valid arguments and based, above all, on some isolated observations that are contradicted by many other conflicting ones, form a body of doctrine that seems made to chase away from the study of geology those who love to reason. This doctrine, propagated by one hundred pens, some good and some bad, has already penetrated into France and England, and now tries to creep into Italy. It is therefore necessary for the Italians to be protected and to be cautious about it and to become accustomed to recognizing what is good in that system, which certainly is a lot, in mineralogy and also in the geology of observation, but abounding what is strange and absurd in its systematic part. (Breislak, 1811a, p. XX–XXII)

In the rest of the book Hutton's ideas are treated as parts of current science and are supported or refuted according to Breislak's own observations and interpretations. On p. 158–166, Breislak gives an excellent sketch of Hutton's theory, specifies what he dislikes in it and presents a defense of it against an objection by Deluc. After having reviewed data on the increase in temperature in mines and in lakes as one descends toward the center of the earth in Europe, Breislak finds that the extreme heat postulated about the interior of the earth today cannot be true, notwithstanding the fiery beginning of the planet that he defends and implies that after a certain increase in temperature as one descends into the earth, the rate of increase must diminish. He continues:

From all that has been said, there appears to be no argument to assume, in the internal parts of the earth, an intense heat sufficient to produce the effects felt [he means at the surface]. The theory of the Englishman [160] Mr. Hutton is founded on this central fire, and it will not be outside our purpose to give an idea of it. The continents that currently exist are destroyed by the action of air, gravity and running water: their materials are transported from the same waters to the great depth of the sea, where they are distributed by the motion of the waves, tides and currents above its bottom. An intense internal heat hardens these compressed materials caused by the mass of water and results in substances similar to those that make up our continents. When the existing continents are so degraded, destroyed and reduced to the level of the sea, the same heat that has hardened the layers above the bottom, raises and upends them. This operation results in the sea being pushed over the destroyed continents. Such events repeat and new continents are generated, also exposed to the action of the air, of the gravity, of the running waters, and then of that of the sea that scatters those materials above its bottom, where heat prepares the layers of newer continents that will again be raised in convenient epochs. To form a clearer idea of this system, we summarise the propositions to which it has been reduced:

1. Our continents are composed of layers formed in the sea;

2. The layers of our continents have been produced by the accumulation of substances from other continents that are gradually destroyed by the actions of the atmosphere and the waters: the materials of these first continents were similar to those which we find on the coasts of our seas;

3. While the fragments of the continents being eroded are dragged by the waters to the bottom of the sea, where the waves, tides and currents distribute them and scatter them on the same bottom;

4. Under the waters of the sea, there is excessive heat, so that the eroded materials that subsequently arrive from the banks [at sea margins], are melted and changed into new rocky layers;

5. When a certain generation of continents is nearly destroyed on our globe, the materials of an older generation, long since arrived at the sea, are consolidated and turned into rocky layers, and then the same heat, which had prepared them to form new continents, raises them and gives them the same character [as the earlier continents];

6. The alternative operations of continent destruction and new continent generation emerging from the sea, have already been repeated innumerable times on our globe, separated by millions of centuries;

7. Our continents are the last in this series of operations that alternately produce the sea and the land in the same portion of the globe; these continents are in a state of degradation, their materials are subsequently scattered first on the low lands where they form a soil, then on the bottom of the ocean, where future continents are going to be prepared from their fusion.

Such is the ingenious opinion of Hutton, supported and defended by Playfair, which provides easy explanations for some geological phenomena, such as the formation of stony strata, their reversals, etc. The same gave the opportunity to Mr. James Hall to undertake a series of experiments on a completely new topic, that is, on the modifications that compression produces in the way heat acts; these experiments enriched the science of new important facts, and have opened up a new field of research for geologists (see the description of these experiences published in Geneva in 1807). However, though this theory is illuminating in its applications, I cannot accept two of the principles it is based on, namely the periodic renewal of the continents and the existence of a very intense heat in the seabed. The first of these two principles has been validly refuted by Mr. G. A. De-Luc in his Elementary Geology Treatise. To his reasons I will add only a reflection, and it is that if ever the state of our globe were to be renewed in certain determined periods, it seems to me that it could not happen if not in the way I indicated above[161], that is by means of the circulation of caloric. The second principle of Hutton is contradicted by all the observations already reported, and which were made to determine the temperature of the sea floor and of the terrestrial depths, which is far less than that required to melt the earthy materials and renew them in compact form. If at the bottom of the ocean and around the globe there was such intense heat, it would have to send some traces to the surface. Admit, however, if you will, this central heat, and suppose it is still sufficient and melt to consolidate incoherent terrestrial materials, deposited on the seabed, you still cannot explain how the fused layers have been melted and overturned. Heat alone fuses the materials, but to uplift them, the action of some other cause is necessary: thus we need to resort to an expandable fluid, whose strength can be conceived to be infinite. But how does this fluid develop under the mass of the molten layers? If it pre-existed at the bottom of the sea, or if its constituents were there, its expandability could not remain suppressed because the central heat has always been there in the globe, according to Hutton; thus the elasticity of that vapor should have prevented the formation of the layers. The materials transported by water could not have collected on a bottom, from which an elastic fluid, with the energy that Hutton supposes, was continuously developing. It could be assumed that this steam develops only from the bottom of the sea after the terrestrial layers have formed and melted; but this new assumption only complicates the theory further. Mr. De-Luc, wishing to disprove this part of Hutton's theory, argued that an expandible fluid, which allegedly raised a mass of layers, like that of our continents, could not have done so, except by means of an extreme density, for, as soon as this fluid breaks the seabed, most of it would escape by the openings; and then it could no longer support the intumescence it supposedly created, so that the uplifted mass would fall to pieces: this would be an inevitable consequence of the initial uprising of the continents. This reasoning by Mr. De-Luc, however, does not seem to me conclusive—as he calls it—against the Huttonian system. An elastic fluid that opens a path through the already consolidated layers, should certainly break them and smash them; but this will not happen if we are dealing with bodies that are in a state of fusion or softness. In this, it seems to me that one of two things may take place: the first is, in case the expansive force of the vapor not being able to break the surface of the body—then it will lift the surface; but, eventually, it will break it; then, the fluid will escape due to the opening, and the raised parts will be able to stay for some time in the situation into which they had been brought by the force of the steam without further breakage, and consolidate in that position, passing to a state of perfect cooling. Anyone who has had occasion to examine the currents of lava, will have been able to observe in them some parts raised by the explosion of some gas, and which have become consolidated, remaining above the general level of the lava.

Breislak quotes Hutton again when discussing the origin of granite. After having pointed out the untenability of the neptunian position about its origin and having quoted Pallas' memoir on the origin of mountains ascribing an igneous origin to granite[162], he wrote:

> ...did not another equally celebrated geologist, Mr. Hutton, reporting (see the British Library, volume VII, p. 256[163]) different observations made in some places in Scotland, alleged that the granite flowed in a fluid state, after having been previously melted in the bowels of the earth? Nor was Hutton the only English mineralogist who supported this opinion.

It was followed by Doctor Beddoës [sic] in his Observations on the affinity between basalt and granite (see Philosophical Transactions 1791[164]). This must be said with the sole purpose of demonstrating that the origin of granite in the humid way is not something so obvious that it should be considered a geological paradox to argue the opposite. It seems to me that on this theoretical point, as several others mentioned above, a certain system of intolerance has been introduced by a mineralogical school, moreover a very famous one in Europe, that is incommensurate with the spirit or with the enlightenment of this century. I will certainly not embrace the idea of Hutton that the granite melted by the underground fires, broke through the rocks and ran like lava, spreading over them and creeping into their fractures; I will only say that I do not see any impossibility in conceiving the molten granite due to a fiery fluidity which then consolidated and crystallized during cooling. (Breislak, 1811a, p. 181–182)

A few pages later, when Breislak is reviewing observations supporting an igneous origin for the granite, he refers to Hutton's description of the graphic granite as a supporting instance for the rock's magmatic origin (p. 185). Concerning the internal structure of the granitic masses, Breislak noted that "According to Hutton we must suppose our granite masses to be without any specific structure except the veins and cracks formed by the withdrawal of the solid mass at the time of consolidation." (p. 221). On p. 249–251, Breislak praises Sir James Hall's experiments and, like Carl Vogt three decades after him, deplores that they are forgotten,

> Perhaps the cause is the preponderance in the minds of geologists of the principles of the Neptunian system, little supported by those experiments. How much physics had to struggle to get rid of the occult qualities. Even the great spirit of Galileo could not get rid of them completely. Probably a similar destiny is reserved in geology for those mysterious dissolvers that we wished to see in the waters during the first formation of the globe. Perhaps even the errors that is recognized in some part of the Huttonian theory have influenced in making us forget the experiments made to support it; but this would not be the first case of important truths, discovered by some erroneous opinion. To the ravings of alchemy we owe some major operations in chemistry, and that beautiful science might not have been born without the vagaries of the alchemists. The hypothesis of Hutton is wrong concerning the present existence of fire under the seabed in such a degree of intensity as to be able to unite in solid and compact formations the fragments of the continent that are transported and distributed by the waters and then uplift and upend them thus producing new continents; it may be true, however, that fire still acted in the first constitution of the globe and on the substances that shared the first period of its existence.

In this quotation Breislak says that the adherence to the neptunian theory has hampered the recognition of the importance of Hutton's theory and led to the neglect of the experiments supporting it, although he himself has important disagreements with some of Hutton's views. He recognizes, however, that errors in current theories may lead to future discoveries. His denouncement of the neptunism is directed, I think, not so much to the content of the theory of neptunism, but to the dogmatism of its supporters. Finally, on p. 288–289, Breislak, while discussing the plutonic nature of what was then called "primitive rocks," once more refers to Hutton's system and mentions Marc-Auguste Pictet's (1752–1825) reference to it by Richard Kirwan's (1733–1812) term "plutonism":

> If we then consider the traps found together with the primeval rocks, this is the creed belonging to the domain of Pluto, since, using the sentence applied by Mr. Pictet [165] to the Hutton system, here it is no longer a question of the furnace of Volcano, but the hearth of the general combustion should be sought lower in the obscure realm of Pluto; and removing the veil of mythology, the primitive rocks, before they were consolidated, seemed to me to have been fluid because of the calorific particles scattered throughout the mass of matter from the surface to the center of the earth.

In Breislak's second volume (1811b), Hutton is mentioned only once, on p. 23, and only in reference to the mention of his description of a septarian nodule from Aberlady in East Lothian in Breislak's v. 1, p. 128 (where Hutton's name is misprinted as Huston).

As in the case of my discussion of Deluc's geology, my purpose here is not an assessment of Breislak's what I think to be rather idiosyncratic geological theory, notwithstanding its obvious merits. For such an assessment one can consult Anonymous[166] (1816) in the *Edinburgh Review*. My account of Breislak's book is to underline the fact that for a prominent geologist in Italy, whose Treatise was translated both into French (in two editions! Breislak, 1812, 1818a, 1818b, 1818c, 1818d[167]) and German (1819, 1820, 1821[168]), years before Lyell wrote anything on geology, Werner's theory had long been *passée* and had been a hindrance rather than a help in advancing geology, whereas Hutton's was the basis of further theorizing on the history and behavior of our planet. Hutton was for him a "celebrated geologist."

After Lyell and Before Geikie

The Italian geologist Carlo Gemmellaro (1787–1866; Fig. 53), one of the founders of volcanology and a friend and advisor to Sir Charles Lyell in matters Sicilian (his son, the geologist Gaetano Giorgio Gemmellaro, 1832–1904, who eventually became even more famous than his father, was Lyell's guide, when Sir Charles visited some of the outcrops at Aci Reale north of Catania in 1857: see Lyell, 1859, p. 711, footnote ✱), published a useful little textbook in 1840, under the title *Elementi di Geologia ad Uso della Regia Universita' degli Studi in Catania* (Elements of geology for use in the Royal University in Catania). This book was written for his lectures at the University in Catania. He had originally used the Belgian administrator and geologist and the inventor of the System name Cretaceous[169], Jean Baptiste Julien Baron d'Omalius d'Halloy's (1783–1875) *Éléments de Géologie* (Elements of Geology) as a basis for his lectures[170]. Gemmellaro then began feeling the need to show his students some local examples of the things he was lecturing about. He also culled information from other authors about the geology and fossils of foreign lands. He says that in geological studies one needs to use the present-day phenomena to understand past events. Having said that, he hastens to add "it should not be required, however, that nature should always operate following the same laws that it today does; while, in view of what can be seen in many ancient terrains, I do not hesitate to remind my readers, that extraordinary and unknown phenomena once had to operate such events, which none of the present would be able, even approximately, to imitate." (p. VIII). Almost in a Delucian fashion, Gemmellaro says that since the universal flood, we can follow geological phenomena in agreement with the currently observed natural laws, but before "the uncertain time of the flood" (p. IX) who can possibly say anything without resorting to hypotheses (but, unlike Deluc, he does not fall back on Biblical mythology). He follows this by a reminder that in d'Halloy's book much doubt is expressed about the relations of various formations, their conditions of superposition etc. This leads to skepticism about older systems and Gemmellaro nicely describes the transition from Wernerism to Huttonism.

> What weight must then be given to the systems of the modern geologists, can well be calculated from the fate that those of many illustrious geologists have encountered in the passing of the years. Who did not say up to the beginning of this century with Werner, that the rocks of the crust of the Globe were nothing but sediments of loose materials suspended in the waters, and that they were deposited according to their specific

Figure 53. Carlo Gemmellaro (in the possession of the Accademia Gioenia di Catania).

weight? Meanwhile, that system then negated the idea of de BUFFON that our Globe, like the rest of the planets, was only portions of a one eighthundredth part of the sun, sparked by the impact of a comet. But Werner's disciples could not long sustain that system; the Huttonians shook him, and the theory of upheavals and the theory of central heat caused them to forget almost everything. Today, another astronomical theory of the condensation of nebulae as the origin of the globular bodies of the universe is taking root. (p. IX–X)

When it comes to discussing the origin of metallic veins Gemmellaro again turns to Hutton:

It was an old opinion of Hutton, developed by the illustrator of his theory, Prof. Playfair of Edinburgh, which showed that veins in general to have been injected from the bottom up in the already formed rocks; against the ideas of the Wernerians, whose supporters wanted veins to be a result of infiltration from top down. Hutton reasoned over the facts: the observation that the metal veins always consisted of materials or substances that were not those of the nearby rocks

…

This opinion is always supported when one reflects on the metallic sublimations that occur in the volcanoes. (p. 342–343)

Finally, Gemmellaro refers on his p. 383, to Gregory Watt's 1804 publication concerning the volcanic character of basalt, in support of Hutton's views (see Watt, 1804).

Seven years after the publication of Gemmellaro's book, a similar textbook written by one of the most extraordinary geologists of the nineteenth century, Giacinto Ottavio Provana Count of Collegno (1794–1856; Fig. 54) was published in Torino under the title *Elementi di Geologia Pratica e Teoretica Destinati Principalmente ad Agevolare lo Studio del Suolo dell'Italia* (Elements of Practical and Theoretical Geology especially for facilitating the study of the ground of Italy). Originally trained as a soldier, Collegno took part in Napoleon's Russian campaign. After the fall of the French Empire, he fought in Spain and in Greece for the liberation of the oppressed. He left the military in 1827 and studied botany and geology, the latter under Élie de Beaumont in Paris in 1835. Later, he became one of the more colorful figures of the *Risorgimento* rising to become the minister of war in the Lombard provisional government of Gabrio Casati (1798–1873). He was a friend of General Giuseppe Garibaldi (1807–1882) and supporter of Count of Cavour (Camillo Paolo Filippo Giulio Benso, Count of Cavour, Isolabela and Leri; 1810–1861), the two grand heroes of the unification of Italy.

The little book reveals that its author, a very busy man with things far outside geology and not exactly conducive to a relaxed lifestyle, was astonishingly well informed and remarkably creative in his thinking on geology. One wonders what he could have done as a scientist, had he not spent most of his life fighting wars as a soldier and being a politician in a particularly volatile episode of the European history.

By the time the Count was writing, Werner's neptunism was long dead and Hutton's view of geology, with various modifications in the hands of different authors, had acquired general acceptance by the geological community worldwide. Count di Collegno's little book reflects this state of affairs nicely.

On p. 352, the Count points out that granite is an intrusive rock penetrating into the country rock as seen in Cornwall, in the Vosges and in other mountains and adds that:

The first authentic example, in which a similar penetration is mentioned, is that observed by the celebrated Hutton in the valley called *Glen Tilt* in Scotland. The granite is covered there with clayey schists, and from the lower mass depart dykes of a certain width which then thin out, to the extent that they penetrate into the clayey schists; then the granitic structure of the injected rock is less evident. (Collegno, 1847, p. 352, italics his)

Let me underline here that the Count calls Hutton "celebrated." When he wrote that, Sir Archibald Geikie was only 12 years old in distant Scotland.

While discussing the origin of schists and marbles, the Count asks whether all schists and marbles are to be viewed as ancient, altered sedimentary deposits: "Thus thought Hutton" he wrote, "the founder of the school of vulcanists, at the end of the eighteenth century; he admitted that the intense internal heat of the earth could, so to speak, cause the matter of the sediments compressed by the water of the sea, to turn into substances analogous to those making up the crystalline rocks" (Collegno, 1847, p. 404).

Count di Collegno then reminded his readers that this interpretation of Hutton had met strong opposition, but, he argued, because he was not properly understood. He explained:

It was objected to Hutton that the observations made in the various seas to determine the temperature were absolutely contrary

Figure 54. Giacinto Ottavio Provana Count of Collegno. Source: Leone Ottolenghi, La vita ei tempi di Giacinto Provana di Collegno. Col Diario dell'assedio di Navarino 1825 che si pubblica la prima volta nell'originale francese, E. Loescher, Torino 1882.

to the hypothesis of intense heat at the bottom of the sea; but the Scottish naturalist did not say that the desired temperature to modify, to metamorphose the sediments eroded from the land existed in the bed of the sea itself; it is under that bed that Hutton placed the great source of the caloric, and certainly no one will want to contend today that starting from the bed of the sea, the temperature must grow with depth, as it happens when we descend below the plane of invariable temperatures below the land. Let us suppose that the terrestrial globe is divided into any number of isothermal zones; those near the center will necessarily have a spheroidal shape, while the external areas will have to conform to the irregularities of the surface of the globe; for, if this were not the case, the bed of the sea should be at a temperature much higher than that, which is shown by direct observation; it is easy, moreover, to understand that water being a better conductor of the caloric compared with rocks, it dissipates more easily in the layers covered by the sea, which is not the case that prevails at the base of the mineral masses of a more or less considerable height. (p. 405)

Figure 55. Narciso Carl'Antonio Stoppani. Photographer unknown. Public domain (PD-US-expired).

This is an important point that no one seem to have considered before Collegno within the context of what Hutton said, with the exception of Joseph Black and possibly Stephan Behlen, but what Black thought about it was never published; he wrote it in the letter he sent to Princess Dashkova that I quoted at length above: "At such depths these fires act in silence [and darkness] during successions of centuries & under the pressure of the immense load of incumbent Earth & water their most general effect is to harden & consolidate into stones of different kinds the materials of former land which have long since been carryed [sic] into the ocean & deposited at the bottom of it in the form of regular but loose stratified matter; the strata of sand are hardened into sandstone." For the purpose of this book, it is immaterial whether Hutton really meant what the Count thought he did (but what Hutton's intimate friend and scientific confidant Black wrote to the Princess makes it very likely that Hutton did indeed mean what the Count understood); what is important is that after more than half a century, Hutton's model gave rise to the suggestion of a very plausible scenario, one we still entertain. That scenario included the idea of isothermal surfaces and their behavior in the underground. This formulation of the isothermal surfaces is commonly attributed to Charles Babbage, which he first published in a short abstract in 1834, then in the note F in his celebrated *Ninth Bridgewater Treatise* (1837; second enlarged edition, 1838) and finally in a full paper in 1847. I am not aware whether the Count knew about Babbage's ideas. When he was writing his book, the only source where he could have learnt them was the *Ninth Bridgewater Treatise*. In any case, Hutton's ideas remained current and not just part of the history of geology.

The great Italian geologist (and Catholic Rosminian priest) Antonio Stoppani (1824–1891; Fig. 55), the inventor of the idea of the Anthropocene (he called it Anthropozoic: Stoppani, 1873, p. 723, paragraph 1327) and the author of the delightful popular book on the natural history and especially geology of the then recently unified Italy, *Il Bel Paese—Conversazioni sulle Bellezze Naturali—La Geologia e la Geografia Fisica d'Italia*

(=The beautiful land—conversations on natural beauty—geology and the physical geography of Italy; 1876), which went through numerous editions subsequently, had earlier published, in the years 1871 and 1873, a much-used, three-volume, treatise on geology, entitled *Corso di Geologia* (A Course of Geology). In this very comprehensive book, there is no section devoted to the history of geology, except for a brief history of paleontology and stratigraphy with emphasis on the latter in the seventh chapter entitled *Principi della Cronologia Stratigrafica Dedotti dalla Stratigrafia e dalla Paleontologia* (Principles of stratigraphic chronology deduced from stratigraphy and paleontology), mainly based on the books by Brocchi and d'Archiac (v. 2, p. 88–106, paragraphs 127–152). It is in that history, Stoppani wrote the following, on p. 99, after having summarized the Abbott Lazzaro Moro's ideas:

> Here are the ideas expressed by Lazzaro Moro in 1740, that is almost half a century before the publication of the famous system of Hutton, who, by some, is considered the founder of modern geology. However, what is true of the system of Lazaro Moro, what he established for the geological science until the end of the eighteenth century, supports those that we have announced as the second true basis of palaeontology; that is to say, it supports the fact that *fossils are relics of organisms, that they lived where their remains are found.* (Stoppani, 1873, p. 99)

Here is a prominent Italian geologist, who was also a Catholic priest, writing some 19 years earlier than the first mention of Hutton by Sir Archibald Geikie about Hutton as "one of the founders of geology." Stoppani clearly approved of what Hutton did, as he cites his compatriot Moro with some pride, as having put forward similar ideas earlier.

In 1875 came out the little booklet entitled *Trattato Storico-Scientifico sull'Origine, su'Progressi, sullo stato della Geologia in Sicilia* (Historical-scientific treatise on the origin, progress and the state of geology in Sicily) by Benedetto Lupi, the learned professor of eloquence and French language and member of the Palermo and Catania literary academies. Lupi wrote:

> *The Theory of the Earth*, published by Hutton in 1785, greatly influenced the progress of geology; this learned scholar rejected some of the hypotheses that ascribed to water the origin of certain rocks, and explained by means of the action of a central fire the formation of a large number of rocks and minerals, but not that of our continents: he was the head of the school of *Vulcanists*. (Lupi, 1875, p. 25–26)

It seems as if Lupi's information about Hutton was second hand; he does not know of the 1788 paper, although he does know its influence. Lupi was no geologist and his reference to Hutton reflects what he learnt from the information available to him in Sicily documenting Hutton's reputation in that remote corner of Italy, which however, was visited by many eminent foreign geologists of the day, such as Baron Leopold von Buch, Élie de Beaumont, Sir Charles Lyell, Sartorius Baron von Waltershausen, and many others. It seems clear that Hutton's name and reputation were well known in Sicily.

Two decades later, Professor Giovanni Omboni (1829–1910; Fig. 56) of Padova, one of the ardent popularizers of geology in Italy in the nineteenth century and one of the founders of the Geological Society of Italy, published his small *Brevi Cenni sulla Storia della Geologia compilati per I suoi Allievi* (Short notes on the history of geology compiled for his students; Omboni, 1894). On Hutton he wrote the following:

> The Scotsman Hutton, after having divided the rocks into sedimentary ones (some dislocated and others still in their original place), ancient (without fossils) and eruptive or volcanic ones, and having claimed that they betrayed the action of the same causes, he admitted the occurrence of upheavals for the production of the continents and mountains, and demonstrated the eruptive or volcanic origin of the basalts and also of the granites, by means of their veins and the metamorphisms caused by these veins on other rocks. And his compatriots Hall and Playfair helped him, the first with special experiments (he fused volcanic rocks and then allowed them to cool and crystallize), and the second with publications and special observations, and by publishing the book entitled *Huttonian Geology* [in Omboni's original text he wrote *La Geologia Huttoniana*], which is, precisely, a clear and complete exposition of Hutton's ideas. The discoveries and theoretical ideas of these geologists were taken into consideration and supported by many others; and so it formed the first nucleus of a phalanx of *Vulcanists*, who accepted the volcanic origin of basalt, of granite and of other rocks; and a fierce struggle, which lasted many years, began between these and the *Neptunists*, leading to strange exaggerations on both sides, and to many interesting discoveries. Many geologists from all countries, as well as several Italians of that time, took part in this struggle, as we will see later. (Omboni, 1894, p. 29–31)

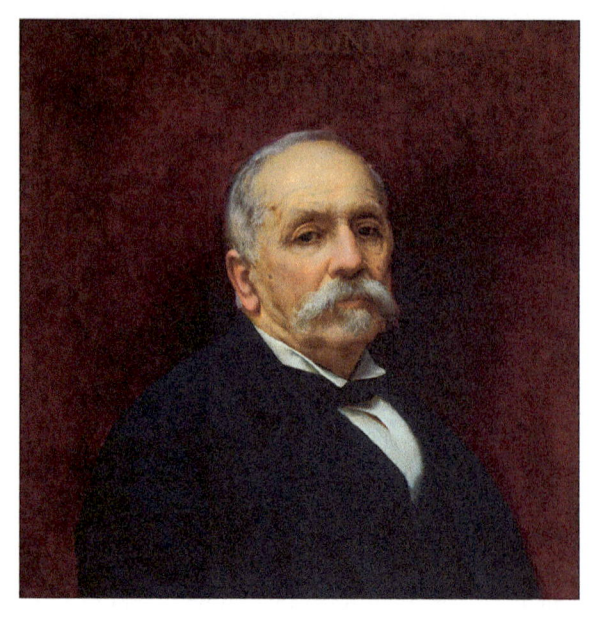

Figure 56. Giovanni Omboni. Phaidra Digital Collections, Sistema Bibliotecario di Ateneo, Universita degli Studi di Padova, CC-BY-NC-SA 4.0.

Later, in a footnote on p. 42 (the final part of the footnote 21 that had started on p. 35!), we read a brief biography of the Count Giuseppe Marzari-Pencati (1779–1836), the Vicenzan geologist, mathematician and botanist, where Omboni points out that after a tour of the central volcanic district of France (in 1804; see Mori, 1975; Ciancio, 2008), the Count abandoned the ideas of Werner on the behavior of the earth and adopted those of Arduino, Hutton etc. Let me quote from the Count's biography published in the *Dizionario Biografico degli Italiani* in 2008:

> The esteem for Marzari-Pencati shown by the Parisian savants—who appreciate the acumen and talent for observation—allowed him to meet personalities such as A. von Humboldt, A.-L. de Jussieu, L. von Buch, J. C. Delamétherie, L. Cordier. His stay in France proved decisive in expanding his skills. However, the contact with scholars, who were mostly supporters of A. G. Werner's geognosy, did not induce him to abandon his plutonist convictions drawn from the works of A. L. Moro, G. Arduino, Fortis and J. Hutton, as confirmed by his association with B. Faujas de Saint-Fond, professor of geology at the Muséum, among the few opponents of triumphant Wernerism. Geological excursions carried out at the request of Faujas in regions of extinct volcanism such as Auvergne (May 1804) and the Vivarese (September 1804) strengthened its non-conformism. (Ciancio, 2008, http://www.treccani.it/enciclopedia/giuseppe-marzari-pencati_(Dizionario-Biografico)/)

Thus, we not only have a description of Hutton's work in an Italian booklet on the history of geology three years before the publication of Sir Archibald's *Founders*, but also a report that, even in pre-Lyellian days, Hutton was read and appreciated in Italy and gained converts to his way of thinking in geology (in fact already in the first decade of the nineteenth century!).

THE UNITED STATES OF AMERICA

Before Lyell

Notwithstanding the presence of such towering intellectuals and polymaths as Benjamin Franklin (1706–1790) and Thomas Jefferson (1743–1826), and such daring expeditions into the wild west shared at the time between the Spanish and the British empires (also the French Empire from 1800 until the Louisiana Purchase in 1803; between 1762 and 1800 the entire area of the future Louisiana Purchase[171] was under Spanish jurisdiction), as those of Lewis and Clark (1804–1806), Zebulon Montgomery Pike (1805–1806 and 1806–1807), Freeman and Custis (the Red River expedition: 1806) and those of Stephen Harriman Long (five expeditions between 1817 and 1823), the fledgling United States was a scientific backwater in the times before the publication of Lyell's *Principles*. For an adequate orientation of the state of science in the United States after the independence to 1830, see Struik (1948), Nye (1960), Greene (1984), Porter (1986), and Daniels (1994); also see the documents published in Reingold (1964). The scarceness of the pre-1830 documents in Reingold corroborates the modest state of science in the new country (also see the new preface in Daniels, 1994[172]). For the development of geology in the young United States, the best summaries are Merrill (1904, 1924), Dana et al. (1918) and Schneer (1979). In his "Translator's Introduction," Edmund Maute Spieker (1972) also gives a succinct account of the earliest geological work undertaken in North America.

Both Ospovat (1967) and Newcomb (1990) touched upon parts of this early episode of American geology from the viewpoint of Werner's and Hutton's influence on it. Ospovat, with some justification, called it the "Wernerian Era" of American geology. Most American writers on geology at that time were aware of both Hutton and Werner. Their preference for the latter was not so much because they thought his interpretation of the nature of the planet superior (though some indeed did think it), but they found his nomenclature of rocks handy for field work, very much like some late twentieth century geologists using Hans Stille's (1876–1966) terminology of "orogenic phases" (see Stille, 1924, 1940) because they thought it convenient, long after plate tectonics showed that it was bereft of either empirical or theoretical basis and thus obsolete. As soon as a better nomenclature emerged on the basis of biostratigraphy, Werner's terms were promptly abandoned.

European publications concerning American geology in the second half of the eighteenth and earliest nineteenth centuries (e.g., Kalm, 1753–1761; Evans, 1760; Schöpf, 1787; Volney, 1803) had essentially zero influence on work done later (see Spieker, 1972). The Scottish businessman and geologist William Maclure (1763–1840) published the second geological map of the then United States in 1809, which he had started working on in 1807[173]. Maclure used the Wernerian nomenclature for his units, namely, Primitive rocks, Transition rocks, Secondary rocks, Alluvial Rocks, Old Red Sandstone, and Salt and Gypsum (separated from the rest by means of a green line drawn as a limit; Maclure, 1809, foldout map). However, he was not as much a Wernerian as Silliman's epithet for him, "American Werner" (Silliman, 1829, p. 5), or Ospovat's (1967) description of his work might lead one to believe. He adopted the Wernerian nomenclature, because:

> Necessity dictates the adoption of some system, so far as respects the classification and arrangement of names the Wernerian appears to be the most suitable. First, because it is the most perfect and extensive in its general outlines and secondly, the nature and relative situation of the minerals in the United States, whilst they are certainly the most extensive of any field yet examined, may perhaps be found to be the most correct elucidation of the general exactitude of that theory, as respects the relative position of the different rock series. (Maclure, 1809, p. 411–412)

But, he said at the end of his explanatory note to his map, that

> In adopting the nomenclature of Werner, I do not mean to enter into the origin or first creation of the different substances, or into the nature and properties of the agents which may have subsequently modified and changed the appearance and form of those substances; I am equally ignorant of the relative periods of time in which those modifications or changes may have taken place; such speculations are beyond my range, and pass the limits of my inquiries. (p. 427)

Maclure published a somewhat enlarged account of his now revised map in 1817. It says nothing different in spirit from what he had already said eight years earlier. In these two publications he appears as a field worker using a convenient terminology to delineate his units, much like a late twentieth century geologist using Stille's orogenic phase terminology as a shorthand for the description of the major unconformities in his area.

Did Maclure believe Werner's theory? He did not and, when in the volcanic district of Auvergne in 1805, he seems to have concluded that the neptunian origin of basalt was wrong and that basalt was indeed a volcanic rock (see Doskey, 1988, p. xxvi, footnote 35 and p. 15). Doskey notes that Maclure struggled long to fit his observations in Europe into Werner's straightjacket, but eventually despaired. Leonard Warren concludes the same about the Wernerism of Maclure, whom he identifies, I believe correctly, as a "Baconian" (Warren, 2009, p. 61). George White thought of Maclure as an "uniformitarian" and not a "real Wernerian" (White, 1970). I think Spieker's (1971) assessment of Maclure's attitude to Werner's ideas is very apposite: after pointing out that Maclure did not attempt an interpretation of his observations in terms of a general theory, he wrote (p. 1333) "The suggestion is here offered that Maclure, who adopted Werner's classification of the stratigraphic succession, was simply shying away from Wernerian interpretation because he did not like the looks of it." (See also Spieker, 1972, p. 16.) Why did he not? I think the best answer to this question so far was provided by Hugh Torrens: "The middle path which Maclure steered between the extremes of Wernerism and Huttonism is closely mirrored by the attitude of Watt junior [i.e., Gregory Watt, James Watt's son; see above], who influenced both Maclure and Humphrey Davy" (Torrens, 2000, p. 87). Maclure was thus also influenced by the ideas generated in Hutton's larger circle in Scotland.

The geological autodidact Parker Cleaveland (1780–1858), long-time professor of mathematics, philosophy, chemistry, and mineralogy in Bowdoin College in Maine, was a convinced Wernerian and produced the first indigenous textbook of mineralogy and geology in the United States in 1816, entitled *An Elementary Treatise on Mineralogy and Geology, being an Introduction to the Study of these Sciences, and designed for the use of pupils,—for persons attending lectures on these subjects,—and as a companion for travelers in the United States of America*. In that stately octavo, 585 pages are devoted to mineralogy and the section between 586 and 641 to geology. Cleaveland's main interest was in mineralogy and he was much admired for having presented for the first time in one place descriptions of American minerals. In the short geology section of his book Cleaveland wrote, after having summarized the general structure of the earth's crust according to Wernerian neptunism, under the sub-heading "Geological Systems":

> It is, perhaps, universally admitted that the *fluid agent*, employed in the formation of minerals, must have been either *water* or *caloric* [i.e., heat]. Hence two geological systems have arisen, according as the principal agency in the production of the mineral kingdom is attributed to water or caloric. Hence the *Neptunian* theory, on the one hand, and the *Vulcanian* theory, on the other. Hence the supporters of these theories are respectively called *Neptunians* and *Vulcanists*, or *Wernerians* and *Huttonians*. (p. 591)

After having thus mentioned the two dominant theories in his day of rock origin, he proceeds to summarize their basic tenets. Two pages are allocated to Werner's views and one-and-a-half to those of Hutton. Cleaveland finds Hutton's interpretation of the origin of sedimentary rocks difficult to follow for the following reasons:

> In regard to the preceding theory we remark, that the materials, of which primitive rocks are composed, have a crystalline and uniform structure, are perfectly distinct from each other, and are frequently few in number. In stratified minerals there is, in general, a remarkably distinct and sudden transition from one stratum to another; and in many cases, contiguous strata are totally unlike each other. Thus beds of shale, sandstone, limestone, coal, clay, &c. alternate with each other, and, in some cases, several times in succession. Indeed, soft strata of clay are sometimes found under beds of limestone, &c. and *loose* sand is sometimes interposed between indurated strata.
>
> It is sufficient to ask, could these facts exist, if minerals had been formed by the fusion of heterogeneous masses of sand and gravel, or consolidated by the injection of melted matter among loose grains of different substances promiscuously mingled? Could, for example, certain varieties of anthracite lose their bitumen by *heat* and yet retain their pyrites, composed in part of *sulphur*?—On the contrary, the facts just stated would probably result from the formation of minerals in an aqueous fluid. (p. 594)

Cleaveland then discusses "veins" cutting through preexisting rock masses. He gives a competent description, but at the end confesses ignorance as to how they might have formed. He gives the opinions of others and says that the neptunians assume they were filled by water from above and that the vulcanists think that molten material welled up into them from below. He is of the opinion that the interpretation that basalt comes out of aqueous solution seems the best available, although he says "difficulties still remain" (p. 286). It is very interesting that to him basalt is a mineral, not a rock, for he does not list it under "Wernerian arrangement of rocks" (p. 597–598). Under that heading a variety of trap-rocks are discussed ("Transition Trap," "Flœtz Trap," "Newest Flœtz Trap") and under "Volcanic productions" trap is implied to be equivalent to hornblende. He is confused about the significance of trap and ends up saying

> It is also obvious, that, although Trap may be a convenient word to designate a certain series of rocks, of which a greater or less number are often associated, it in fact conveys no definite idea of any one species or sort of rocks. (p. 609)

Cleaveland's book was in such high demand that by 1822 a second edition was called for. Cleaveland enlarged his book mainly in the mineralogical part, but the geology remained the same. What he said in the first edition about Hutton's ideas is repeated on p. 725–726 with the same verdict. The same applies to his views on veins. Although there was high public demand for, and occasional claims in the literature about the existence with a date of 1856, of a third edition, it regrettably never materialized as "he was unable to respond to the call, having turned his thoughts and efforts in new directions" (Brush, 1882, p. 801).

Amos Eaton (1776–1842; Fig. 57), who became one of the pioneers of American geology after a most unfortunate career in law (he was in prison for almost five years as a consequence of claims of forgery, which he repeatedly denied) he turned to botany and geology, taught these subjects first in Williams College, then in the Lyceum of Natural History in New York City (ancestor of the New York Academy of Sciences) and finally in the Rensselaer School, which he co-founded with Stephen Van

Rensselaer III (1764–1839) and in which he ended his scientific career as the "senior professor." Eaton published in 1818 the first edition of his *Index to the Geology of the Northern States, with a Transverse Section from Catskill Mountain to the Atlantic* as a textbook and field trip guide for his students (presumably his hearers in the Lyceum). In the preface addressed to his pupils, Eaton explains the purpose of his booklet. It is meant to assist instruction in the field:

> It was observed by a late mineralogical writer; that, until the days of Werner, mineralogy was a traditionary science. Is it not still in a great measure traditionary, notwithstanding all the improvements of Werner, Haüy and Brongniart? Who is able, with all the assistance of those celebrated naturalists, to become acquainted with minerals without being taught by specimens? (Eaton, 1818, p. 6)

Accordingly, the book is almost exclusively descriptive, outlining the geology along two cross-sections across eastern New York near Albany. In the descriptive sections Werner is mentioned each time an objection had been made to his sequence or the characterization of a rock type by Bakewell, and Eaton almost always defends Werner's original statements. He may have been aware of Hutton too, however, because on p. 35 he refers to the melting experiments by Sir James Hall and Gregory Watt, but he seems to know of them second hand, via Bakewell. On p. 42 commences a section entitled "Conjectures respecting the formation of the earth." This section reads almost as a commentary on the Old Testament with frequent references to it and to Moses with a view to telling the students that earth history was exactly as it is related in the Bible, with the proviso that "six days cannot be supposed to have been equal to six apparent diurnal revolutions of the sun, as no such regulation was then made. During several of the first days the greater light [i.e., the Sun] was not appointed to rule the day, nor the lesser light [i.e., the Moon] to rule the night. Consequently, time could not have been measured as at this day. But with the Lord a thousand years are as one day, and one day as a thousand years" (Eaton, 1818, p. 44).

In two years, another edition Eaton's book was called for and the tenor of the book changed considerably. The book is again dominantly descriptive but the last section on the history of the earth has been reduced to six pages; although Eaton still wishes to see the *Bible* vindicated, his prose no longer reads like a *Bible* commentary. He refers to Kirwan and De Luc in support of his religious reading of the geological record, but he has a lot more confidence in Cuvier's discoveries. In the long preface, he indicates his preference for Werner's theoretical views, but confesses that he had to make some changes on them:

> With respect to the theoretical part, as far as I have given any theory, it is to that of Werner, with the improvements of Cuvier and Bakewell[174]. But I hope I have no where compelled fact

Figure 57. Amos Eaton. From original engraving by A.H. Richie. Popular Science Monthly, v. 38, 1890–1891.

> to bend to *any* theory. I expect to be censured for rejecting so many of the strata, which are given by European geologists. My reply is a short one—*I do not insert them, because I cannot find them*[175]. It is certainly most advisable to begin with those only, which have been clearly ascertained, and calculate to increase our list, when we shall be authorized by facts. (Eaton, 1820, p. vi–vii; italics Eaton's)

Yet, he cannot do without the central heat and the Huttonian theory of heat-driven uplift, in flat contradiction to Wernerian interpretation of earth behavior:

> That in due time some force (perhaps the expansion of steam by the agency of the internal heat of the earth) was exerted beneath the granite, sufficient to overcome the weight and strength of the granite and of all the rocks outside of it. That this force, extending around the earth between the granite and the unknown stratum next beneath it, at length burst forth driving granite through the outer strata in several places, and thereby raising up their broken edges into a highly inclined position. By this process islands, or even continents, of rocks, and perhaps supporting some submarine soils, were raised out of the waters hitherto enveloping the earth. (Eaton, 1820, p. 15)

He then outlines the importance of the external phenomena entirely in a Huttonian spirit and ends by stating:

> Whether this theory accords with the real origin of the present state of things or not, is immaterial. It is introduced solely for the purpose of aiding the memory in studying the strata, which we know do exist. It is presented for his use; because it is more simple and seems to accord with more phenomena, than any theory hitherto suggested. (p. 17)

One cannot tell from which source Eaton obtained this somewhat modified outline of Hutton's theory. Hutton is nowhere mentioned in this book and Eaton seems not to have realized that "his" theory contradicts the Scripture-based evolution he outlines in his last chapter.

We finally find out in his *Geological Text-Book* published in 1830 that he had indeed become aware of Hutton and realized that Werner had to be given up:

> Hutton had done much in the life time of Werner. But his theoretical views were at variance with those of Werner; and much overheated controversy existed between their respective pupils. Werner ascribed most geological phenomena to the agency of water, while Hutton referred the same to the action of internal heat. Hutton died in an unhappy state of mind, at a time, when his views were almost universally rejected; but Werner died amidst the universal plaudits of his favorite views. Scarcely had that earth, which he had studied with such interest and applause, received his manes, when his theory was reviewed, and began to be received with less approbation.
>
> At length the illustrious Cuvier came before us. He balanced the two theories with a giant hand. He demonstrated the Huttonian theory in its application to many phenomena; but left much to be explained upon the Wernerian. (Eaton, 1830, p. 13)

Eaton's last paragraph is quite astonishing. Few at the time, or even today, would think that Cuvier had "balanced" the two theories, but I believe there is much truth in that; however, Eaton does not seem to have realized that Hutton's theory accounted for the external phenomena even better than that of Werner. What Eaton considers as Werner's "theory" is really his rock sequence and the terminology he invented for it. Cuvier's biostratigraphy very quickly made it obsolete. Be as it may, Eaton's slow conversion to Hutton's views was a characteristic of the time and is paralleled in a remarkable way by that of Leopold von Buch in Germany at almost exactly the same time.

Despite the enthusiasm of Cleaveland for the neptunian theory and his admiration for Werner, and Eaton's slow abandonment of Werner's views, others in the United States were not so disposed. Only three years after the publication of the second edition of Cleveland's textbook, the medical doctor and geologist Dr. Jeremiah van Rensselaer III (1793–1871), the first American to climb Mont Blanc in the Alps, published his *Lectures on Geology; Being Outlines of the Science*, which he had delivered, in his capacity as associate and lecturer on geology to the New York Athenaeum (van Rensselaer, 1825). On p. 35–37, van Rensselaer gives a short account of Hutton's theory and says that it is ably supported by Playfair's book which, van Rensselaer says, "continues to be the text book of the best English Geologists" (p. 37). Dennis Dean, in a paper he read in the session S14 entitled "Hutton, Lyell, Logan—And Their Influence in North America" at the Geological Society of America Annual Meeting in Toronto in 1998, said that van Rensselaer's book contained the earliest evidence of a first-hand knowledge of Hutton's theory in the United States of America (see Oldroyd, 1999, p. 11)—although he did point out that allusions to Hutton's theory had been published in 1798 and 1803 (Dean, 1998). Van Rensselaer criticizes the idea of an eternally burning central fire in the earth, but emphasizes the importance of subterranean fires. He sees in Hutton's theory, "a happy union of the agency of both fire and water; the one collecting and depositing, the other consolidating and elevating." (p. 36). On p. 207 we read further that "when Hutton allowed a certain agency to water, and a certain power to the action of subterranean heat, he combined the good of both theories, and appears above the rank of a *mere theorist*" (Italics van Rensselaer's). Van Rensselaer also reviews Werner's interpretations, emphasizes his love of order and method, acknowledges the reality of the Deluge (but he ascribes it to supernatural powers), but cannot follow him in the interpretation of magmatic phenomena in which he appears to side with Hutton. Even in some nomenclatorial issues he deviates from Werner:

> In retaining the term Transition, I do violence to my own opinions. The division of rocks in which that term is adopted is unnecessary and perplexing. I have elsewhere given reasons for not employing it … I shall however, retain this name, as I do not consider my own authority sufficient to banish it, …. (van Rensselaer, 1825, p. 200)

Van Rensselaer mentions the uplift of the Alpine chains by means of an expansive heat below, but cannot understand why this heat issued from a conduit everywhere parallel with the original inclination of the superjacent beds. Neither can he agree that massive granite can uplift mountains (p. 161–162), although he is happy that volcanoes were uplifted by "subterranean heat" (p. 121). He is by no means a pure Huttonian, but, as Ospovat noted (1967, p. 242), leans toward Hutton's interpretations as what he says on p. 207 shows. Ospovat (1967) says that van Rensselaer gives a tabular view of Werner's arrangement of rocks, but omits to say that it is in reality Jameson's rendering with modifications (van Rensselaer, 1825, p. 355).

One year after the publication of van Rensselaer's lectures at the Athanaeum, one of the great pioneers of American geology and the man who named the Adirondacks in 1838, Ebenezer Emmons (1799–1863; Fig. 58), a one-time student of Amos

Figure 58. Ebenezer Emmons. Appleton's Popular Science Monthly, v. 48, p. 319 (1896).

Eaton in the Rensselaer School, published his small *Manual of Mineralogy and Geology* as a textbook. There is no theory in it, but on p. 180 and 181, we read

> It appears to be admitted by all geologists that the materials of the earth must have been at some former period in a fluid state, and that they must have been brought and held in this state either by caloric [i.e., heat] or water: that minerals were deposited from this fluid state, and that the waters of the ocean once covered the tops of the high mountains. But it is not clearly settled which of these agents was the solvent. Hence there have arisen two geological systems, according as the principal agency in the production of the mineral kingdom is attributed to water or caloric—the one called the *Neptunian* theory, the other the *Volcanic*; or more properly, after their respective supporters, as Wernerian and Huttonian. The latter at the present day seems more likely to prevail, though with some essential modifications. (Emmons, 1826, p. 180–181)

This statement was written almost a decade earlier than the birth of Sir Archibald Geikie, on the other side of the Atlantic!

Benjamin Silliman (1779–1864; Fig. 59), chemist and geologist, professor of geology at Yale and the founder of the *American Journal of Science and Arts* (so titled until 1880; *American Journal of Science* afterwards), the oldest continuously published science journal in the United States, edited an American printing of the third English edition of Robert Bakewell's (1768–1843) *An Introduction to Geology* (1839), a famous textbook of the first half of the nineteenth century in the English-speaking world, to which Sir Charles Lyell owed his awakening enthusiasm in geology. To this book, Silliman not only added a "Preface to the American Edition" summarizing why the book was being reprinted in America, but also an appendix entitled "Outline of the Course of Geological lectures given in Yale College" to summarize his own views in the subject "although with some repetition of ideas contained in the author's text" (p. xii). It is in the first paragraph of that appendix that we read the following revelatory sentence:

> As it is the *fashion* of the day, to attribute almost every thing in the earth to igneous agency, I shall probably be thought to be behind the present state of opinion, while I maintain, that the chemical affinities, through the medium of aqueous solutions of the great chemical agents—as well as of water itself, have also produced important effects in the early arrangement of the planet. (Silliman, 1829, p. 3)

He further says on p. 4

> It has become fashionable do decry Werner; but, without being his blind admirer, I may be permitted to ask, who has done more for geology, and who has done it better?

Nevertheless,

> So far as the following arrangement is founded upon the Wernerian plan, it is one of convenience merely, and therefore there is no hesitation in deviating from it, or in substituting other views, when they appear preferable.

> Had Werner lived till this time, he would probably have admitted that the difference between the trap rocks and the lavas have become evanescent, and that it is certainly possible, if not probable, that they may have had a similar origin. (p. 5)

The rest of Silliman's appendix is a mixture of Huttonian and Wernerian interpretations. But this promiscuous attitude did not last long as Silliman tells us himself with great candor in the appendix he attached to the third American edition of Bakewell's book (corresponding to the fifth English edition) that came out in 1839, as I quote at some length below.

After Lyell and Before Geikie

In 1832, Maclure published two small pamphlets: one on the origin of rocks (Maclure, 1832a) and the other on the geology of the Lesser Antilles from Barbados to Santa Cruz (Maclure, 1832b), based on his travels there in 1816 with the French naturalist Charles Alexandre Lesueur (1778–1846), both privately published in the utopian settlement of New Harmony founded by Maclure in Indiana, when Lesueur was also living there (for the history of New Harmony, see Warren, 2009). These pamphlets clearly show that in the meantime he had embraced the essence of Hutton's view of the planet as clearly shown by the following remarks:

Figure 59. Benjamin Silliman. Engraving by William G. Jackman between 1841 and 1860. Public domain (PD-US-expired).

1) When we pretend to limit the operations of nature, to suit our contracted ideas, we most probably do her injustice. To proceed from the known, which we see daily forming, towards the unknown, through a chain of reasoning strictly analogous, is perhaps all that our present knowledge will permit us to do. (Maclure, 1832a, p. 10)

2) How the origin of basalt could be doubtful with the Wernerians, can only be accounted for, by Werner having at first put the detached masses of basalt, found in Saxony, into the Neptunian origin, and that his disciples have since persevered in the arrangement. (Maclure, 1832a, p. 45)

3) The islands were probably thrown up from the bottom of the ocean (Maclure, 1832b, p. 17)

This last statement is based on Hutton's theory of magmatic uplift, but by 1832 it had been widely adopted internationally.

In the same year, Ebenezer Emmons brought out the second, completely revised edition of his textbook, but the main revision pertained to the longer mineralogical part. In the second, geological part, under the section subtitled "History of Geology," he wrote:

> Geology is a modern science. Its foundation was laid by Lehman [sic], the German, about the middle of the last century. He was followed by Mitchell and Whitehurst in England, and Werner in Saxony, all of whom have left monuments of their industry and ability. The latter especially has given character and great interest to this department of science. The name of Werner always brings to mind that of Hutton, from the fact that they respectively advanced and supported theories diametrically opposed to each other. Which deserves the meed of having done the most for geology, it is not for partial critics to say. Hutton, however, aided by the happy illustrations of Playfair, seems to have ultimately triumphed, though many of Werner's views are as unshaken as the rocks of his own country. (Emmons, 1832, p. 260)

In 1833 William Williams Mather (1804,1859), officer of the United States Army and geologist, published his small textbook entitled *Elements of Geology, for the Use of Schools*, in which he compared the Wernerian and the Huttonian theories and showed a preference for the latter. In his brief account of the history of geology at the end of his book, he wrote:

> In 1788, Hutton published his theory of the Earth, and this is a work which has exerted a lasting influence over geologists. Hutton has the merit of having first directed the attention of geologists to the important fact of granite veins issuing apparently from beds and strata of granite, and traversing all the surrounding rocks.
>
> He also brought forward in a striking manner, the circumstances that seem to shew the igneous origin of the trap rocks; but the wildness of some of his theoretical views, may well go to counterbalance the utility of the facts he gathered from observations. (p. 119)

About Werner's theory, Mather's verdict was much more severe:

> His *theory* must *now* appear to almost all, as among the most unphilosophical and unsuccessful yet framed, and his few remaining adherents are one by one abandoning his most characteristic opinions. (p. 120, italics Mather's)

In the penultimate chapter entitled "Recapitulation," Mather explained what he meant by Hutton's "wild" theories:

> The Wernerian theory is, in many parts, founded on a close observation of facts; but it is so interwoven with hypothetical points, which are not confirmed by succeeding observations in other parts of the world, that it cannot be received. It is not all true, and it does not account for all the facts.
>
> The Huttonian theory consists of inferences drawn from facts, but the deductions are sometimes carried farther than the facts warrant. It supposes the present continents derived from the wrecks of older ones, because, we find a large portion of the rocks upon the surface of the earth formed of the fragments of others cemented together, and often filled with the remains of animals that once inhabited the sea. But whether volcanic agency has caused those lands to be elevated, which have evidently been for a long period under the surface of the ocean, is a point which is not conceded by all. It is however rendered highly probable by some facts which have been mentioned under the head of volcanos and earthquakes. The Huttonian theory supposes the secondary rocks, after they were formed in the bed of the ocean, to have been indurated by subterranean heat, before they were raised to become islands and continents. Some experiments by Sir James Hall, which have been already mentioned, render this supposition probable.
>
> The Huttonian theory supposes the earth to have once been in a melted state, and to have cooled gradually. (p. 109)

After having related Sir James' and Gregory Watt's experiments about the formation of "trap rocks" (p. 84–85), Mather comes to the matter of the origin of marble, in part quoting Bakewell:

> Dr. Hutton had advanced the opinion, that the beds of limestone were formed of the shells and exuviae of marine animals, which have been melted by central fire, and crystallized. It was however objected to this theory, that the well known action of fire on limestone rocks would expel the carbonic acid, and make them soft and pulverulent, as we see in making lime from limestone. This objection was answered, by saying, that these beds of shells, &c. were heated under the pressure of

the ocean which for so long time covered the land, and that this pressure would prevent the escape of the carbonic acid, and render the limestone fusible. This was regarded as mere hypothesis for some time, but Sir James Hall determined to try its validity by experiments. Having calculated the resistance which a column of water, 1500 feet, or any given depth, would present to the escape of the carbonic acid of the limestone, he inclosed a quantity of powdered chalk in a gunbarrel, and confined it in such a manner as to present an equal degree of resistance. He subjected the powdered chalk, thus confined, for some time to the action of a furnace; it was then drawn out and cooled, and was found converted into crystalline limestone or marble; and in one instance where the chalk inclosed a shell, the shell had acquired a crystalline texture without losing its form. Hence, when chalk or limestone is found to have a crystalline texture in contact with trap rocks, we may, with a high degree of probability, infer that the limestone had been fused by the trap. (p. 85–86; I deleted Mather's quotation marks to ease legibility)

In his final chapter, Mather reviewed the history of geology. What he says about Hutton's and Werner's theories is most instructive:

Werner and Hutton, both of whom raised a new and very superior class of geologists. Whatever may have been the errors of their theories, it is certain that their influence on the minds of men has been of much importance in causing the advance of the science. Hutton's doctrine of the consolidation of earthy materials by heat and pressure, and Werner's theory of universal formations, were brought to the tests of experiment and observation. The one was confirmed more satisfactorily than could have been expected under factitious circumstances; the other was shewn to have originated in its utmost generality in the narrow views of its ingenious but untravelled author. One was passed by in silent neglect, the other attracted crowds of disciples from all parts of Europe. The former of these men died when his theory had hardly attracted notice, the latter, in the full career of glory. Like the fabled Phoenix, Hutton's theory arose from its ashes, but from the time Werner was laid in his grave, his was found wanting in the generality which he assigned to it. (p. 122)

Mather's second edition in 1838, repeated these same words verbatim (p. 254).

Only a year after the first edition of Mather's textbook came out the first edition of John Lee Comstock's (1789–1858) potboiler, *Outlines of Geology*. After having reviewed the current theories of the earth at the time, Dr. Comstock concludes:

It is generally acknowledged at the present day, that Hutton's theory will account for a much greater number of phenomena than Werner's. It is impossible, for instance, to account for the present situation of stratified rocks containing sea shells, unless we suppose, either that the sea occupied the earth for ages, or that these strata were formed under the ocean, and elevated by some mighty force; and as we know that islands are thrown up from the sea, by volcanic force, at the present day, it is reasonable to attribute the same effect to the same cause anciently. (Comstock, 1834 p. 14)

In the third edition of Comstock's book, published in 1839, we encounter precisely the same paragraph on p. 30.

The most illuminating account of the conversion of one prominent American geologist from Wernerian geology to Huttonian geology was provided by Benjamin Silliman in 1839 in his appendix to Bakewell's third American edition as mentioned above. Below I quote him at length, because of the important points he makes about how he had not originally realized that he had been carried away by Wernerian propagandists and how subsequent accounts of worldwide observations had opened his eyes and the importance of evidence in science:

The outline of my lectures, annexed to the first American edition of 1829, does not present a correct view of the courses which I now give. Fifteen years have elapsed since that outline was first sketched, and ten since it was published. Within those periods geology has made great advances, particularly in the proofs of igneous action, in all ages, ancient and modern; and perhaps my own admissions of its agency were not commensurate with the proofs that existed in 1833. The powerful direction early given to my mind, towards the Wernerian theory, by the captivating eloquence of the late Dr. John Murray of Edinburgh, whose lectures on geology as well as chemistry I attended, was not, at the time, fully appreciated by myself.

I was also a deeply interested listener in the lectures of Dr. Hope [Silliman added a footnote here: "The distinguished Professor of Chemistry, &c. in the University of Edinb." His p. 462. Thomas Charles Hope, 1766–1844, discoverer of strontium; he was an assistant to Joseph Black, Hutton's close friend.] based on the Huttonian theory, and I was a careful student of Playfair's splendid illustrations of that theory; while Playfair himself, with many other eminent men of that school of geology, as well as its rival [another footnote by Silliman here: "Professor Jameson, then recently returned from Germany, where he had studied under Werner, had not at that time entered on his public duties"; his p. 462], was then in full vigor and activity at Edinburgh. It was delightful to listen to their eloquent statements and acute reasonings. In this way I heard both sides of the question fully vindicated, while, from my youth and inexperience, I endeavored to sustain a neutral position, and reserved the liberty to decide ultimately, with an unbiased mind. Still, I was, to a degree, incredulous in regard to the fundamental postulates of the Huttonian geologists, and could not perceive that they made out their case, as to the extent and energy of internal fire. The powerful arguments in favor of great igneous action, contained in Mr. Bakewell's Geology, were supported by Dr. Daubeny's fine Treatise on Volcanoes, and this by the full and exact work Mr. Scrope on the same subject, with particular reference to the extinct volcanos of France, illustrated also by an ample atlas of volcanic regions. The more recent exhibition of proofs by Mr. Lyell, as to the extent, persistence and energy of igneous action; the satisfactory evidence accumulating every

Figure 60. Portrait of Edward Hitchcock, circa 1863. Courtesy of Amherst College Archives & Special Collections, Edward and Orra White Hitchcock Papers.

Thus, in the United States too, the tide had turned completely and Werner was abandoned, even within what Ospovat (1967) had called the "Wernerian era" of American geology, especially after biostratigraphy had reached the shores of the New World. Edward Hitchcock (1793–1864; Fig. 60), professor of chemistry and natural history (1825–1845) and afterwards professor of natural theology and geology (1845–1864) in the Amherst College wrote possibly the most successful geology textbook in the United States of America (31 editions between 1840 and 1868!). I quote here from his first edition, which I happen to have in my library:

> Nearly at the same time, a Scotch geologist, by the name of Hutton, published a "Theory of the Earth," opposed in most respects to the doctrines of Werner. He supposes that the rocks which form our present continents were derived from the ruins of former continents; which were abraded and carried into the sea by the agency of running water; just as the same agency is now spreading over the bottom of the ocean, deposits of mud, sand, and gravel. Afterwards, the unstratified rocks, in a melted state, were protruded through these deposits, by which they were consolidated, rendered more or less crystalline, and elevated into their present condition. Many of the fissures also, were filled with metallic and other matter, injected from beneath. When our present continents are nearly all worn down, Hutton supposes this process of consolidation and elevation may be repeated. Indeed, in these changes he sees "no traces of a beginning, no prospect of an end." Professor Playfair, however, in his illustration of his theory, endeavors to show that Hutton did not mean by such language that the world is eternal: but only geology, like astronomy, does not disclose to us the time when this series of changes commenced.
>
> These rival hypotheses [i.e., Werner's and Hutton's] excited a great deal of discussion, both on the European continent and in England, for a great number of years. The final result is, that the theory of Werner has been almost universally abandoned; while that of Hutton, denominated also the Plutonian theory, has, in its essential principles, been adopted by most geologists of the present day. (Hitchcock, 1840, p. 280)

In the eighth edition (Hitchcock, 1847), which I also happen to have, he repeats these same paragraphs on p. 306. In the later editions, Hitchcock no longer recounted this by then well-known story, but referred to Hutton for specific innovations, specifically on veins. In the last edition, the 31st, published in 1868, for example, Hutton is approvingly mentioned with respect to his interpretation of how metallic veins are filled (on p. 396).

Alonzo Gray (1808–1860), the famous American educator, published in 1853, together with the well-known conchologist Charles Baker Adams (1814–1853; Fig. 61), the first state geologist of Vermont, from 1838 to 1847 the Burr Professor of Chemistry and Natural History at Middlebury College, and then professor of astronomy, zoology and natural history in Amherst College and a member of the American Academy of Arts and Sciences, the very popular textbook *Elements of Geology*.

year respecting the increasing internal heat as we descend into the earth; the decisive influence of galvanic power in mineral veins, as ascertained by Mr. Fox [Robert Were Fox, the younger, 1789–1877; the work referred to here is Fox, 1830], its efficiency in producing mineral crystallization and its power even in rousing into life the long latent eggs and germs of insects, as established by Mr. Crosse [Andrew Crosse, 1784–1855; for the experiments here referred to, see Crosse, 1857, p. 353–360]—with the splendid proofs which our galvanic and electro-magnetic machines now afford of an igneous energy inherent in the earth—an energy which knows no limits—attended also by magnetic and decomposing power, equivalent to all which geology demands; these and many other considerations that may be stated, have removed my doubts, and I have been for a series of years in a condition to do full justice to the internal agency of fire, as my various classes in the university and elsewhere can attest.

It is of little importance to occupy the reader's attention, even for a moment, with my own personal views and opinions, nor would I have ventured to do so, were it not of some importance that the science may not suffer by any apparent, although not real, caprice of opinion in those who teach it to others. I have therefore thought it but honest to make this frank declaration, my *amende honorable*, of the change in my views, and of the grounds of it; and perhaps it may not be entirely without utility, as an exhibition of the effect of progressive development and accumulation of evidence upon one mind, inasmuch as other minds may, by similar means, be led to the same result. (Silliman, 1839, p. 462–463)

Figure 61. Charles Baker Adams. Courtesy of Amherst College Archives & Special Collections.

Figure 62. David Ames Wells. Popular Science Monthly, v. 32.

According to WorldCat, this book had 27 editions (http://worldcat.org/identities/lccn-n85252755/, last visited on 28 May 2019)! Already in the first edition, on p. 4, Gray and Adams (1853) wrote:

> Werner was the father of Mineralogy; but the title of the father of Geology belongs to Hutton, a Scotch geologist, who soon after proposed the Plutonian theory. According to Hutton's theory, which has been established, the unstratified rocks are of igneous origin, like the lavas of the present epoch, and the stratified rocks were originally sand, clay, mud, gravel, &c., like the aqueous deposits of the present time. Hutton supposed that these deposits were derived from the abrasion of ancient continents; that some of them were rendered crystalline by the heat of the protruded igneous rocks; and that in such a series of changes Geology can discover no proof of a beginning nor prospect of an end.

I own two further "editions" of this book: those of 1857 and 1860. In both, the above passage occurs verbatim and on the same page. The successive "editions" of this book are simply reprints of the first edition (Adams passed away in St. Thomas of the Virgin Islands of malaria in the year of the first edition; Gray may not have felt comfortable in changing the contents after his co-author's untimely demise at age 39!). There are no further references to Hutton by name in Gray and Adams' book, but the entire book is constructed on Huttonian principles, as they themselves admit by pointing out that Hutton's theory, "has been established."

The American scientist, engineer, journalist, and economist David Ames Wells (1828–1898; Fig. 62) wrote a number of popular textbooks on natural sciences. Among those is *Wells's First Principles of Geology. A Text-Book for Schools, Academies and Colleges* (Wells, 1861). In it, Wells followed Hitchcock almost to a word (without reference), which I reprint here despite the repetition thus occasioned to emphasize the prevailing sentiment:

> Nearly at the same time, a Scotch geologist, by the name of Hutton, published a "Theory of the Earth," opposed in most respects to the doctrines of Werner. He taught, that the rocks which form our present continents were derived from the ruins of former continents, which were abraded and carried into the sea by the agency of running water; just as the same agency is now spreading over the bottom of the ocean deposits of mud, sand, and gravel. Granite and trap rocks, he asserted, were of igneous origin, and have intruded in a melted state into fissures in the earth's crust. The crystalline stratified rocks, included in the classes "primary" and "transition" of Werner, he regarded as merely sedimentary strata, altered by heat—a supposition which accords very well with the views at present entertained respecting the origin of the metamorphic rocks.
>
> These rival theories excited controversy among the scientific men of Europe, which for years was carried on with a bitterness and animosity almost unprecedented in the history of such disputes; and all geologists allied themselves to various schools or sects, under the name of "Wernerians" or "Neptunists;" "Huttonians" or "Plutonists;" "Cosmogonists;" "Diluvialists;" "Fossilists," etc. etc. The final result was, that the views of Werner were almost universally abandoned, and those of Hutton, or their essential features, adopted. (p. 172)

Beginning with 1863, the geological education in American universities was dominated by two famous textbooks. The earlier was Dana's *Manual of Geology* which had four editions, the last being in 1895. With unnumbered revised and corrected editions published by different publishers this famous book had six different editions, all by Dana himself alone. None of them says anything about Hutton, because they make no reference to the historical development of the science of geology or any of its branches. The later of the two widely used textbooks was Joseph Le Conte's *Elements of Geology* first published in 1879, which also had four editions until 1896 with a reprint of the last edition in 1899. Le Conte does not talk about Hutton either and for the same reasons as Dana. But there were other books by prominent American geologists. I here cite only the great geochemist Thomas Sterry Hunt's (1826–1892; Fig. 63) *Chemical and Geological Essays* of 1875 (second edition in 1878) and his *Mineral Physiology and Physiography*, published in 1886, because Hunt was not a blind follower of Hutton and whose views on the origin of crystalline rocks may be regarded as a sort of neo-neptunism (his "crenitic" hypothesis: Hunt, 1886, p. 132). Bearing his sympathies in mind, his testimony about Hutton's views prevalent during his professional life are of the greatest interest.

In the first of these books Hunt stated (the statement and the page on which it occurs are the same in both editions):

> I accept in its widest sense the view of Hutton and Bouë [sic], that all the crystalline stratified rocks have been produced by the alteration of mechanical and chemical sediments. (Hunt, 1875, p. 24)

When discussing the origin of the schists in the Alps, Hunt referred again to Hutton (once more, both the statement and the page numbers are identical in the second edition):

> Hutton, as early as 1788, had taught that what he called the primary schists were sediments, the ruins of earlier rocks altered by heat, but it does not appear that he attempted to fix the relative age of any such altered rocks. In fact, the notion of geological periods, based upon the study of fossils, was not as yet fully recognized. The suggestions of Bergmann and Bertrand, that the crystalline rocks of the Alps are newer than the fossiliferous limestones which pass beneath them, seems to have been the first attempt to give to Hutton's view a definite and special application.… (Hunt, 1875, p. 338–339)

In *Mineral Physiology and Physiography*, Hunt no longer appears as enthusiastic about Hutton's interpretation of metamorphism, because he thinks the layering in the Laurentian gneiss is sedimentary. This interpretation led him, beginning with 1847, to be "perplexed by the difficulties of the Huttonian tradition (then and for many years generally accepted in America), that the mineral characters of these rocks was in no obvious way related to their age and geologic sequence, but that the strata of paleozoic and even cenozoic times might take on the forms of the so called azoic rocks" (p. 112). From this erroneous deduction, he develops a complete neptunian chemical theory of all what we now call (and what Hutton's later followers have called) metamorphic rocks, because "Convinced of the essential truth of the principles laid down by Werner, and embodied in his distinctions of Primitive, Transition, and Secondary rocks, I sought, during many years, to define and classify the rocks of the first two of these classes, and by extended studies in Europe, as well as North America, succeeded in establishing an order, a succession, and a nomenclature…" (p. 113). This hypothesis derives granites "by aqueous secretion from a primary igneous and quartzless mass, it would follow that the highly basic compound, assumed by Bunsen[176] to represent the typical pyroxenic or basaltic rock … would be the … insoluble residuum" (p. 188–189). "Werner argued, and, as we shall endeavor to show, correctly, from their analogies with concretionary granitic veins, that all granitic rocks were deposited from water..." (p. 73). It is not here the place to discuss this extraordinary hypothesis of neptunian origin of granites by a great geologist in the latter half of the nineteenth century; nevertheless, it is important to underline that Hunt could not accept an "original" Werner:

> As regards the … hypothesis of Werner, according to which the whole of the materials of the crystalline rocks were originally dissolved in a primeval sea,—its chemical difficulties are evident to the modern student. That the ocean could have ever held at no time in solution, under any conceivable conditions, the elements of the whole vast series of crystalline rocks, and could have deposited them successively, in that orderly manner which we observe in the earth's crust, was seen to be incredible. This argument, successfully urged by Playfair and his followers, contributed, with others, to the discredit which, as we have seen, soon fell upon the Wernerian hypothesis. (p. 109)

Figure 63. Thomas Sterry Hunt (from Popular Science Monthly, v. 8, November 1875 to April 1876; public domain).

Despite his dislike of Hutton's metamorphism, Hunt could not do without Hutton's other ideas: "A second place in the history of exoplutonic rocks [i.e., heat-indurated supracrustals by injection of magma according to Hutton], already foreseen by the Huttonians, here presents itself for our consideration" (p. 186). Hunt supposes that the internal heat of the planet combined with the water included in the buried sediments would lead them to suffer "a softening which permitted them, as a result of subsequent movements of the crust, to appear again at the earth's surface, as exoplutonic or exotic rocks of the trachytic or granitic type" (p. 186). Hunt was fighting a losing battle against the idea of regional metamorphism, which he ascribed to Hutton, and because of that not only extensively discussed Hutton's interpretations, which he tells us were widely accepted, but also borrowed from his ideas ranging from the internal heat of the planet to the heat induration and mobilization of buried sediments to temper his neptunism.

We know to what kind of an impasse Hunt's train of thinking led during the brief geosyncline controversy in the nineteenth century America. Unaware of the private criticism Joseph Henry (1797–1878) had directed against the paleontologist James Hall's (1811–1898) views on mountain-building (see Clarke, 1921, p. 327–328), Dana criticized them years later in 1873 severely, as a consequence of a defense published by Hunt (1873) in Hall's favor, declaring that the foot-per-foot subsidence under the weight of the accumulating sediments and the folding allegedly resulting from the subsidence were "physical impossibilities" (Dana, 1873, p. 349). Le Conte, in a paper in the same issue of *The American Journal of Science and Arts* as Dana's paper, pointed out that Hall and Hunt "leave the sediments just after the whole preparation had been made, but before the actual mountain formation had taken place" (Le Conte, 1873, p. 450). To Hunt's charge that his and Hall's statements had been misunderstood, Le Conte replied that "neither he [i.e., Hunt], nor Hall ever produced any theory of mountain formations at all, but only a return to the views of de Buffon and Montlosier, that 'mountains are fragments of denuded continents'"(ibid.). Between his lines, Le Conte was saying that they remained static neptunians and did not appreciate the full significance of Hutton's dynamic earth theory!

With Hunt's 1886 book we are only a decade away from Geikie's 1897 account, which we defined as our younger limit for the compass of this book.

DISCUSSION

The purpose of this book is to document what the late eighteenth and nineteenth century geologists thought of Hutton's legacy before the 1897 publication of Sir Archibald Geikie's *The Founders of Geology*. Did they think he was the founder of modern geology as we know it? Was what he published something new? How widely was he later remembered by geologists? In addition to his general theory of the earth, did they think Hutton founded any one particular branch of modern geology, one that is still *worked* on? The section before this provides lengthy answers to these questions. I have selected out and list below the contents of what I think direct, unequivocal statements by the representatives of the geological world of their times that answer these questions in a straightforward manner and thus, in a way, summarize the previous section. To these I only added William Henry Fitton's quote. I do not cite him in the previous section, because Dean (1992) deals with him adequately.

Hutton Regarded as the Founder of Modern Geology

Let us begin with answers to the first question, namely "Did they think he [i.e., Hutton] was the founder of modern geology as we know it?" The answer to that question automatically answers also the second, i.e., how widely Hutton was remembered. I present the answers in chronological order.

1829

It is, however, unnecessary that I should here recapitulate the Plutonian and Neptunian evidence. The former, as modified by Dr. Hutton and his commentators, is that which most plausibly accounts for the phenomena, is least at variance with facts, is least hypothetical, and best entitled to the appellation of a theory of the earth (Brande, 1829, p. 233).

1830

The first volume of Sir Charles Lyell's *Principles of Geology* is published.

1834

In short, the prevailing geological theory is that of Hutton. (Boase, 1834, p. 217)

1834 (Published 1838)

These views make Hutton without a doubt the creator of the ideas now generally accepted in our science. We therefore can argue that Scotland is the cradle of the new geology. (Hoffmann, 1838, p. 208–210)

1835 (28 December)

Sir Archibald Geikie is born.

1838

...it is but an act of justice to the memory of an illustrious philosopher, and of his able illustrator, Professor Playfair, to state that this theory, corrected and elucidated by the light which modern discoveries have shed upon the physical history of our planet, is now embraced by the most distinguished geologists. (Mantell, 1838a, p. 79–80)

1839

We are convinced that nobody will partake in our satisfaction more cordially than himself [i.e., Charles Lyell], if we shall have succeeded in proving that Dr Hutton was really the founder of the TRUE THEORY OF THE EARTH. (Fitton, 1839, p. 62 {466}, capital letters his)

1840

These rival hypotheses [i.e., Werner's and Hutton's] excited a great deal of discussion, both on the European continent and in England, for a great number of years. The final result is, that the theory of Werner has been almost universally abandoned; while that of Hutton, denominated also the Plutonian theory, has, in its essential principles, been adopted by most geologists of the present day. (Hitchcock, 1840, p. 280)

The history of the earth in time is now the task of the present and the future and James Hutton is the Copernicus, Leopold von Buch is the Galileo of the new episode of research. (Hauff, 1840, p. 410)

Hutton, is however, the father of modern geology. (Hauff, 1840, p. 470)

1842

...celebrated system of Dr. Hutton; a system which has not only supplanted that of Werner, but has formed the foundation of the researches and writings of our most enlightened observers, and is justly regarded as the basis of all sound Geology at the present day. (Richardson, 1842, p. 41–42)

1844

... the molten initial condition of the earth, Hutton proved with irrefutable facts. Although the volcanic origin of certain rock formations (traps, basalts) had been shown before him, it is to his credit alone to have recognized and proved the same of granite, this basal foundation of the crust of the earth. He also strictly excluded from geology or the evolutionary history of the earth all geogenetic hypotheses (i.e., those about the origin of the earth). His "Theory of the Earth" appeared in 1788.

Thus the basis, on which a scientific geology could be constructed, was laid down by Werner and Hutton towards the end of the eighteenth century. (von Maack, 1844, p. 19–20)

1845 (Also in 1857)

The Scotsman James Hutton has to be regarded as the father of modern geology, who, in 1795, brought his *Theory of the Earth* in circulation. (Wagner, 1845, p. 8; also 1857, p. 23)

1849

James Hutton must be regarded as the founder of modern geology, without belittling the services of Alexander von Humboldt, Leopold von Buch, Cuvier, Élie de Beaumont, Murchison and a few other luminaries of our times. (von Leonhard, 1849, p. 8)

1853

Werner was the father of Mineralogy; but the title of the father of Geology belongs to Hutton, a Scotch geologist. (Gray and Adams, 1853, p. 4)

1860

The importance of the work of Hutton and Playfair, in which the fundamentals of modern geology are established and developed for the first time, ... requires that I here summarise its principal propositions.... (Daubrée, 1860, p. 9)

Like most of the men of genius who opened new paths, Hutton indeed exaggerated the scope of the ideas that he had conceived. One cannot, however, consider without admiration with what profound penetration and what rigor of induction this farsighted man imagined, at a time when accurate observations were still very few, a system which, for the first time, called for the simultaneous competition of water and heat and embracing the entire history of the globe. He put forward principles that are today universally accepted, at least in fundamental issues. (Daubrée, 1860, p. 12–13)

1861

The final result was, that the views of Werner were almost universally abandoned, and those of Hutton, or their essential features, adopted. (Wells, 1861, p. 172)

1868

The English geologist Hutton entered the fray in 1788 and by numerous examples taken from Nature demonstrated that some rocks are of igneous origin, others of aqueous origin and that some rocks of sedimentary origin had been altered profoundly, modified and metamorphosed at their contacts with the molten igneous rocks.

These discussions established in an irrefutable manner many important points that form the basis of current geology. (Chevalier, 1868, p. 13–14)

1873

...famous system of Hutton, who, by some, is considered the founder of modern geology.... (Stoppani, 1873, p. 99)

1876

[Hutton] may be fairly looked upon as the veritable father of the science. (Green, 1876, p. 4)

1877

...we may go further and call him the founder of Geology as a whole. (Green, 1877, p. 4)

1887

...it can be said that Hutton gave geology its direction in many decades to come and even in its present form the influence of his teaching is not inconsiderable. The next success was the general victory of fire over water that came about after Werner's death and after a long and bitter struggle between the neptunists and the plutonists, i.e., Wernerians and Huttonians. (Neumayr, 1887, p. 23–24)

1888

Sir Archibald Geikie mentions Hutton as one of the founders of modern geology for the first time.

1897

Sir Archibald Geikie publishes *Founders of Geology*, first edition.

Now, let me remind my readers that the above quotes are only those absolutely unequivocal, direct statements made to identify Hutton as the founder of modern geology and they are all before Sir Archibald Geikie published anything saying the same (some had been stated even before he was born!). *Therefore, the statements in the revisionist literature that Sir Archibald made Hutton into the founder of modern geology or that it was he who established a tradition of saying so are clearly fallacious.* I cannot understand what they are based on, short of the reasons Gibbon listed against Tacitus plus plain ignorance. Geikie was simply following a common and unambiguously expressed conviction in the entire world of geology since the end of the third decade of nineteenth century! That he was also proud of it as a Scot goes without saying and cannot be held against him. Until about the end of the fourth decade of the nineteenth century there were still opinions here and there to be heard disputing Hutton's prominence. Even these dwindled to a total whisper by the middle of the century and eventually ceased altogether until revived in the last quarter of the twentieth century by some historians of geology whose familiarity with the literature of geology seems inadequate.

Hutton's Theory Regarded as Something New

The next question I listed above is "Was what Hutton published something new in the eyes of his contemporaries and successors who wrote before Geikie's 1897 book?" The quotes that follow are those, also selected from the previous section, that designate unequivocally Hutton's theory as something novel, something that had not been seen earlier. When the reader reads the entire text in the previous section, he or she will find numerous other statements implying the same. Not all of the authors who thought Hutton's ideas original agreed with what he said. But, nevertheless, they recognized its novelty, something some of the recent revisionists among the historians of geology unaccountably deny.

1787

In this System of Dr. Hutton there is a grandeur & sublimity by which it far surpasses any that has been offered. ... Other authors have observed a great part of the phenomena on which this system is built, and have perceived some of the truths deducible from them; but none has perceived them in such a comprehensive manner, or employed them so ably to explain the whole of these subjects. The Paper he has published in our Transactions is but a specimen; he is pping [i.e., preparing] a larger work. (Black, 1787; Edinburgh University ref: Gen. 873/III/36–39)

An III (1794–1795)

Although in the preceding notices we have confined ourselves to a review the works of deceased savants, we believe we must deviate from this position with respect to those of that Scottish savant, because the English naturalists have dealt with them extensively, both as supporters and as opposers. One cannot really regard the new views that Dr. Hutton developed and the new questions that he discussed with indifference. They are of interest to those who, by their researches, have occupied themselves to enrich the natural history of the globe. (Desmarest, *An III* (1794–1795), p. 732)

1795

Now, if I am to compare that which I have given as a theory of the earth, with the theories given by others under that denomination, I find so little similarity, in the things to be compared, that no other judgement could hence be formed, perhaps that they had little or no resemblance. I see certain treatises named Theories of the Earth; but, I find not any thing that entitles them to be considered as such.... (Hutton, 1795, p. 270)

1815

As the science of geology is closely related to all the mathematical and physical knowledge forming the subject of your studies, I believe that I would be useful to you by translating into our own tongue two new theories of the earth that are based on researches and observations made with great care. (Basset, 1815, p. iii)

1819

...at the end of the last century, Hutton in Edinburgh published a geological system entirely different from those hitherto appeared. Supported and commented upon by Mr. Playfair, distinguished as mathematician and author, it is generally accepted in Scotland and even in England. The beautiful experiments of Hall on the effects of heat on objects under great pressure seem to give it a new degree of probability and were greatly in its favour. (d'Aubuisson de Voisins, 1819, p. xxii–xxiij)

1820

Everybody knows that there are only two dominant geological theories: one, the neptunian theory, the other volcanic and that this latter comprises a special geological system due to the genius of Dr. Hutton. (Boué, [1820], p. 394; italics Boué's)

1837

Hutton was dead, leaving behind him an ingenious theory of the formation of the earth, based only on some facts. Playfair supported it with much energy and talent and the celebrated Hall rendered to the principal consequences of the Huttonian system a high degree of probability with his experiments on the influence of heat and pressure. (Rozet, 1837, p. x)

1840

Doctor James Hutton (born 1726, died 1797), a private gentleman in Edinburgh, observed with great industry and clear eye the conditions in England and Scotland. On their basis he erected a distinctive history of the formation of the earth in his brilliant Theorie [sic] of the Earth (in the Transact. of the royal Soc. Of Edinburgh 1, 1788) and as an independent work in 1795, of which a German extract was published in Voigt's Magazin der Physik in 1789. (Keferstein, 1840, p. 80)

Whereas Werner infinitely reduced the activity of vulcanism and the pyric rocks, Hutton accorded them so much a larger field of activity. He was the first to prove definitely the intimate connexion of all massive, crystalline rocks positing them as contrasts to the stratified rocks and completely undermined the idea of "primordial rocks," thus introducing into geology completely new ideas. (Keferstein, 1840, p. 83)

1860

The importance of the work of Hutton and Playfair, in which the fundamentals of modern geology are established and developed for the first time, ... requires that I here summarise its principal propositions.... (Daubrée, 1860, p. 9)

So Hutton is the founder of the fecund principle of the transformation of sedimentary rocks under the action of heat. (Daubrée, 1860, p. 13)

1866

It is especially to the observations and theoretical ideas of Hutton, then to the experimental researches of Sir J. Hall, that the Scottish school owes its originality and its greatest brilliance. (d'Archiac, 1866, p. 236)

1876

Hutton was the first clearly to enunciate the laws of denudation, which he had learned by observation (Green, 1876, p. 116)

1878

As I said in finishing the last lesson, Hutton also presented a complete theory of the earth from this viewpoint, which played an important rôle in the aetiological school of geology. But he clearly distinguished himself from all his predecessors and, one can add, all those who followed him in this order of considerations, in that the facts that he cites and knows, almost all as eye-witness, are admirably observed and the deductions made from them generally betray an excellent logic. (Sainte-Claire Deville, 1878, p. 145).

Of this account it clearly emerges, I think, that the founder of the Scottish school of geology was at once an excellent observer, a great logician and that in a large number of questions he proved to have an admirable perspicacity that allowed him to foresee and announce many of the discoveries made after him. (Sainte-Claire Deville, 1878, p. 174–175)

Hutton Regarded as the Founder of Metamorphic Geology

Apart from being the creator of the modern conception of geology, Hutton is regarded by the various practitioners of the various branches of geology as their founder. However, in the theory of metamorphism, most metamorphic geologists of the nineteenth century saw him as their first predecessor.

The crystalline schists he [i.e., Hutton] regarded as ordinary layered rocks altered by the internal heat of the earth, i.e., reworked and recrystallized long after their deposition, and thus became the founder of the theory of metamorphism that has been and still is controversial. (Neumayr, 1887, p. 23)

Evaluation of the Historical Record

The quotations above, summarizing the long account in the last section, are representative of the prevailing view during the latest eighteenth and the entire nineteenth century as well as in our own day among geologists about Hutton's place in the history of geology on the basis of what he contributed to our understanding of the structure and the history of our planet. I wrote this book without leaving my personal library and with a minimum of internet support (I used the *Wikipedia* mostly for the dates of nativity and demise of the people mentioned in this book and some of the portraits, which I otherwise could not find so quickly). Experience tells me that had I cast my net wider, I would have collected a much richer harvest of statements corroborating what I have here.

I repeat what I said in the introduction: I am surprised that such a review as published herein was even necessary. I had been under the impression that everybody interested in geology was aware of Hutton's unique, pioneering contribution and that nobody would even think of attributing his great fame to an alleged nationalistic propaganda by one geologist. The international tradition of considering Hutton's theory of the earth as a novel view of our planet's behavior and Hutton himself as the father of modern geology was firmly established already during the first half of the nineteenth century.

Very astonishing has been the rather quiet reaction from geologists against the revisionist histories putting forward grossly erroneous claims with no substance whatever. Some prominent geologists even wrote laudatory reviews for the publications containing such claims. Perhaps they have not read the books they reviewed in adequate detail. A part of this lack of attention from the side of the geologists may have been because many have not taken the new revisions seriously. A prominent Scottish geologist, to whom I complained about the revisions, wrote back: "Silly to waste time belittling Hutton because anything … write in this vein will have little influence" (12 December 2018). I had thought so too, but some of the reviews by some geologists I highly respect of the publications of the revisionists have led me to change my mind and made me think that a well-documented rebuttal was necessary. Let us recall that at least one reviewer (Baker, 2008) even recommended that the textbook writers in geology today take cognizance of what is said in one such revisionist account; I would recommend exactly the opposite! I hope this essay has sufficiently documented that what the revisionists say about Hutton's having had a limited contribution to geology or that his fame had been created by Sir Archibald Geikie's little book are entirely untrue and are fabrications of their authors rather than revelations of a careful reading of the geological literature post-dating Hutton and pre-dating Sir Archibald's book.

But why these fabrications? It needs emphasizing that the revisions have been mostly the work of professional historians rather than scientist-historians. In the works that I have been able to read by the revisionists, what struck me as the principal causes of the misinterpretations/misrepresentations were (1) a confusion between methods and theory in the sciences, something that Sergei Tomkeieff also pointed out, and (2) a paucity of knowledge of, or attention to, the geological literature of the times studied. Both can be traced back to lack of experience in actually doing geology. If one has never grappled with actual geological problems, especially of a theoretical nature, it is exceedingly difficult to understand how it is done. Geology is a very complex science dealing with historical problems using the methods of all the other natural sciences plus its own peculiar methods that have developed mainly from the Renaissance to the present day. To understand *why* a certain geologist does *what*, it is important to know (1) what question(s) that geologist was trying to answer; (2) what prompted the question(s); (3) what method(s) the geologist employed to tackle the question(s); (4) had the method(s) been already in existence, or did the particular geologist (or group of geologists) had to invent it/them; (5) what was the common scientific wisdom with respect to the question(s) that needed answering at the time; and (6) were the means of the geologist(s) attempting to answer the question(s) adequate to do so? Answering these questions would completely define why a certain geologist did what in the past *scientifically* and allow the historian to begin evaluating it. Notice that no social question needs answering here to understand the geological situation. A geologist may indeed wish to tackle a problem to earn fame, or to receive a promotion, or to increase his wealth, or to embarrass a colleague (the infamous Murchison-Sedgwick, Cope-Marsh, or Bittner-von Mojsisovics feuds come to mind) etc. However, none of these sources of extra-scientific *motivation* would give us any idea about how the geological question had originated, what methods were available to tackle it or what common knowledge was at hand to be used in answering it. The scientific aims of, and the methods used by, either Murchison or Sedgwick, or either Cope or Marsh would have been exactly the same even if they had not been trying to ruin each other's reputations and research efforts. Even if Alexander Bittner had not so profoundly hated the Austrian "establishment" in paleontology and stratigraphy, he would still have made the same objections to what he (rightly) thought were erroneous observations and conclusions about the Triassic stratigraphy. That need for coal led to the foundation of geological surveys in the civilized countries and in their colonies during the nineteenth century and that this situation immensely furthered geology is a social phenomenon, but a study of the societal needs and reactions will contribute nothing to our understanding of through what steps of problem identification, hypothesis generation, and then testing used by the geologists in these surveys and other state institutions responsible for finding and extracting coal, developed.

Let us consider a much more limited non-geological example: the barometer was invented by Evangelista Torricelli (1608–1647) in 1643, because the pump makers in Florence were failing to suck water into heights more than 10 m. This troubled the Grand Duke of Tuscany, Ferdinando II de'Medici (1610–1670; reigned 1621–1670), and he asked Toricelli to see what the problem was. So far social events. But the problem of the weight of air was entirely independent of the Grand Duke's worry or his request and was in fact as old as Aristotle. Earlier, the scholarly Genoese physicist, engineer, and astronomer Giovanni Battista Baliani (1582–1666) had asked Galileo in a letter dated 1630 why water would not rise through a siphon he had designed to carry water over a 21-m-high hill. Galileo could not solve the problem and passed it over to his disciple Torricelli. That Torricelli's *solution* (not the reason why he considered that problem when he did) and *the way he went about to reach it* also had nothing to do with the social circumstances prevailing in Florence is shown by the fact that both Descartes and Gasparo Berti (ca. 1600–1643; Berti lived in Rome) had earlier come up with similar solutions, but did not publish them. Since "the weight of the air" had long been a problem, Galileo's and the Duke's requests only prompted Torricelli to consider it at a certain time. That it became widely known by the public was because it solved also a practical problem of concern for the society. Thus the extra-scientific reasons of motivation are totally irrelevant for the origin of the scientific problem(s) to be solved and the thinking that goes into their solution. A scientist may be inspired to study a problem or consider a solution to a problem by multifarious sources: religion, poetry, chance encounter with a physical object, or even a pleasantry made by a friend. None are parts of the history of science. What is a part of the history of science is what the scientist does with the inspiration thus received.

Returning to geological examples, the biblical flood is a myth born of the social circumstances, possibly in Mesopotamia, most likely inspired by a Meghalayan-age (0.0042 Ma to the present) local storm surge of the kind that still devastates the shores of the Persian Gulf under the influence of the Shamal winds, although at the time the sea level was probably higher by a few meters, exacerbating the flood conditions in the Sumerian settlements (e.g., El-Sabh and Murty, 1989; Lambeck, 1996; for a short summary the local geological evidence, see Woolley, 1953). That the Sumerian myth, taken over by the holy books of the Abrahamic religions (Woolley, 1953), gave an impetus in Christian Europe (taken as a scientific theory) to global stratigraphic correlation (as a method) is, however, a purely scientific fact, because a universal flood seemed a *causa vera* for the possible correlatability of strata around the world. It was tested scientifically and abandoned scientifically when it failed the tests to which it was put. The theory of the universal deluge failed, but not the method of lithostratigraphic correlation. There is no point here in insisting that "religion was helpful for the development of science" because any idea that seems plausible can and has served science as a source of inspiration for erecting theories in the past and still does so. Greek mythology can be seen as the "cause" of the rise of science in Miletus more than 2600 years ago, because, as an explanatory theory, it could not provide satisfactory answers to the questions put to it, such as the origin of tempests or earthquakes, and was consequently abandoned for a better theory by Thales, which in turn was dethroned by Anaximander's yet better theory. Accidental discoveries in science are dime a dozen and they all originate from investigating unexpected appearances. Social events do not come into question in their origin and development.

It is certainly true, for example, that the great demand on the mines in Saxony to sustain the kingdom's crumbling economy after the Seven Years' War (1756–1763) leading to widespread state support for mining-related activities was a social phenomenon, but the continued survival, to our own day, the method of lithological correlation despite Werner's choice of neptunism and its failure as a geological theory, have nothing to do with that social event. To understand the roots and evolution of neptunism, one has to understand above all the invention and evolution of the concept of a "bed," i.e., a "stratum," as I showed elsewhere (Şengör 2009a, 2016) and a whole scientific tradition from Georgius Agricola to Füchsel and Lehmann. Laudan (1987) thought what she called the Becher-Stahl cosmogonic tradition played an important role in developing Werner's neptunism (its theory or its methods?) by influencing the training of miners in the German states. It did not, simply because the practical Werner had little interest in chemistry. His world view was shaped by the descriptive mineralogical-geognostical tradition, mainly of methods, developed by such mining-based geologists as Georgius Agricola (1494–1555); Johann Mathesius (1504–1565); Petrus Albinus (1543–1598), who wrote the *Meißnische Land- und Berg Chronica…* (=Mining chronic of Meissen) published in Dresden in 1589; Georg Engelhardt von Löhneyß (1552–1622), the author of the famous *Bericht vom Bergwerck* (1617: Report of the Mine; and many subsequent printings till well into the eighteenth century, although much of its geology and metallurgy was plagiarized from Agricola and Lazarus Ercker's {1528 or 1530–1594} *Beschreibung der Allerfürnemsten Mineralischen Erzt- und Bergwerksarten* of 1574: Description of all the best mineral and ore mine types); Alvaro Alonso Barba (1569–1662), the author of *Arte de los Metales* (=Art of Metals, published in 1640 in Madrid and translated into German by Johann Lange in 1676 with the title *Berg-Büchlein, darinnen von der Metallen und Mineralien Generalia und Ursprung*, i.e., Mining booklet, in which [are contained] generalities and origin of metals and minerals; despite Werner's criticism of it); Balthasar Rößler (1605–1673); Johann Gottschalk Wallerius (1709–1785); Johann Gottlob Lehmann (1719–1767); Georg Christian Füchsel (1722–1773); Johann Ehrenreich von Fichtel (1732–1795); and mainly the mostly bibliolatrous theories, developed by the naturalists Carl von Linné (1707–1778) and Torbern Bergman (1735–1784) in Sweden and the Scheuchzer brothers in Switzerland (Johann Jacob: 1672–1733, and Johannes 1684–1738). From the literature Laudan cites, I obtained the impression that she is not familiar with the mining tradition of central Europe, in which Werner had grown up. Her statement that she finds it unclear why Wernerianism had such a romantic appeal as it did (Laudan, 1987, p. 111), shows further her unfamiliarity with the rich literature on romanticism and natural sciences in Germany (e.g., Snelders, 1970) and her insufficient knowledge about the rich mining folklore from Scandinavia to the Alpine countries in Europe in which Werner had been raised and from the lore of which his school had been nourished (see Şengör, 2001)[177].

Reasons to engage in science, apart from personal curiosity, are naturally abundant in the history of science, but true scientists are those who go after their personal desire to find things out. In fact, Cornford suggested that the difference between the science of our days and the science of those who actually invented it, i.e., the Ancient Greeks, was that science in its earliest infancy, "was suffered to remain as part of the pursuit of peaceful wisdom and of a happiness independent of wealth and even of material comfort. The fruits it gathered from the Tree of Knowledge were not the Baconian fruits of utility and progress" (Cornford, 1950, p. 94). For science to remain science, it has to seek *understanding of Nature* and nothing else, which was the cause of its birth as a distinct human activity. That its findings, once obtained, may be put to multifarious uses is not a part of the definition of science. During the reign of the Third Reich in Germany (1933–1945), for example, a number of geologists sought personal furtherment by joining the NSDAP (*Nationalsozialistische Deutsche Arbeiterpartei*; National Socialist German Workers' Party) and published opinions in support of Nazi policies. This, however, did not affect the scientific questions the good ones tackled, the methods they used to answer them, or the answers they eventually came up with, most of which have since become respectable ideas in the intellectual treasury of geology. Edwin Hennig (1882–1977), the great German geologist of mainly East African fame, was certainly a Nazi sympathizer, but look what he said about the goal of science at the height of the Nazi domination in Germany:

> Science is not a means to any old goal, but a drive, an expression of life, like breathing. Where it is not, where it serves only ambition and livelihood, however high it may reach in learning, it would not lead us farther. (Hennig, 1938, p. 113)

The Hennig student and convinced Nazi Karl Beuerlen's (1901–1985) support of the anti-uniformitarian ideas and the orthogenetic evolution theory (named by Theodor Eimer in 1898, although the theory had many predecessors) has been ascribed to his being influenced by the racist Nazi ideology. However, not only his, but other very reputable paleontologists', such as Othenio Abel's (1875–1946; also a Nazi sympathizer) or Otto Schindewolf's (1896–1971; not a Nazi sympathizer), support of orthogenesis was based on science, which turned out to be wrong, and not on politics. Beuerlen lost his job in Munich as a result of the denazification policy of the invading Allied armies and went to Brazil where he did some superb geology (I still use his *Geologie von Brasilien*, 1970, as a sourcebook). His emigration was a social event, but not his interest in the problems of Brazilian geology or his becoming a supporter of Wegener's theory (which the Nazi establishment never liked).

Such pseudo-scientific theories as Hanns Hörbiger's (1860–1931) *Glacial-Cosmogonie* (or *Welteislehre*; Fauth, 1913) were not taken seriously by the scientific world and had zero effect on the development of geology in the twentieth century, despite strong support from Hitler himself and Himmler. Trofim Denisovich Lysenko's (1898–1976) influence on the agricultural science and genetics in the Soviet Union was not a scientific, but a political aberration; he harmed biology in the Soviet Union not by any science, but by brute political force of Stalin, who outlawed any scientific criticism of what was called "Lysenkoism" (Graham, 1993).

Even in a country like Turkey, where there has been no scientific tradition until the twentieth century, political pressure could not succeed in derailing science: Mustafa Kemâl Atatürk (1881–1938), the creator of modern Turkey and certainly one of the more enlightened leaders in the twentieth century, proposed a theory, with a view to boosting the self-confidence of his people, claiming that after the desiccation of Central Asia, the Turks living there, who supposedly had attained a high degree of civilization, had been dispersed and carried civilization to the four corners of the world. He tried to push his theory and there were certainly many sycophants, even in the university, supporting it. Yet, in the face of determined resistance by serious scholars, such as Mehmed Fuad Köprülü (1890–1966) and Zeki Velidi Togan (1890–1970), he eventually had to abandon it (see Şengör, 2014b). Thus, no social circumstance could make nonsense triumph or hide a good explanation, unless science itself is banned or abandoned, as it happened in the European Middle Ages, or in the Soviet Union in the case of Lysenkoism; or is restricted, as in some countries today.

Sivin (1988, p. 52) pointed out that some astronomical and alchemical practices in China were religious in character and purpose, and therefore religion should be taken into account in the history of science. I disagree entirely. Sivin says "I have argued that although alchemists applied chemical knowledge, their goals were not cognitive but religious or oriented toward healing." Anything that has no cognitive aim is not a science (but it may be an art or a craft) and therefore cannot come under the compass of the history of science. No illusionist, employing for his trade the techniques developed for scientific purposes, would be considered a scientist. Similarly, no priest or shaman concocting hallucinogenic drugs for religious ceremonies or prescribing certain natural products discovered by chance against certain ailments (such as the *Cinchona calisaya* WEDDELL 1849, used by the Quechua Indians in the Central Andes as a muscle relaxant) can be regarded as a chemist or a pharmacologist, although such concoctions or natural products, when discovered by a scientist, may indeed be the starting point of a scientific investigation (as in the case of *Cinchona* leading to the proper understanding of how it functions, i.e., not only *what* it does, but *how* it does what it does, and the consequent development of quinine by European scientists). Neither can the pharmacist in our district in İstanbul, who owns a laboratory to prepare some of our prescription drugs, be considered a scientist or his activity a subject for a future history of science treatise. Even if he discovers by chance a more potent drug than the prescribed one, it would not be science until a pharmacologist investigates it to *understand* how it works.

The paragraphs above depict elements of the training and the scientific milieu of a particular scientist in the past, of the kind that one must take into account in evaluating a scientist's work. Let me emphasize again: it is not the social milieu or personal extra-scientific ambitions that determine the origin and the manner of handling of scientific questions. If it were the social milieu, it would be difficult to understand the widely differing attitudes toward geology in such a small town as Edinburgh or indeed in Werner's circle of students and colleagues. Neither the great veneration toward Werner's person, nor his dogmatism, could stop the development of chemical mineralogy, or the rise of the vulcanist sympathies, among his students, albeit some of his students felt that they needed to publish their deviant thoughts anonymously, so as not to hurt their teacher's feelings. But publish them, they did (see also Gaudant, 2013, p. 231, for the case of d'Aubuisson de Voisins)!

We should also add to his or her scientific milieu the scientist's profession. As Kristeller (1979b, p. 25) rightly observed, "the achievements of a given nation or period in particular branches of culture depend not only on individual talents but also on the available professional channels and tasks into which these talents can be drawn and for which they are trained. This is a subject to which cultural and social historians apparently have not yet paid sufficient attention." Werner, for example, was not only a miner, but also a professor in a mining academy. That is why his work was more systematization than discovery. His primary question was to find means of making mining easier by sorting out the geological structure of the mined areas and by generating a terminology that would facilitate communication amongst miners. How that structure came about was incidental and clearly Werner did not give much thought to it. His approach was descriptive and not explanatory. He developed methods and nomenclature that helped description and spatial prediction ("where is the next vein of ore?").

Hutton's entry into geology was via a completely different approach. He was surprised that continuous erosion and consequent denudation do not level all land area to sea-level in a short time. His entire geological enquiry and the theory that developed around this simple question were to generate *not descriptions* but *explanations*. Hutton was not out to train miners who would need to obtain a good knowledge of the methods of mining to go out and find and run more mines. He was trying to understand how this planet of ours functioned.

The second problem with the revisionists' accounts of the history of Hutton is their inadequate attention to the primary literature. A common characteristic of their accounts is that a number of standard works in the time interval under study are cited and then the social milieu in which they originated is related followed by a peremptory statement not supported by any evidence whatever, but with the insinuation that the before-said is the basis of the final statement. In order to pass judgement on Hutton's influence on the posterity and the evolution of his reputation, something that revisionists do so inadequately, one needs to present evidence from the time of his death to the time when one thinks his memory was "resurrected" and allegedly made material for nationalistic propaganda from the primary literature of that interval. This evidence should show, to make the revisionists' claims tenable, that nobody before Sir Archibald Geikie had ever referred to Hutton as the father of modern geology; or at least this had not become commonplace, a "tradition."

Even there, one needs to proceed cautiously because negative evidence may turn out to be entirely misleading. In our case, however, there is abundant *positive* evidence, even before Sir Archibald was born, that Hutton had been hailed as the founder of modern geology and it happened right across Europe, from St. Petersburg to Paris and even beyond to New York. Greene (1982), who wrote that Hutton's originality was restricted to Great Britain and to geomorphology, seems also to have been a victim of neglecting a careful study of the primary sources and relying mostly on modern historiography of geology. How else could he have misunderstood what Davies (1969) had said or could have cited H.H. Read to support a position exactly the opposite of what he intended it to? The late David Oldroyd once pointed out that he had been a victim of the secondary literature concerning Werner, generated by the "wave of scholarship" that Laudan (1987) refers to (Oldroyd, 2003, p. 430), because he could not read the primary literature if it was not in English or French (D. Oldroyd, personal commun., 2002). More flagrant cases of inattention of primary sources are Laudan (1987) and Claudia Schweizer's 2009 paper on three religious Swiss scientists who dealt with geology. Laudan cites Balthasar Rößler's (1605–1673) book as "Balthzar Roesler, *Speculum Metallurgie*," whereas the correct orthography of the author's name and the full title of the book are: "Rößler, Balthasar, 1700, *Speculum Metallurgiæ Politissimum oder: Hell-Polierter Berg-Bau-Spiegel*." She further cites the classic book on the history of mineralogy by Franz von Kobell as "Kobell, Franz von, 1863, *Geschichte der Mineralogie von 1650–1850*. Munich: Merhoff;" but the correct reference would have been "Franz von Kobell, 1864, *Geschichte der Mineralogie von 1650–1860*. Munich, J. Cotta." Claudia Schweizer thinks that in all three cases of the authors she studied, a religious belief proved to benefit scientific knowledge. In the case of Scheuchzer, her own paper belies this claim (and she herself says that "religion had only in part positively motivated science...," her p. 98; religion was indeed partly Scheuchzer's motivation, but it massively misled him as indeed Schweizer mentions). Von Haller argued that Church's fear of science was not justified. And for Deluc, Schweizer used only a single source: Rudwick's 2005 book! (Although she refers to a book by Deluc, which, by title, is the one I referred to in my footnote 16 above, but allegedly in seven volumes! The title she cites is in reality a single volume work. What she probably meant is the five volume—not seven—*Lettres Physiques et Morales sur l'Histoire de la Terre et de l'Homme adressées à la Reine de la Grande Bretagne*, but not one argument out of it is mentioned by her). Her information is entirely second hand and from a source not particularly reliable about Deluc (see also Oldroyd, 2009, about Rudwick on Deluc).

I wonder whether these two authors have even seen the books they cite? If they have, whence the sloppiness?

All this reinforces my point that strong claims in the history of science must be based on quotations from the primary literature, not even on paraphrases by the author who is putting forward the claims. As Laplace said in the second edition of his

Théorie Analytique des Probabilités (1814, p. xij): "the more something is extraordinary, the more it would need to be supported by strong proofs"[178]. And strong proofs in history are to be gathered from original documents and monuments, and especially not from secondary sources with an additional hidden agenda outside historiography.

That an historian ought to use primary source materials goes without saying, as Edward Gibbon stressed in the second motto of this book. It is a part of the historian's most important responsibility, although one frequently not observed with due diligence so that good historians often feel the need of reemphasizing it. The great German codicologist and authority on the history of Renaissance philosophy and culture, Paul Oskar Kristeller (1905–1999), who, from 1939, taught at Columbia University in New York, felt that the following needed emphasizing in the introduction to a book of some of his collected writings:

> …in spite of many objections, I have not been driven to change my basic views. This may be due to…my conviction…that such matters do not, or should not, depend on opinions or preferences, but on the testimony of the texts and documents. History, and especially the history of thought and philosophy, has its own "scientific method." A historical interpretation derives its validity, not from the authority of the scholars who endorse it, or from its agreement with conventional views (which may be wrong) or with current fashions (which constantly change), but from its agreement with the primary sources, that is with the original texts and documents of the past. I tried to formulate my views on the basis of such primary sources, and I am ready to change them (or to see them changed by others) when they are confronted with primary sources that contradict them. (Kristeller, 1979a, p. 4)

Every historian of geology should take to heart the advice of the great Russian petrologist and historian of petrology Franz Yulyevich Loewinson-Lessing (1861–1939) to the young petrologists interested in the history of their subject:

> I earnestly advise my readers, in particular young petrologists, to acquaint themselves with the classical monographs of leading petrologists and so derive their knowledge direct from the fountainhead rather than through the medium of textbooks. (Loewinson-Lessing, 1954, p. viii)

But isn't this precisely the same in the case of the natural sciences? Hawking and Mlodinow (2010, p. 72), in their magnificent popular account of the universe, wrote, when describing the weird nature of the quantum theory "Though that is distasteful to some, scientists must accept theories that agree with experiment, not their own preconceived notions."

Gould's (1987) very popular book is yet another case in point about ignoring the primary documents: he first claims that Hutton made his first geological observation after he worked out his theory. Rudwick, in his various books (e.g., 2018, p. 70) says precisely the same thing. In principle, there is nothing wrong in what they claim Hutton did as far as scientific method is concerned: did Darwin not formulate his theory of coral reefs before he ever got to see one (Darwin, 1887)? *However, what Gould and Rudwick say about Hutton is manifestly not true.* We have documentary evidence that Hutton had started making geological observations already in 1753, if not a bit earlier (Playfair, 1805, p. 44). We have specific cases in 1764 and in 1765 of his excursions that I cited above. Jones et al. (1994) documented, using Hutton's correspondence with James Watt and George Clerk-Maxwell, his geological excursions through Wales (during which Hutton found fossils "in what may be called primitive mountains"), the Midlands and the southwest of England in 1774, before he published a word on geology. Jones et al. (1994, p. 641) wrote: "The principal importance of the letters, however, lies in the picture they give of Hutton in the field. For most of his life he was a relentless traveler and observer. We know that between 1750 and 1788 he journeyed through nearly every area of Britain, except northwest Scotland and the Hebrides. These are facts that modern commentators who brand him as a mere theorist have chosen to ignore." Jones et al. (1995) add further letters containing more field work and sending of rock samples to Hutton, but the letters in Part II of Jones et al. (1995) deal mostly about engineering and commercial subjects about Watt's steam engine.

Gould then claims that Hutton did not have a historical view of geology and that his world revolved and revolved around endless cycles repeating everything over and over again, to infinity. But this is also clearly not true, as Hutton himself pointed out as quoted above. Hutton was not an eternalist either, as Playfair bears testimony: he simply pointed out that the *available evidence* indicated no vestige of a beginning and no prospect of an end—and, at the time, he was perfectly right. How can anybody write a book, the claims of which fly in the face of historical evidence, is something I have great difficulty understanding.

Among the revisionists about Hutton's place in the history of geology, Martin Rudwick takes a special place, because not only he had been trained as a geologist in the superb environment of Cambridge with its rich library, museum, and archival resources, but worked, for some time, both as a research geologist and an academic teacher in Cambridge and, by all accounts, he was very good at what he did (except for an unfortunate philosophical encounter with wing-flapping pterosaurs: Rudwick, 1964). I cannot believe that he does not appreciate the value of the primary sources, abundantly stocked in Cambridge, Oxford, and in the great institutions in London, such as the Royal School of Mines of the Imperial College, University College London, the Geological Society, the Linnean Society, the Royal Geographical Society, and the Natural History Museum including its neighbor the Geological Museum. So, what can be the source of his misrepresentation of Hutton's position in the face of such positive evidence as recounted above in his account the ostensible purpose of which is to relate the rise of geohistory in the eighteenth and early nineteenth centuries?

I believe the answer lies in what some of the reviewers of his two recent books (Şengör, 2009c; Wilson, 2009; Oldroyd, 2010) and, in a special analysis of the influence of the creeds of historians on their interpretations of the past (Oldroyd, 2009), have pointed out: his desire to exonerate religion from its sins against geology by arguing that some religious savants, as he calls them, had a better understanding of geology than Hutton and in fact religion not only did not adversely affect their work, but instead helped them by giving them a unique historical perspective. By contrast, so he contends, the deist Hutton, with his allegedly ahistorical and anachronistic *Theory of the Earth*, really did not have much of a contribution and influence on geology until his countryman Geikie saved him from oblivion. When I pointed this out in my essay review (Şengör, 2009c) of his *Bursting the Limits of Time* (Rudwick, 2005), Rudwick retorted:

> So, finally, Şengör concludes his review by stating that "Rudwick's historical interpretations seem more than a little colored by his own religious commitment." I could respond, quoting the apt English proverb, that "what's sauce for the goose is sauce for the gander": that is, I could retort that Şengör's historical interpretations seem more than a little colored by his own atheistic commitment. (Rudwick, 2009, p. 141)

That response, however, is clearly wide of the mark, because atheism is not an article of faith; it is not just another religion: that I do not believe that our earth was created and shaped by Snow White and the seven dwarfs (and possibly also by the Prince Charming), does not imply that I have a special commitment not to believe that our Earth was created and shaped by Snow White and the seven dwarfs (and possibly also by the Prince Charming); it simply implies that I refuse to believe in nonsense. Should somebody, one day, show in an intersubjectively testable way, by different groups of sentient beings not in communication with one another, i.e., as independent investigators, that our earth was really created by Snow White and the seven dwarfs (and possibly also by the Prince Charming), I would have no hesitation in proclaiming that our Earth was indeed created and shaped by Snow White and the seven dwarfs (and possibly also by the Prince Charming).

My attitude toward religion, as a historian of geology, is not because I view religions from the viewpoint of the present day: as we saw above, science arose because certain individuals refused to believe in the tales of the ancient Greek religion already in the sixth century BCE[179]. Anaxagoras was chased out of Athens because he thought that the belief that the Sun was a god was nonsense. Titus Lucretius Carus, the great Roman philosopher and poet of the first century BCE, thought all religion and belief in god or gods nonsense as he elaborated in his immortal *De Rerum Natura*. Both Celsus (second century CE) and Porphyry of Tyre (his real name was Malchus: third century CE) published books nearly two millennia ago exposing what they considered to be follies and absurdities in the Christian creed (Hoffmann, 1987 and 1994). Giardano Bruno was burned at stake in Rome in 1600 by the Catholic Inquisition, for, among other offenses, his conviction that a number of Christian dogmas expressed nonsense.

In his most recent book, summarizing his earlier efforts, Rudwick wrote:

> Hutton's system was not ignored or neglected; and living in one of the cultural centers of the Enlightenment it was of course unthinkable that he would have been persecuted for his opinions. But by the end of the 18th century the genre of "Theory of the Earth" was generally regarded by savants as having outlasted its usefulness. Hutton's example of it, like Buffon's, was widely considered too speculative to be taken seriously. Although some of his detailed observations were accepted as valuable, his theory might well have been forgotten, along with other 18th-century works in the same genre, had it not been repackaged, after Hutton's death, to suit the scientific tastes of a new century. (Rudwick, 2014, p. 73)

How can anybody familiar with the literature I cited above —and many others I could not so as not to inflate the size of this book—throughout the late eighteenth and the whole of the nineteenth century write such a statement that contradicts so flagrantly the available documentation is extremely difficult to comprehend, unless its author had a specific agenda in mind to belittle Hutton's contributions to geology. What does Rudwick mean by Hutton's theory having been repackaged, after Hutton's death, to suit the scientific tastes of a new century? I wish he told us what the "repackaging" entailed. Not only is a comparison between de Buffon's theory of the earth with Hutton's without any merit of analogy, but Rudwick had to ignore (or distort: "repackaging") almost the entire geological literature of more than a century, to come to his extraordinary conclusion. The only reason that he did so, I think together with some other outstanding historians of geology such as Leonard Wilson and David Oldroyd, is to be found in his commitment to exonerate his religion from its pernicious rôle against science. How else could he call Deluc, whom others have identified as a religious bigot (Hübner, 2011, p. 31ff.) and possibly somewhat insane (Heilbron, 2011b, p. 186), "one of the most perceptive" of Hutton's critics (Rudwick, 2018, p. 73)? Whatever Deluc wrote in geology was mostly nonsense if it was original with him, and whatever was right in his geological writings, had already been said by other serious geologists. As one of the greatest geologists in the history of our science and still its best historian, Karl Alfred von Zittel (1899, p. 109), wrote as I already quoted above and repeat here, "it is not necessary to consider any more closely the views of this assiduous but fleeting observer and prolific author of phantastic writings; Deluc's publications have been deservedly forgotten and his strong attacks on Hutton and Playfair became pointless." As the forthright Georg Christoph Lichtenberg said already in the late eighteenth century, all such bibliolatrous theories of the earth as that by Deluc were not contributions to the history of the earth, but contributions to the history of the aberrations of the human mind (see above).

CONCLUSIONS

Hutton's position in the development of geology as a science has been questioned during the past four decades by some historians, who, to use the editorially imposed title of Victor Baker's *Nature* review, have generated a history of the science of geology "turned upside down" (Baker, 2008, p. 406). They have claimed that Hutton's theory was an anachronistic representative of the old seventeenth and eighteenth century "Theories of the Earth" genre and that it was an ahistorical view of our planet and its influence in the future development of geology was limited. The purpose of this book is to document that neither Hutton's contemporaries, nor his successors thought so, and there is no need to turn the history of geology upside-down. In the beginning, until about the end of the third decade of the nineteenth century, those who disagreed with Hutton's theory of the earth, nevertheless acknowledged that what he had said was novel. That certain components of Hutton's theory had been proposed earlier by others and some, like Moro, came very close to what Hutton later said, does not change the originality of his total model of earth evolution. Already in the third decade, people began acknowledging the fact that modern geology was moving along the lines Hutton had sketched and geological theory was evolving in dialogue with Hutton's concepts. Although many of his interpretations, such as the heat induration of submarine sedimentary layers, were given up (but see Count Collegno's point quoted above), the newer arguments always moved within the basic framework he had erected, namely, a heat-driven internal dynamics constantly and irregularly repairing the asperities of the planet's surface, which are under perennial attack by the external agencies of air, water, and ice fueled by the energy of the Sun, which seek to reduce it all to sea-level. All events are local, but this limitation may imply an entire continent or an ocean. Hutton shut out all global events from geology and the argument, whether they exist, continues to our own day. Geology neither retained Werner's neptunism, nor Cuvier's catastrophism, nor indeed the bibliolatrous geology of Deluc. Many of Werner's methods and terms in mineralogy and lithostratigraphy have certainly remained, but neither had originated in his thinking anyway. Soon, chemical classifications, promulgated by his own students (initially anonymously, lest the master's feelings be hurt), came to dominate his favorite subject and left him behind. In stratigraphy, he had simply elaborated his predecessors' interpretations, but did not necessarily improve them. Igneous and metamorphic agencies, defined and advocated by Hutton, assumed places in geological theory that Werner could have never dreamt of. Cuvier's biostratigraphy, supported by the independent discoveries by Smith in England, had become the most important pillar of geology, but not because of the causes Cuvier had imagined; his friend Alexander von Humboldt had already drawn his attention to the fact that a globally valid stratigraphy could not consist of natural independent units with global lateral extent, i.e., Werner's universal formations, thus implicitly supporting Hutton's provincialism. So far, only one of the five great extinctions in the geological past, namely that of the Cretaceous/Paleocene boundary event, turned out to be a real Cuvierian catastrophe, but that would have hardly surprised Hutton (except perhaps by its global impact), who himself had retrodicted the Messinian salinity crisis:

Let us but suppose a rock placed across the gut of Gibraltar, (a case nowise unnatural), and the bottom of the Mediterranean would be certainly filled with salt, because the evaporation from the surface of that sea exceeds the measure of its supply. (Hutton, 1788, p. 242)

By the end of the third decade of the nineteenth century, most leaders in geological thinking in Europe and America began saying that Hutton was the creator, the father of modern geology[180]. Sir Archibald Geikie simply followed that general opinion in his later writings and was no doubt very proud to have a countryman of his regarded as the modern architect of his favorite science. Modern geology continues to roll on the rails Hutton had put in place and there seems no indication at present that it will be any different in the future.

The strange turn of events concerning Hutton's legacy in the recent historiography of geology appear to have two main reasons. (1) A confusion of what makes up science: its methods or its theories? Some historians without experience in geological research appear to confuse methods with theories as the main core of science. (2) Insufficient attention to the primary literature. This may in part be because of difficulty of access, in part because of language barriers, and in part because of the difficulty in grasping the problem situations of past times. Some cases may be simply examples of carelessness in note-taking, which does not inspire confidence in the reliability of their authors. In only one case, I have found that religious convictions stood in the way of an impartial interpretation of the historical data. David Oldroyd once wrote, in a tolerant mode, that religion has given us superb art and that it might also be the spring-head of great historiography (Oldroyd, 2010). I have difficulty sharing this particular opinion of my late, lamented friend.

There may yet be another source for the desire to debunk Hutton: the recent tendency in the western culture of derogating heroes and the great cultural edifices they created and the condemnation of the science-based, secular civilization that Europe has produced. In his remarkable book, *In Defense of Elitism*, the late William A. Henry III observed that

universities, ...entrusted with the curatorship of our heritage... have caved in almost entirely to the rigorous dogma of what might be termed "special pleading studies." These purportedly scholarly undertakings are really intended to redress historic grievances, sometimes by willfully misunderstanding and reinventing the past.... (Henry, 1994, p. 4)

In some of the revisionist publications on the history of geology, I do sense the presence of some of what Henry complained about.

Let me end this book with a plea that I made in my acceptance speech during the conferral ceremony of the Mary C. Rabbitt Award of the History and Philosophy of Geology Division of the Geological Society of America during the Society's Annual Meeting in 2017 in Seattle, Washington, USA:

The great cosmologist Edward Harrison wrote a commentary in *Nature* in 1987 entitled "Whigs, prigs and historians of science" in which he pointed out that writing history of science without a retrospect from today would make it impossible to appreciate the significance of past science. I entirely agree with him. I do history of science to understand better where I stand now, but looking at history of science from my present vantage point also enables me to understand the problems and the struggles of my predecessors much better than somebody without a grasp of present science. The march of science through time is itself whiggish; its progress can only be interrupted by abandoning it entirely. That scientific progress includes many dead alleys is no argument against the continuity of its development. Dead alleys are a part of hypothesis testing and therefore of scientific method. Doing history of science without a sound scientific knowledge is a recipe for ship wreckage. Some historians of science such as Steve Woolgar even advocated the view that it is better for the historians of science not to know any science at all. Such a view I can only describe as complete nonsense. Neither is it true that hypothesis selection is dependent on the whim of the individual scientist as the late Thomas Kuhn wanted us to believe. His claim that in science the highest court of appeal is the consent of the majority of the concerned parties is also untrue. If it were true, neither Copernicus, nor Galileo, nor Eduard Suess, nor Einstein, nor Émile Argand would have caused the progress they did. That is why "sociology of science" is not a helpful pursuit to understand the development of science. As Goethe once said "the history of science is the science itself" and therefore those unwilling to learn science have no business doing its history. (Şengör, 2017, https://www.geosociety.org/awards/17speeches/rabbitt.htm)

In other words, there is little reason, and indeed grave danger for our civilization, for us to turn into prigs in the history of science.

REFERENCES CITED

Abel, C., 1818, *Narrative of a Journey in the Interior of China, and of a Voyage to and from that Country, in the Years 1816 and 1817; containing an account of the most interesting transactions of Lord Amherst's Embassy to the Court of Pekin, and Observations on the Countries which it visited*: Longman, Hurst, Rees, Orme, and Brown, London, xvi + 420 p. + 1 foldout map + unnumbered colored plates.

Abrahamse, J.E., and Feiken, R., 2019, Driftig veen en onderaards bos: Nicolaas Witsen, het landschap van Amstelland en de grondbeginselen van de moderne geologie: *Bulletin Knob*, v. 118, p. 33–54.

Adıvar, A. A., 1943, *Osmanlı Türklerinde İlim* (İkinci Tabı): Maarif Matbaası, İstanbul, 225 p. + 3 colored plates.

Agassiz, L., 1840, *Études sur les Glaciers*: Jent et Gassmann, Soleure, V + 237 p.

Agassiz, L., 1841, *Untersuchungen über die Gletscher*: Jent & Gaßmann, Solothurn, XII + 326 p. + 1 errata page.

d'Alembert, J. le R., 1749, *Recherches sur la Précession des Equinoxes et sur la Nutation de l'Axe de la Terre dans le Système Newtonien*: David L'aîné, Paris, xxxviij + [ii] + 184 p. + 4 plates.

Allchin, D., 1994, James Hutton and phlogiston: *Annals of Science*, v. 51, p. 615–635, https://doi.org/10.1080/00033799400200461.

Ampère, A.M., 1834, *Essai sur la Philosophie des Sciences, ou Exposition Analytique d'une Classification Naturelle de Toutes les Connaissances Humaines*: Bachelier, Paris, lxx + 272 p. + 3 foldout tables.

Anderson, D.L., 1989, *Theory of the Earth*: Blackwell, Boston, xiii + 366 p.

Anderson, D.L., 2007, *New Theory of the Earth*: Cambridge University Press, Cambridge, xv + 384 p.

Anderson, G.A., 1988, The Cosmic Mountain—Eden and its early interpreters in Syriac Christianity, *in* Robbin, G.A., ed., *Genesis 1–3 in the History of Exegesis—Intrigue in the Garden*: Studies in Women and Religion, v. 27, Edwin Mellen, Lewinston/Queenston, p. 187–224.

Anderson, R.G.W., and Jones, J., eds., 2012, *The Correspondence of Joseph Black*, volume one (xiv + 803 p.) and volume two (p. 805–1564): Ashgate, Farnham, Surrey.

Andrée, K., 1938, Rezente und fossile Sedimente. Erdgeschichte mit oder ohne Aktualitätslehre?: *Geologische Rundschau*, v. 29, p. 147–167, https://doi.org/10.1007/BF01806879.

Anonymous [Füchsel, G.C.], 1773, *Entwurf zu der ältesten Erd- und Menschengeschichte, nebst einem Versuch, den Ursprung der Sprache zu finden*: publisher not specified (privately printed) Frankfurt und Leipzig, [II] + 273 p.

Anonymous [François Dominique de Reynaud, Comte de Montlosier], 1789, *Essai sur la Théorie des Volcans d'Auvergne*: No publisher, no place of publication, xi + 134 + [vii] p.

Anonymous [John Playfair?], 1816, Introduzione alla Geologia, di Scipione Breislak, Amministratore ed Ispettore de' Nitri e delle Polveri del Regno d'Italia. 2 tom. 8vo. Milano. 1811: *Edinburgh Review*, v. 27, p. 144–163.

Anonymous [Nose, K.W.], 1821, *Kritik der Geologischen Theorie besonders der von Breislak und Jeder Ähnlichen*: Eduard Weber, Bonn, 79 + [I] p.

Anonymous [Nose, K.W.], 1822, *Fortgesezte Kritik der Geologischen Theorie*: Eduard Weber, Bonn, 65 + [III] p.

Anonymous, 1874, *Meyers Hand-Lexikon des Allgemeinen Wissens in einem Band*. Erste Hälfte A bis Gyromantie—Seite 1 bis 740. Bibliographisches Institut, Hilburghausen, 740 p. + 24 unnumbered Nachträge und Berichtigungen zur ersten Hälfte + 52 plates (including a geographic atlas).

Anonymous, 1965, *Bergakademie Freiberg—Festschrift zu Ihrer Zweihundertjahrfeier am 13. November 1965*, herausgegeben von Rektor und Senat der Bergakademie Freiberg: VEB Deutscher Verlag für Grundstoffindustrie, Leipzig, v. I., 414 p.

Ansted, D.T., 1844, *Geology, Introductory, Descriptive & Practical*: John van Voorst, London, v. 2, viii+[ii]+572 p.

Appleby, J.H., 1985, John Grieve's correspondence with Joseph Black and some contemporaneous Russo-Scottish medical intercommunication: *Medical History*, v. 29, p. 401–413, https://doi.org/10.1017/S0025727300044719.

D'Archiac [Desmier de Saint-Simon, Vicomte d'Archiac, É.J.A.], 1866, *Géologie et Paléontologie*: F. Savy, Paris, XVI + 776 p.

Ashworth, W.B., Jr., 2003, Faujas-de-Saint-Fond visits the Herschels at Datchet: *Journal for the History of Astronomy*, v. 34, p. 321–324, https://doi.org/10.1177/002182860303400305.

Ashworth, W.B., Jr., and Bradley, B., 1984, *Theories of the Earth 1644–1830—The History of a Genre*: Linda Hall Library, Kansas City, 68 p.

d'Aubuisson de Voisins, J.F., 1819, *Traité de Géognosie, ou Exposé des Connaisances Actuelles sur la Constitution Physique et Minérale du Globe Terrestre*: F.G. Levrault, Strasbourg and Paris, tome premier, ljix + 496 p. + 1 foldout table and 1 foldout plate; tome second, 665 p. + 1 foldout plate.

d'Aubuisson de Voisins, J.F., 1828, *Traité de Géognosie, ou Exposé des Connaisances Actuelles sur la Constitution Physique et Minérale du Globe Terrestre*, nouvelle édition revue et corrigée: F.G. Levrault, Paris and Strasbourg, xlvij + 524 p. + 1 foldout plate.

Aydın, A.E. (Chorepsikopos), 2008, *Der Heilige Ephrem—Leben–Werk–Texte*: place and name of publisher not indicated, [I] + 233 + 26 pp + 1 frontispiece.

Babbage, C., 1834, Observations on the Temple of Serapis, at Pozzuoli, near Naples, with remarks on certain causes which may produce geological cycles of great extent: *Proceedings of the Geological Society*, v. 2, p. 72–76.

Babbage, C., 1837, *The Ninth Bridgewater Treatise. A Fragment*: John Murray, London, xxii + 240 p.

Babbage, C., 1838, *The Ninth Bridgewater Treatise. A Fragment*, second edition: John Murray, London, vii + [i] + xxii + 270 p.

Babbage, C., 1847, Observations on the Temple of Serapis, at Pozzuoli, near Naples, with remarks on certain causes which may produce geological cycles of great extent: *The Quarterly Journal of the Geological Society of London*, v. 3, part 1 (Proceedings of the Society), p. 186–240 + 1 foldout plate.

Bailey, E.B., (Sir), 1967, *James Hutton—the Founder of Modern Geology*: Elsevier, Amsterdam, xii + 161 p.

Baker, V.R., 2008, Geological History Turned Upside Down: *Nature*, v. 454, p. 406–407, https://doi.org/10.1038/454406b.

Baker, V.R., 2010, Channeled Scablands: A megaflood landscape, *in* Migon, P., ed., *Geomorphological Landscapes of the World*, Springer, p. 21–28.

Bakewell, R., 1839, *An Introduction to Geology: Intended to Convey a Practical Knowledge of the Science, and Comprising the Most Important Recent Discoveries; with Explanations of the Facts and Phenomena Which Serve to Confirm or Invalidate Various Geological Theories*, third American from the fifth London edition, edited with an appendix by Prof. B. Silliman: B. & W. Noyes, New Haven, xxxvi + 596 p. + 8 foldout plates.

Balderston, W., 1961, Sir William Herschel and his place in the history of science: *The Journal of the Royal Astronomical Society of Canada*, Royal Astronomical Society of Canada, v. 55, p. 1–8.

Ballantyne, [J.], 1820, *Catalogue of the Library of the late John Playfair, esq. F. R. S. E. & E. &c. &c. Professor of Natural Philosophy in the University of Edinburgh Comprising a Valuable Collection of Mathematical, Philosophical and Miscellaneous Books, Maps, &c. &c. with a few Philosophical Instruments To Be Sold by Auction, Without Reserve by Mr. Ballantyne*: James Ballantyne and Co., Edinburgh, [iii] + 106 p.

Bartlett, R.A., 1962, *Great Surveys of the American West*: University of Oklahoma Press, Norman, xxiii + 408 p.

Basset, C.-A., 1815, *Explication de Playfair sur la Theorie de la Terre par Hutton et Examen Comparatif des Systèmes Géologiques Fondés sur la feu et sur l'eau, par M. Murray en réponse à l'explication de Playfair*: Bossange et Masson, Paris, Bossange et Masson et Leblanc, Londres, xxiv + [2 pages of errata] + 424 + iv + 190 p. + 2 colored plates.

Becher, J. J., 1733, *Physica Subterranea Profundam Subterraneourm Genesin, e Principiis Hucusque Ignotis...*editio novissima: Weidmanniana, Lipsiae, 504 p. + unpaginated index pages.

Becker, U., 2015, *Exegese des Alten Testaments—Ein Methoden-und Arbeitsbuch* 4., überarbeitete Auflage: Mohr Siebeck, Tübingen, X + 236 p.

Beddoes, T., 1791, Observations on the affinity between basalts and granites: *Philosophical Transactions of the Royal Society of London*, v. 81, p. 48–70.

Behlen, S., 1826, *Lehrbuch der Gebirgs- und Bodenkunde in Beziehung auf das Forstwesen*, *in* Bechstein, J.M. and Laurov, C.P., eds,, *Die Forst- und Jagdwissenschaft nach allen ihren Theilen für Angehende und Ausübende Forstmänner und Jäger*: Hennig'sche Buchhandlung, Erfurt und Gotha, part 8, v. 4, [II] + 251 p. + 1 plate.

Berger, J.F., 1811, Observations on the physical structure of Devonshire and Cornwall: *Transactions of the Geological Society*, v. S1-1, p. 93–184, https://doi.org/10.1144/transgsla.1.93.

Bertrand, A., 1839, *Lettres sur les Révolutions du Globe, cinquième edition, revue, corrigée et considérablement augmentée, enrichie de nouvelles notes par MM. Arago, Élie de Beaumont, A. Brongniart, etc.*: Just Tessier, Paris, VII + 501 p. + 3 plates.

Bertrand, M., 1884, Rapports de structure des Alpes de Glaris et du bassin houiller du Nord: *Bulletin de la Société Géologique de France*, série 3, v. 12, p. 318–330.

Berzelius, J.J., (Baron), 1840, Neptunische Ansichten von der Bildung der Urgebirge, *in* Berzelius, J., ed., *Jahres-Bericht über die Fortschritte der Physischen Wissenschaften*, im Deutschen herausgegeben von F. Wöhler, 19. Jahrgang, p. 736–744.

Beudant, F.S., 1841, *Minéralogie.—Géologie*, *in* Beudant, F.S., de Jussieu, A., and Edwards, M., eds., *Cours Élémentaire d'Histoire Naturelle A l'Usage des Collèges et des Maisons d'Education*: Langlois et Leclercq and Fortin Masson et Cie, Paris, 303 + 248 p.

Beuerlen, K., 1970, *Geologie von Brasilien*: Beiträge zur Regionalen Geologie der Erde, v. 9, Gebrüder Borntraeger, Stuttgart, VIII + 444 p. + 2 foldout plates.

von Beust, F.C., (Baron), 1840, *Kritische Beleuchtung der Werner'sche Gangtheorie aus dem Gegenwärtigen Standpunkte der Geognosie*: J.G. Engelhardt, Freiberg, 135 p.

Bischof, G., 1837, *Die Wärmelehre des Innern unseres Erdkörpers ein Inbegriff aller mit der Wärme in Beziehung Stehender Erscheinungen in und auf der Erde nach physikalischen, chemischen und geologischen Untersuchungen*: Johann Ambrosius Barth, Leipzig, XXIV + 312 p.

Bischof, G., 1847, *Lehrbuch der Chemischen und Physikalischen Geologie*, erster Band: Adolph Marcus, Bonn, XXXIV + 989 p. + 2 plates.

Bischof, G., 1851, *Lehrbuch der Chemischen und Physikalischen Geologie*, zweiten Bandes erste Abtheilung: Adolph Marcus, Bonn, XX+844 p.

Bischof, G., 1854, *Elements of Chemical and Physical Geology*, translated from the manuscript of the author by B.H. Paul and J. Drummond, v. 1: Cavendish Society, London, xxiii + 455 p.

Bischof, G., 1855, *Lehrbuch der Chemischen und Physikalischen Geologie*, zweiten Bandes zweite Abtheilung: Adolph Marcus, Bonn, XXXVI p. + p. 845–1666.

Bischof, G., 1866, *Lehrbuch der Chemischen und Physikalischen Geologie*, zweite gänzlich umgearbeitete Auflage, dritter Band: Adolph Marcus, Bonn, XVI + 974 p.

Blackadder, A., 1826, On the superficial strata of the Forth District [+ plate XI.]: *Memoirs of the Wernerian Natural History Society*, v. 5, part II, p. 424–439.

Blei, W., 1977, War der Aktualismus HUTTONS und LYELLS ahistorisch?: *Zeitschrift der Geologischen Wissenschaften*, v. 5, p. 537–541.

Blei, W., 1981, *Erkenntniswege zur Erd- und Lebensgeschichte. Ein Abriß*: Wissenschaftliche Taschenbücher. Texte und Studien. Akademie-Verlag, Berlin, 433 p.

Blumenberg, H., 1987, *Das Lachen der Thrakerin—Eine Urgeschichte der Theorie*, 4th edition: Suhrkamp Taschenbuch Wissenschaft, Berlin, 162 p.

Boase, H.S., 1834, *A Treatise on Primary Geology; being an Examination, both Practical and Theoretical, of the Older Formations*: Longman, Rees, Orme, Brown, Green, & Longman, London, xi + 399 p.

Du Bois-Aymé, J.-M., 1812, *Mémoire sur les Anciennes Branches du Nil et ses Embouchures dans la Mer*: Jean Marenigh, Livourne, 30 p. + 3 plates.

Bölsche, W., 1894, *Entwicklungsgeschichte der Natur*: Th. Knaur Nachf., Berlin and Leipzig, v. 1, [I] + 806 p. + unnumbered colored plates.

Bonnaire-Mansuy, J.-S., 1824, *Cosmogonie, ou de la Formation de la Terre et de l'Origine des Pétrifications, Nouveaux Principes de Géologie d'après lesquels l'Antiquité incontestable des matériaux dont notre monde est formé se concilie naturellement avec l'époque récente de la creation indiquée dans la Genèse*: Librairie Ecclésiastique de Rusand, Paris, Rusand, Lyon, xxiv + 236 p. + 3 p. of errata.

De Boucheporn, F., 1844, *Études sur l'Histoire de la Terre et sur les Causes des Révolutions de sa Surface*: Carilian-Goeury and Langlois et Leclercq, Paris, 394 + iv + [iv] p. + 3 foldout plates.

Boué, A., [1820], *Essai Géologique sur l'Écosse*: Mme. Ve. Courcier, Paris, 519 p. + 4 p. errata + 2 maps + 7 foldout plates.

Boué, A., 1822, On the geognosy of Germany, with observations on the igneous origin of trap: *Memoirs of the Wernerian Natural History Society*, v. 4, part I, p. 91–108.

Boué, A., 1832, *Mémoires Géologiques et Paléontologiques*, v. 1 (all published): Chez l'Auteur, F.-G. Levrault, Paris and Strasbourg and Librairie Parisienne, Bruxelles, xvj + 362 p. + 4 plates.

Brande, W.T., 1817, *Outlines of Geology; being the Substance of a Course of Lectures Delivered in the Theatre of the Royal Institution in the Year 1816*: John Murray, London, viii + 144 p. + 1 foldout colored plate.

Brande, W.T., 1829, *Outlines of Geology*: John Murray, London, xiv + 234 p. + 1 frontispiece + 1 foldout colored plate.

Breislak, S., 1811a, *Introduzione alla Geologia*: Stamperia Reale, Milano, parte prima, XXVIII + 1 page of errata + 27 + 367 + [I] p.

Breislak, S., 1811b, *Introduzione alla Geologia*: Stamperia Reale, Milano, parte II, 490 p.

Breislak, S., 1812, *Introduction à la géologie ou à l'Histoire Naturelle de la Terre*, par Scip. Breislak, traduit de l'italien par J.-J.-B. Bernard: J. Klostermann, Paris, X + 595 p.

Breislak, S., 1818a, *Institutions Géologiques* traduites du manuscrit italien en français par P.J.L. Campmas, tome premier, Imprimerie Impériale et Royale, Milan, XXIX + 1 p. of errata + 468 p.

Breislak, S., 1818b, *Institutions Géologiques* traduites du manuscrit italien en français par P.J.L. Campmas, tome second, Imprimerie Impériale et Royale, Milan, 550 p. + 1 p. of errata.

Breislak, S., 1818c, *Institutions Géologiques* traduites du manuscrit italien en français par P.J.L. Campmas, tome troisième, Imprimerie Impériale et Royale, Milan, 557 p. + 1 p. of errata.

Breislak, S., 1818d, *Institutions Géologiques* traduites du manuscrit italien en français par P.J.L. Campmas, Atlas, Imprimerie Impériale et Royale, Milan, 56 plates.

Breislak, S., 1819, *Scipio Breislak's Lehrbuch der Geologie* nach den zweiten umgearbeiteten französischen Ausgabe mit stäter Vergleichung der ersten italiänischen, übersetzt und mit Anmerkungen begleitet von F.K. von Strombeck, erster Band: Friedrich Vieweg, Braunschweig, XXII + 658 p.

Breislak, S., 1820, *Scipio Breislak's Lehrbuch der Geologie* nach den zweiten umgearbeiteten französischen Ausgabe mit stäter Vergleichung der ersten italiänischen, übersetzt und mit Anmerkungen begleitet von F.K. von Strombeck, zweiter Band: Friedrich Vieweg, Braunschweig, VIII + 703 + 1 plate.

Breislak, S., 1821, *Scipio Breislak's Lehrbuch der Geologie* nach den zweiten umgearbeiteten französischen Ausgabe mit stäter Vergleichung der ersten italiänischen, übersetzt und mit Anmerkungen begleitet von F.K. von Strombeck, dritter und letzter Band: Friedrich Vieweg, Braunschweig, XII + 762 p.

Brock, S., (introduction and translation), 1990, *Saint Ephrem Hymns on Paradise*: St Vladimir's Seminary Press, Crestwod, 240 p.

Brush, G.J., 1882, The progress of American mineralogy: The *Popular Science Monthly*, v. 21, p. 795–809.

von Buch, L., 1818–1819[1877], Ueber die Zusammensetzung der basaltischen Inseln und über Erhebungs-Kratere: *Abhandlungen der physikalischen Klasse der Akademie der Wissenschaften zu Berlin aus den Jahren 1818–1819*, p. 51–86 (reprinted in Ewald, J., Roth, J., and Dames, W., eds., *Leopold von Buch's Gesammelte Schriften*, v. III: G. Reimer, Berlin, p. 3–19).

von Buch, L., 1824, [1877], Ueber geognostische Erscheinungen im Fassathal. Ein Schreiben an den Gehemrath von Leonhard: *von Leonhard's Mineralogisches Taschenbuch für das Jahr*, v. 1824, p. 343–396 (reprinted in Ewald, J., Roth, J., and Dames, W., eds., *Leopold von Buch's Gesammelte Schriften*, v. III: G. Reimer, Berlin, p. 141–166).

von Buch, L., 1827, [1877], Ueber die Verbreitung grosser Alpengeschiebe: *Poggendorff's Annalen der Physik und Chemie*, v. 85, p. 575–588, https://doi.org/10.1002/andp.18270850407 (reprinted in Ewald, J., Roth, J., and Dames, W., eds., Leopold von Buch's Gesammelte Schriften, v. III: G. Reimer, Berlin, p. 659–668).

de Buffon, [G.-L.L.], (Comte), 1749, *Histoire Naturelle Générale et Particulière, avec la Description du Cabinet du Roi*, t. 1: Imprimerie Royale, Paris, [iii] + 612 p. + 2 plates.

de Buffon, [G.-L.L.], (Comte), 1778, *Histoire Naturelle Générale et Particulière, Supplement*: L'Imprimerie Royale, Paris, v. 5, p. viii + 615 + xxviii p.

de Buffon, [G.-L.L.], (Comte), 2018, *The Epochs of Nature*, translated and edited by Jan Zalasiewicz, Anne-Sophie Milon, and Mateusz Zalasiewicz: The University of Chicago Press, Chicago and London, xxxiv + 190 p.

Bunsen, R., 1851, Ueber die Processe der vulkanischen Gesteinsbildungen Islands: *Annalen der Physik und Chemie*, v. 159, p. 197–272, https://doi.org/10.1002/andp.18511590602.

Burke, K., 1977, Aulacogens and continental breakup: *Annual Review of Earth and Planetary Sciences*, v. 5, p. 371–396.

Burke, K., Dewey, J.F., and Kidd, W.S.F., 1976, Dominance of horizontal movements, arc and microcontinental collisions during the later permobile regime, *in* Windley, B.F., ed., *The Early History of the Earth*: John Wiley & Sons, London and New York, p. 113–129.

Bury, J.B., [1910], Introduction, *in* Gibbon, E., ed., *The History of the Decline and Fall of the Roman Empire*, edited in seven volumes ... by Bury, J.B.: Methuen & Co., London, v. 1, lxviii + 464 p. + 1 foldout map.

de Bury, R., 1903, *The Love of Books The Philobiblon of Richard de Bury newly translated into English by E.C. Thomas*: The King's Classics under the general editorship of Israel Gollancz, and Alexander Moring: The de la More Press, London, XXI + 148 p. + frontispiece.

Cailleux, A., 1979, The Geological Map of North America (1752) of J.-E. Guettard, *in* Schneer, C.J., ed., *Two Hundred Years of Geology in America—Proceedings of the New Hampshire Bicentennial Conference on the History of Geology*: University Press of New England, Hanover, p. 43–52.

Carey, S.W., 1988, *Theories of the Earth and Universe: A History of Dogma in the Earth Sciences:* Stanford University Press, Stanford, xviii + 413 p.

Carozzi, A.V., 1962, Editor's introduction, *in* Werner, A.G., ed., *On the External Characters of Minerals*, translated by Albert V. Carozzi: University of Illinois Press, Urbana, p. ix–xvi.

Carozzi, A.V., 1965, Lavoisier's fundamental contribution to stratigraphy: *The Ohio Journal of Science*, v. 65, p. 71–85.

Cavendish, H., 1798, Experiments to determine the density of the earth: *Philosophical Transactions of the Royal Society of London*, part II, v. 88, p. 469–526.

Chang, K.-M., (Kevin), 2007, From vitalistic cosmos to materialistic world—The lineage of Johann Joachim Becher and Georg Ernst Stahl and the shift of early modern chymical cosmology, *in* Principe, L.M., ed., *Chymists and Chymistry—Studies in the History of Alchemy and Early Modern Chemistry*: Chemical Heritage foundation and Science History Publications, USA, a division of Watson Publishing International LLC, Sagamore Beach, p. 215–225.

de Charpentier, J., 1841, *Essai sur les Glaciers et sur le Terrain Erratique du Bassin du Rhône:* Marc Ducloux, Lausanne, X + 363 p. + 1 foldout map + 8 plates.

Chenevix, R., 1808, Réflexions sur quelques méthodes minéralogiques: *Annales de Chimie, ou Recueil de Mémoires Concernant la Chimie et les Arts qui en Dependent et Spécifement la Pharmacie*, v. 65, p. 5–43, 113–160, 225–276.

de Chesnel, A., 1849, *Dictionnaire de Géologie suivi d'Esquisses Géologiques et Géographiques*, *in* Migne, J.-P. (Abbé), ed., *Encyclopédie Théologique, ou Série des Dictionnaires sur Toutes les Parties de la Science Religieuse, offrant en Français la plus claire, la plus facile, la plus commode, la plus variée et la plus complète des théologies*, v. cinquantième: chez l'Éditeur, aux Ateliers Catholiques du Petit-Montrouge, Paris, 736 columns.

Chevalier, C., abbot, 1868, *Géologie Contemporaine—Histoire des Phénomènes Actuels du Globe Appliquée à l'Intirprétation des Phénomènes Anciens*: Alfred mame et Fils, Tours, 383 p. + 1 frontispiece.

Chladni, E.F.F., 1794, *Über den Ursprung der von Pallas Gefundenen und Anderer ihr Ähnlicher Eisenmassen und über Einige Damit in Verbindung Stehende Naturerscheinungen*: Johann Friedrich Hartknoch, Riga, II + 63 p.

Chorley, R.J., Dunn, A.J., and Beckinsale, R.P., 1964, *The History of Landforms or The Development of Geomorphology*, volume one: *Geomorphology Before Davis*: Methuen & Co. and John Wiley & Sons, London, xvi + 678 p.

Ciancio, L., 2008, MARZARI PENCATI, Giuseppe: in *Dizionario Biografico degli Italiani*: Istituto della Enciclopedia Italiana, Roma, v. 71, http://www.treccani.it/enciclopedia/giuseppe-marzari-pencati_(Dizionario-Biografico)/.

Clarke, J.M., 1921, *James Hall of Albany—Geologist and Palaeontologist*: (publisher not indicated), Albany, 565 p.

Clarke, M.L., 1945, *Greek Studies in England 1700–1830*: University Press, Cambridge, 255 p.

Cleaveland, P., 1816, *An Elementary Treatise on Mineralogy and Geology, being an Introduction to the Study of these Sciences, and designed for the use of pupils—for persons attending lectures on these subjects—and as a companion for travelers in the United States of America*: Cummings and Hilliard, Boston, xii + 668 p. + 4 plates + 1 colored foldout map.

Cleaveland, P., 1822, *An Elementary Treatise on Mineralogy and Geology, being an Introduction to the Study of these Sciences, and designed for the use of pupils—for persons attending lectures on these subjects—and as a companion for travelers in the United States of America*, second edition: Cummings and Hilliard, Boston, v. 2, x + 818 p. + 1 errata page + 6 plates (including the foldout geological map).

Clifford, R.J., (S.J.), 1972, *The Cosmic Mountain in Canaan and the Old Testament*: Harvard University Press, Cambridge, [vi] + 221 p.

Coleman, W., 1964, *Georges Cuvier, Zoologist—A Study in the History of Evolution Theory:* Harvard University Press, Cambridge, 212 p, https://doi.org/10.4159/harvard.9780674283701.

Collegno, G., 1847, *Elementi di Geologia Pratica e Teorica Destinati Pricipalmente ad Agevolare lo Studio del Suolo dell'Italia*: G. Pomba e Comp., Torino, XVI + 446 p. + 1 p. of errata.

Comstock, J.L., 1834, *Outlines of Geology: intended as a Popular Treatise on the Most Interesting Parts of the Science together with an examination of the question, whether the days of creation were indefinite periods*: D.F. Robinson & Co., Hartford, xii + 336 p.

Comstock, J.L., 1839, *Outlines of Geology: intended as a Popular Treatise on the Most Interesting Parts of the Science together with an examination of the question, whether the days of creation were indefinite periods*, third edition: Robinson, Pratt & Co., New York, 384 p.

Comte, P., 1949, Aperçu sur l'œuvre géologique de Lavoisier: Annales de la Société Géologique du Nord, v. 49, p. 369–375.

Conybeare, W.D., 1835, Report on the progress, actual state, and ulterior prospects of geological science: *Report of the First and Second Meetings of the British Association for the Advancement of Science at York in 1831 and at Oxford in 1832*, second edition: John Murray, London, p. 365–414 + 1 foldout plate.

Conybeare, W. D. and Phillips, W., 1822, *Outlines of the Geology of England and Wales with an Introductory Compendium of the General Principles of that Science and Comparative Views of the Structure of Foreign Countries*, Part I: William Phillips, London, lxi + 470 p. + 3 foldout plates.

Cordier, P., 1827, Essai sur la température de l'intérieur de la terre: *Mémoires de l'Academie des Sciences* pour l'année 1827, v. 7, p. 473–556.

Cordier, L., 1828, *An Essay on the Temperature of the Interior of the Earth*, translated from the French by the junior class in Amherst College, John S. & Charles Adams, Amherst, 94 p.

Cornford, F.M., 1932, *Before and After Socrates*: At the University Press, Cambridge, X + 113 p.

Cornford, F.M., 1950, *The Unwritten Philosophy and Other Essays*, edited with an introductory memoir by Guthrie, W.K.C.: At the University Press, Cambridge, xx + 139 p.

Cotta, B., 1842, *Anleitung zum Studium der Geognosie und Geologie, besonders für Deutsche Forstwirthe, Landwirthe und Techniker*: Arnoldischen Buchhandlung, Dresden and Leipzig, XX + 584 p. + 1 plate + 2 foldout tables.

Cotta, B., 1846, *Grundriß der Geognosie und Geologie*: Arnoldischen Buchhandlung, Dresden and Leipzig, XII + 428 p.

Cotta, B., 1848, *Briefe über Alexander von Humboldt's Kosmos—Ein Commentar zu diesem Werke für gebildete Laien*: T.O. Weigel, Leipzig, X +356 p. + 1 p. errata.

von Cotta, B., 1866, *Geologie der Gegenwart—Dargestellt und Beleuchtet*: J.J. Weber, Leipzig, XLVI + 424 p.

Craig, G.Y., ed., 1978, *James Hutton's Theory of the Earth: The Lost Drawings*: Scottish Academic Press in association with the Royal Society of Edinburgh and the Geological Society of London, Edinburgh, [iii] + 67 p. + portfolio of 27 facsimiles.

Critchley, E., 2010, *Dinosaur Doctor—The life and Work of Gideon Mantell*: Amberley Publishing, Stroud, 256 p.

Crook, T., 1933, *History of the Theory of Ore Deposits with a Chapter on the Rise of Petrology*: Thomas Murby & Co., London; D. van Nostrand Co. Inc., New York, 163 p.

Crosse, C.A.H., editor, 1857, *Memorials, Scientific and Literary of Andrew Crosse, The Electrician*: Longman, Brown, Green, Longmans & Roberts, London, ix + [i] + 360 p.

Cunningham, R.J.H., 1838, Geology of the Lothians [+ XVI plates.]: *Memoirs of the Wernerian Natural History Society*, v. 7, p. 3–160.

Curwen, E.C., 1940, *The Journal of Gideon Mantell Surgeon and Geologist—Covering the Years 1818–1852*: Oxford University Press, London, xii + 315 + frontispiece.

Cuvier, G., 1812, *Recherches sur les Ossemens Fossiles de Quadrupeds où l'on rétablit les caractères de plusieurs espèces d'animaux que les révolutions du globe paroissent avoir détruites*, v. 1: Deterville, Paris, not consecutively paginated.

Cuvier, [G. (Baron)], 1813, *Essay on the Theory of the Earth translated...by Robert Kerr with Mineralogical Notes and an Account of Cuvier's Geological Discoveries by Professor Jameson*: William Blackwood, Edinburgh; John Murray and Robert Baldwin, London, xiii + 265 p. + 2 plates.

Cuvier, G., 1825, *Discours sur les Révolutions de la Surface du Globe et sur les Changements Qu'Elles Ont Produits Dans le Régne Animal*: G. Dufour et Ed. D'Ocagne, Paris, ij + 400 p. + 6 plates.

Dana, E.S., Schuchert, C., Gregory, H.E., Barrell, J., Smith, G.O., Lull, R.S., Pirsson, L.V., Ford, W.E., Sosman, R.B., Wells, H.W., Foote, H.W., Page, L., Coe, W.R., and Goodale, G.L., 1918, *A Century of Science in America with special reference to the American Journal of Science, 1818–1918*: Yale University Press, New Haven, 458 p., https://doi.org/10.5962/bhl.title.29430.

Dana, J.D., 1837, *A System of Mineralogy: including an extended treatise on crystallography: with an appendix, containing the application of mathematics to crystallographic investigation, and a mineralogical bibliography*: Durrie & Peck and Herrick & Noyes, New Haven, xiv + 452 p. + 4 plates.

Dana, J., 1848, *Manual of Mineralogy, Including Observations on Mines, Rocks, Reduction of Ores, and the Applications of the Science to the Arts*: Durrie & Peck New Haven; xii + 13–430 p.

Dana, J.D., 1863, *Manual of Geology—Treating of the Principles of the Science with special reference to American Geological History, for the use of colleges, academies and schools of science*: Theodore Bliss & Co., Philadelphia, xvi + 798 p.

Dana, J. D., 1873, On the origin of mountains: *American Journal of Science and Arts*, third series, v. 5, p. 347–350.

Daniels, G.H., 1994, *American Science in the Age of Jackson*: History of American Science and Technology Series, The University of Alabama Press, Tuscaloosa and London, xx + 282 p.

Darwin, C.R., 1887, Autobiography, *in* Darwin, F., ed., *The Life and Letters of Charles Darwin including an Autobiographical Chapter*, John Murray, London, v. I, p. 26–107, https://doi.org/10.5962/bhl.title.50683.

Dashkowa (Fürstin), 1970, *Erinnerungen—Katharina die Große und ihre Zeit*: Winkler, München, 281 + [I] p.

Daubrée, [G.-A.], 1860, *Études et Expériences Synthétiques sur la Métamorphisme et sur la Formation des Roches Cristalliens*: in v. 17, Mémoires de l'Académie des Sciences, Paris, Imprimerie Impériale, Paris, VII + 127 p.

Daubrée, [G.-]A., 1867, *Rapport sur les Progrès de la Géologie Expérimentale*: Recueil de Rapports sur les Progrès des Lettres et des Sciences en France, Imprimerie Impériale, Paris, 142 p.

Daubrée, [G.-]A., 1879, *Études Synthétiques de Géologie Expérimentale*: Dunod, Paris, III + 828 p. + 7 plates.

Davies, G.L., 1967, George Hoggart Toulmin and the Huttonian theory of the earth: *Geological Society of America Bulletin*, v. 78, p. 121–124, https://doi.org/10.1130/0016-7606(1967)78[121:GHTATH]2.0.CO;2.

Davies, G.L., 1968, Another forgotten pioneer of the glacial theory James Hutton (1726–97): *Journal of Glaciology*, v. 7, p. 115–116, https://doi.org/10.1017/S0022143000020451.

Davies, G.L., [Herries], 1969, *The Earth in Decay—A History of British Geomorphology*: Macdonald Technical and Scientific, London, xvi + 390 p.

Davies, G.L. Herries, 1985, James Hutton and the study of landforms: *Progress in Physical Geography*, v. 9, p. 382–389.

Dean, D.R., 1992, *James Hutton and the History of Geology*: Cornell University Press, Ithaca, xiii + [iii] + 303 p.

Dean, D.R., 1997, *James Hutton in the Field and in the Study: Being an Augmented Reprinting of Vol. III of Hutton's Theory of the Earth as first published by Sir Archibald Geikie (1899)—A Bicentenary Tribute to the Father of Modern Geology*: Scholars' Facsimiles & Reprints, Delmar, New York, irregularly paginated.

Dean, D.R., 1998, Hutton and North America: *Geological Society of America Abstracts with Programs*, v. 30, no. 7, p. A-101.

Delamétherie, J.-C., *An III* (1795 *vieux style*), *Théorie de la Terre*, tome troisième: Maradan, Paris, viij + 471 p. + plates V–VII.

Delamétherie, J.-C., *An V* (1797), *Théorie de la Terre*, seconde edition, corrige, et augmentée d'une Minéralogie, tome cinquième: Maradan, Paris, ix + 1 page of errata + 535 p. + plates VI–VIII.

Delamétherie, J.-C., 1816, *Leçons de Géologie données au Collège de France*, tome troisième: V. Courcier, Paris, 364 p.

De Luc, J.A., 1778, *Lettres Physiques et Morales, sur les Montagnes et sur l'Histoire de la Terre et de l'Homme*: Libraires Associés, En Suisse, XXIV + 224 p.

Deluc, J.-A., 1809a, *An Elementary Treatise on Geology: Determining Fundamental Points in that Science, and Containing some Modern Geological Systems and Particularly of the Huttonian Theory of the Earth*. Translated from the French Manuscript By Rev. Henry De La Fite: F.C. and J. Rivington, London, xxvii + 415 p. + 1 errata page.

Deluc, J.-A., 1809b, *Traité Élémentaire de Géologie*: Courcier, Paris, 395 p.

Deluc, J.-A., 1810, *Geological Travels*, v. I, *Travels in the North of Europe, containing Observations on some parts of the Coasts of the Baltic, and the North Sea*. Translated from the French Manuscript: F. C. and J. Rivington, London, 407 p. + unpaginated Index + 1 foldout map + 1 plate with explanatory text.

Deluc, J.-A., jr., 1816, De la matière première des laves: *Journal de Physique, de Chimie et d'Histoire Naturelle*, v. 82, p. 465–471.

Desmarest, [N.], 1757, Géographie physique, *in* Diderot, [D.] and d'Alembert, [J.-B. le Rond], eds., *Encyclopédie, ou Dictionnaire Raisonné des Sciences, des Arts et des Métiers*, v. 7: Briasson, David, Le Brenton, Durand, Paris, p. 613–626.

Desmarest, [N.], 1779, Extrait d'un mémoire sur la determination de quelques époques de la nature par les produits des volcans, & sur l'usage de ces époques dans l'étude des volcans: *Observations sur la Physique, sur l'Histoire Naturelle et sur les Arts*, v. 13, p. 115–126. (Published separately in the same year as *Précis d'un Mémoire sur la détermination de quelques époques de la Nature par les produits des Volcans, & sur l'usage de ces époques dans l'étude des Volcans*: Clousier, Paris, 24 p.)

Desmarest, [N.], *An III* (1794–1795),Géographie-Physique, v. I: in *Encyclopédie Méthodique*, H. Agasse, Paris, 857 + [I] p.

Desmarest, [N.], 1806, Mémoire sur la determination de trois époques de la nature par les produits des volcans, et sur l'usage qu'on veut faire de ces époques dans l'étude des volcans: *Mémoires de l'Institut des Sciences, lettres et Arts de Paris, Sciences Mathematiques et de Physiques*, v. 4, p. 219–289 + plates 6–9.

Desmarest, [N.], 1828, Géographie-Physique, v. 5, continued by Bory de Saint-Vincent, [J.-B.-G.-M.], Doin, G.-T. and Huot, [J.-J.-N.]: in *Encyclopédie Méthodique*, veuve Agasse, Paris, viij + 958 p. + 2 foldout tables.

Deutsch, D., 2011, *The Beginning of Infinity—Explanations that Transform the World*: Penguin, New York, [iv] + 487 p.

Dixon, T., 2008, *Science and Religion—A Very Short Introduction*: Oxford University Press, Oxford, xvi + 150 p.

Doskey, J.S., ed., 1988, *The European Journals of William Maclure*: American Philosophical Society, Philadelphia, xlx + 815 p.

Dott, R.H., Jr., 1969, James Hutton and the concept of a dynamic earth, *in* Schneer, C.J., ed., *Toward A History of Geology*: The M.I.T. Press, Cambridge, p. 122–141.

Drummond, E.E., 1826, Notice regarding fossil bones of a whale discovered in the district of Monteith: *Memoirs of the Wernerian Natural History Society*, v. 5, part II, p. 440–441.

Duhem, P., 1906[1984], *Études sur Léonard de Vinci—Ceux qu'il a lus et Ceux qui l'ont Lu, première série*: A. Hermann et Fils, Paris; reprinted by Éditions des Archives Contemporaines, (no place of publication), VII + 355 p.

Duhem, P., 1958, *Le Système du Monde—Histoire des Doctrines Cosmologiques de Platon a Copernic*, v. 9 *(Cinquième Partie: La Physique Parisienne au XIVe Siècle (suit))*: Hermann, Paris, 442 p.

Eaton, A., 1818, *An Index to the Geology of the Northern States, with a Traverse Section from Catskill Mountain to the Atlantic*: Hori Brown, Leicester, 52 p. + 1 section.

Eaton, A., 1820, *An Index to the Geology of the Northern States, with Traverse Sections extending from Susquehanna River to the Atlantic, crossing Catskill Mountains to which prefixed A Geological Grammar*, second edition: Wm. S. Parker, Troy, xi + 286 p. + 2 plates.

Eaton, A., 1830, *Geological Text-Book, prepared for Popular Lectures on North American Geology; with application to Agriculture and the Arts*: Websters and Skinners, Albany, 68 + [i] p.

Eberhard, J.A., 1788, *Allgemeine Geschichte der Philosophie zum Gebrauch Academischer Vorlesungen*: Hemmerdesche Buchhandlung, Halle, [X] + 308 + [IX] p.

Einstein, A., 1914, Antrittsrede: *Sitzungsberichte der Königlich Preußischen Akademie der Wissenschaften*, Jahrgang 1914, zweiter Halbband, p. 739–742.

El-Sabh, M.I., and Murty, T.S., 1989, Storm surges in the Arabian Gulf: *Natural Hazards*, v. 1, p. 371–385, https://doi.org/10.1007/BF00134834.

Eliade, M., 1961, *Myths, Dreams, and Mysteries—The Encounter Between Contemporary Faiths and Archaic Realities*: The Library of Religion and Culture, Harper, New York, 256 p.

Eliade, M., 1965(1974), *The Myth of the Eternal return or, Cosmos and History*, translated from the French by Willard R. Trask: Bollingen Series 46, Princeton University Press, Princeton, xv + 195 p.

Élie de Beaumont, L., 1829–1830, Recherches sur quelques-unes des Révolutions de la surface du globe, présentant différens exemples de coïncidence entre le redressement des couches de certains systèmes de montagnes, et les changements soudains qui ont produit les lignes de démarcation qu'on observe entre certains étages consécutifs des terrains de sédiment: *Annales des Sciences Naturelles*, v. 18, p. 5–25, 284–417, v. 19, p. 5–99, 177–240.

Élie de Beaumont, L., 1830, Recherches sur quelques-unes des Révolutions de la surface du globe, présentant différens exemples de coïncidence entre le redressement des couches de certains systèmes de montagnes, et les changemens soudains qui ont produit les lignes de démarcation qu'on observe entre certains étages consécutifs des terrains de sédiment: *Revista Fronteiras*, no. 15, p. 1–58.

Élie de Beaumont, L., 1831, Researches on some of the Revolutions which have taken place on the Surface of the Globe; presenting various Examples of the coincidence between the Elevation of Beds in certain Systems of Mountains, and the sudden Changes which have produced the Lines of Demarcation observable in certain Stages of the Sedimentary Deposits [New Series]: *The Philosophical Magazine and Annals of Philosophy*, v. 10, p. 241–264, https://doi.org/10.1080/14786443108675531.

Élie de Beaumont, L., 1832, Fragmens géologiques tirés de Stenon, de Kazwini, de Strabon et du Boun Dehesch: *Annales des Sciences Naturelles*, v. 25, p. 337–395.

Élie de Beaumont, L., 1833, Recherches sur quelques-unes des Révolutions de la surface du globe, présentant différens exemples de coïncidence entre le redressement des couches de certains systèmes de montagnes, et les changemens soudains qui ont produit les lignes de démarcation qu'on observe entre certains étages consécutifs des terrains de sédiment: in *Manuel Géologique par Henry T. De La Beche*, seconde édition, traduction française revue et publiée par A.J.M. Brochant de Villiers: F.-G. Levrault, Paris, ss. 616–665.

Élie de Beaumont, L., 1845, *Leçons de Géologie Pratique professées au Collège de France pendant l'année scolaire 1843–1844*, tome premier: P. Bertrand, Paris, 555 p.

Ellenberger, F., 1972, La métaphysique de James Hutton (1726–1797) et le drame écologique du XXᵉ siècle: *Revue de Sythèse*, 3rd series, nos. 67–68, p. 267–283.

Emmerling, L.A., 1799–1802, *Lehrbuch der Mineralogie*, 2. Ganz umgearbeitete, durchaus vermehrte und verbesserte Auflage:, Heyer, Gießen, v. 1/1, XLVI + 499 p., v. 1/2, XXXII + 928 p.

Emmons, E., 1826, *Manual of Mineralogy and Geology designed for the use of Schools; and for Persons Attending Lectures on these Subjects as also a Convenient Pocket Companion for Travellers in the United States of America*: Websters and Skinners, Albany, xxiii + 229 + 1 page of Addenda.

Emmons, E., 1832, *Manual of Mineralogy and Geology*, second edition: Webster and Skinner, Albany, xii + 299 p.

Enfield, W., 1792, *The History of Philosophy from the Earliest Times to the Beginning of the Present Century; drawn up from Brucker's Historia Critica Philosophiæ*: P. Wogan, P. Byrne, A. Grueber, W. M'Kenzie, J. Moore, J. Jones, R. M'Allister, W. Jones, J. Rice, R. White and G. Draper, Dublin, v. I, xxvii + 470 p.

von Engelhardt, W. (Baron), 1986/1987, Lichtenbergs Gedanken zur Entstehung und Bildung unserer Erde zu ihrer gegenwärtigen Gestalt: *Photorin*, v. 11/12, p. 19–35.

Euler, L., 1749, 1751, Recherches sur la précession des equinoxes et sur la nutation de l'axe de la terre: *Mémoires de l'Académie des Sciences de Berlin*, v. 5, p. 289–325.

Eyles, V.A., 1966, The history of geology: suggestions for further research: *History of Science*, v. 5, p. 77–86, https://doi.org/10.1177/007327536600500104.

Eyles, V.A., 1970, Introduction, *in* White, G.W., ed., *James Hutton's System of the Earth, 1875; Theory of the Earth, 1788; Observations on Granite, 1794 together with Playfair's Biography of Hutton*, Contributions to the History of Geology, v. 5, Hafner, New York, p. xi–xxiii.

Fantuzzi, G., 1774, *Memorie della Vita di Ulisse Aldrovandi Medico e Filosofo Bolognese con alcune Lettere d'Uomini eruditi a lui scritte, e coll'Indice delle sue Opere Mss., che si conservano nella Biblioteca dell'Instituto*: Lelio dalla Volpe, Bologna, vi + 263 p. + 2 unnumbered plates.

Faujas de Saint-Fond, B., 1778, *Recherches sur les Volcans Éteints du Vivarais et du Velay; avec un Discours sur les Volcans brûlans. Des Mémoires analytiques sur les Schorls, la Zéolites, le Basalte, la Pouzzolane, les Laves & les différentes Substances qui s'y trouvent engageés, &c.*: Joseph Couchet, Grenoble, Nyon et Née et Masquelier, Paris, vxiij + [ii] + 460 p. + 4 p. of subscribers + numerous unnumbered plates.

Faujas de Saint-Fond, B., 1797, *Voyage en Angleterre, en Écosse et aux Îles Hébrides; ayant pour objet Les Sciences, les Arts, l'Histoire naturelle et les Mœrs; avec La Description du pays de Newcastle, des montagnes de Derbyshire, des environs d'Édinburgh, de Glasgow, de Perth, de St. Andrews, du duché d'Inveray et de grotte de Fingal*: H.J. Jansen, Paris, tome premier, 430 p. + 1 p. of errata + 3 plates; tome deuxième, 434 p. + 1 p. of errata + 4 foldout plates.

Faujas Saint-Fond, B., 1799, *Travels in England, Scotland and the Hebrides; undertaken for the purpose of examining the state of the Arts, the Sciences, natural History and Manners in Great Britain: containing Mineralogical Descriptions of the Country round Newcastle; of the Mountains of Derbyshire; of the Environs of Edinburgh, Glasgow, Perth, and St. Andrews; of Inveray, and other Parts of Argyleshire; and of the Cave of Fingal*: James Ridgeway, London, v. 1, iii + 361 p. + 3 plates; v. 2, [1 p. of errata] + viii + 352 p. + 4 plates.

Faujas-Saint-Fond, B., 1803, *Essai de Géologie, ou Mémoires pour servir à l'Histoire Naturelle du Globe*, tome premier: C.F. Patris, 493 p. + 17 plates.

Faujas-Saint-Fond, B., 1809a, *Essai de Géologie, ou Mémoires pour servir à l'Histoire Naturelle du Globe*, tome second, première partie, Mineraux, 400 p. + [ii] + 21 plates.

Faujas Saint-Fond, B., 1809b, *Essai de Géologie, ou Mémoires pour server à l'Histoire Naturelle du Globe*, tome second, deuxième partie, Volcans, p. 401–731 p. + plates 22–29.

Faust, B.C., 1789, Nachricht von dem auf dem Meißner in Hessen über Steinkohlen und bituminösen Holze liegenden Basalte—aus dem Augustmonate des 1784er Jahrgangs des Journals von und für Deutschland entlehnt, und mit einer Vorbemerkung, wie auch einigen erläuternden Anmerkungen versehen, von A.G. Werner: *Bergmännisches Journal*, v. 1, p. 261–295.

Fauth, P., ed., 1913, *Hörbigers Glacial-Kosmogonie—Eine Neue Entwickelungsgeschichte des Weltalls und des Sonnensystems auf Grund der Erkenntnis des Widerstreits eines Kosmischen Neptunismus mit einem Ebenso Universellen Plutonismus nach den Neuesten Ergebnissen sämtlicher Exakter Forschungszweige*: Hermann Kayser, Kaiserslautern, XXVII + 772 p.

Ferguson, [A.], 1805, Minutes of the life and character of Joseph Black, M.D. Addressed to the Royal Society of Edinburgh: *Transactions of the Royal Society of Edinburgh*, v. 5, p. 102–117.

Feyerabend, P., 1988, *Against Method*, revised edition: Verso, London, viii + 296 p.

Fitton, W.H., 1839, A review of Mr. Lyell's "Elements of Gelogy;" with Observations on the Progress of the Huttonian Theory of the Earth: *Edinburgh Review*, no. 140, v. 69, p. 406–466.

Forbes, I., 2015, Wesgarth Forster's Strata—a reappraisal of a geological pioneer: *Proceedings of the Yorkshire Geological Society*, v. 60, p. 157–177, https://doi.org/10.1144/pygs2015-333.

Forster, W., 1809, *A Treatise on a Section of the Strata commencing near Newcastle upon Tyne and concluding on the west side of the Mountain of Cross-Fell with remarks on Mineral Veins in General, and Engraved Figures of Some of the different species of these Productions. To which are added Tables of the Strata in Yorkshire and Derbyshire. The whole intended to amuse the Mineralogist and assist the Miner in his Professional Researches*: Preston and Heaton, Newcastle, viii + 156 p.

Forster, W., 1821, *A Treatise on a Section of the Strata, from Newcastle-upon-Tyne to the Mountain of Cross Fell, in Cumberland; with Remarks on Mineral Veins in General.—also Tables of the Strata, in Yorkshire, Derbyshire, &c.—To which is added, A Treatise on the Discovery, the Opening, and the Working of Lead Mines; with the Dressing and Smelting of Lead Ores, second edition, greatly enlarged*: privately printed, Alston and Newcastle-upon-Tyne, xii + x + 422 + xvi + 7 pages of index and glossary + 11 pages of subscribers + 1 page of errata.

Forster, W., 1883, *A Treatise on a Section of the Strata, from Newcastle-upon-Tyne to Cross Fell, with Remarks on Mineral Veins; also Tables of the Strata, in Yorkshire, Derbyshire, West Cumberland, and Fifeshire; to which is added, A Treatise on the Discovery, the Opening, and the Working of Lead Mines and the Dressing and Smelting of Lead Ores, third edition, revised and corrected to the present time by the Rev. W. Nall, with a memoir of the author's life*: Andrew Reid, Newcastle-upon-Tyne, Edward Stanford, London, lvi + 208 p. + 14 foldout plates.

Fortenbaugh, W.W., Huby, P.M., Sharples, R.W., and Gutas, D., eds., 1993a, *Theophrastus of Eresus—Sources for His Life, Writings, Thought & Influence, Part One Life, Writings, Various Reports, Logic, Physics Metaphysics, Theology, Mathematics*: E.J. Brill, Leiden, viii + 465 p.

Fortenbaugh, W.W., Huby, P.M., Sharples, R.W., and Gutas, D., eds., 1993b, *Theophrastus of Eresus—Sources for His Life, Writings, Thought & Influence, Part Two Psychology, Human Physiology, Living Creatures, Botany, Ethics, Religion, Politics, Rhetoric and Poetics, Music, Miscellanea*: E.J. Brill, Leiden, vii + 705 p.

Fortis, A., l'An X (1802), *Mémoirs Pour Servir à l'Histoire Naturelle et Principalement à l'Oryctographie de l'Italie, et des Pays Adjacens*, tome second: J.J. Fuchs, Paris, 362 + 5 p. + 1 page of errata + 5 foldout plates.

Fox, R.W., 1830, On the Electro-Magnetic Properties of Metalliferous Veins in the Mines of Cornwall: *Philosophical Transactions of the Royal Society of London*, v. 120, p. 399–414, https://doi.org/10.1098/rstl.1830.0027.

Francis, E.H., 1991, Carboniferous-Permian igneous rocks, *in* Craig, G.Y., ed., *Geology of Scotland*, 3rd edition: The Geological Society, London, p. 393–420.

Frängsmyr, T., 1994, Linnaeus as a geologist, *in* Frängsmyr, T., ed., *Linnaeus—The Man and His Work*: Science History Publications/USA, Canton, p. 110–155.

French, W.J., 2005, The granite controversy ends: an introduction to J.D. Clemens paper "Granites and granitic magmas" and accompanying invited discussion: *Proceedings of the Geologists' Association*, v. 116, p. 5–7.

Fritscher, B., 1987, Johann Nepomuk Fuchs' "Theory of the Earth": *Sudhoff's Archive*, v. 71, no. 2, p. 141–156.

von Fuchs, J.N., 1844, *Ueber die Theorien der Erde, den Amorphismus Fester Körper und den Gegenseitigen Einfluß der Chemie und Mineralogie—Nebst einer kurzen Inhaltsanzeige aller übrigen Schriften des Verfassers, zu dessen 70ster Geburtsfeier von einigen seiner Freunde herausgegeben*: C.A. Fleischmann, München, VIII + 88 p.

Füchsel, G. C. 1761. Historia terrae et maris, ex historia Thuringiae, per montium descriptionem eruta: *Acta Academiae Electoralis Moguntinae Scientiarum Utilium quae Erfurti*, v. 2, p. 209–254.

Fyfe, W.S., 1988, Granites and a wet convecting ultramafic planet: *Transactions of The Royal Society of Edinburgh*: Earth and Environmental Science, v. 79, p. 339–346.

Gamauf, G., 1818, *Erinnerungen aus Lichtenbergs Vorlesungen über die Physikalische Geographie. Nebst einem Anhang über Barometrische Höhenmessen*: Geistinger, Wien und Triest, 582 p. + 1 foldout plate.

Gamauf, G., 2008, *Georg Christoph Lichtenberg Vorlesungen zur Naturlehre—Gottlieb Gamauf: «Erinnerungen aus Lichtenbergs Vorlesungen». Die Nachschrift eines Hörers*: Krayer, A. and Lieb, K.-P., eds., Wallstein, Göttingen, LXXXII + 1048 p. + non-consecutively numbered 14 foldout plates.

Gantner, J., 1958, *Leonardos Visionen—Von der Sintflut und vom Untergang der Welt*: Francke, Bern, 264 p.

Gaudant, J., 2013, Jean-François d'Aubuisson de Voisins (1769–1841) et le basalte: de la Saxe à l'Auvergne: *Travaux du Comité Français d'Histoire de la Géologie (COFRHIGÉO)*, troisième série, v: XXVII, no. 8, p. 221–232.

Gehler, I.C., 1757, *De Characteribus Fossilium Externis*: Ex Officina Langenhemiana, Lipsiae, 36 p. + 1 plate.

Geikie, A., (Sir), 1858, *The Story of a Boulder or Gleanings from the Note-Book of a Field Geologist*: Thomas Constable and Co., Edinburgh, xvi + 263 p.

Geikie, A., (Sir), 1871a, *The Scottish School of Geology*: Edmonston and Douglas, Edinburgh, 27 p.

Geikie, A., (Sir), 1871b, The Scottish school of geology: *Nature*, v. 5, p. 37–38 and 52–55.

Geikie, A. (Sir), 1874, Earth sculpture and the Huttonian school of geology: *Transactions of the Edinburgh Geological Society*, v. 2, p. 247–267, https://doi.org/10.1144/transed.2.3.247.

Geikie, A., (Sir) 1875, *Life of Sir Roderick I. Murchison...based on his journals and letters with notices of his scientific contemporaries and a sketch of the rise and growth of Palæozoic geology in Britain*: John Murray, London, v. I, xii + [i] + 387 p. + 11 plates; v. 2, vi + [i] + 375 p. + 11 plates.

Geikie, A., (Sir), 1882a, *Geological Sketches at Home and Abroad*: Macmillan and Co., London, x + 382 p.

Geikie, A., (Sir), 1882b, *Text-Book of Geology*: Macmillan and Co., London, xi + 971 p. + 1 foldout plate.

Geikie, A., (Sir), 1888, Hutton, James (1726–1797): in *The Encyclopædia Britannica—A Dictionary of Arts, Sciences and General Literature*, ninth edition, v. 12, p. 414–415.

Geikie, A. (Sir), 1892, Inaugural Address by Sir Archibald Geikie, LL.D., D.Sc., For. Sec. R.S., F.R.S.E., F.G.S., Director-General of the Geological Survey of the United Kingdom, President: *Nature*, v. 46, p. 317–323.

Geikie, A., (Sir), 1893a, Address: in *Report of the Sixty-Second Meeting of the British Association for the Advancement of Science held at Edinburgh in August 1892*, John Murray, London, p. 3–26.

Geikie, A., (Sir), 1893b, *Text-Book of Geology*, third edition, revised and enlarged: Macmillan and Co., London, xvi + 1147 p. + 1 foldout plate.

Geikie, A., (Sir), 1897, *The Founders of Geology*: Macmillan and Co., London, x + 297 p.

Geikie, A., (Sir), 1905, *Landscape in History and Other Essays*: Macmillan and Co., London, viii + 352 p.

Gemmellaro, C., 1840, *Elementi di Geologia ad uso della Regia Universita' degli Studi in Catania*: Salvatore Sciuto, Catania, XII + 432 p. + 3 plates.

Gibbon, E., 1776 [1910], *The History of the Decline and Fall of the Roman Empire*, edited in seven volumes...by Bury, J. B., Methuen & Co., London, v. 1, lxviii + 464 p. + 1 foldout map.

Gillispie, C.C., 1951, *Genesis and Geology—A Study in the Relations of Scientific Thought, Natural Theology, and Social Opinion in Great Britain, 1790–1850*: Harvard University Press, Cambridge, xiii + [i] + 315 p.

Gilluly, J., Chairman, 1948, *The Origin of Granite (Conference at Meeting of The Geological Society of America held in Ottawa, Canada, December 30, 1947)*: The Geological Society of America Memoir 28, [ii] + 139 p., https://doi.org/10.1130/MEM28.

Gilpin, W., 1792, *Observations Relative Chiefly to Picturesque Beauty Made in the Year 1776 on Several Parts of Great Britain; particularly the High-Lands of Scotland*, second edition: R. Blamire, London, 195 + xvi p.

Gohau, G., 1994a, Lavoisier, un géologue méconnu: *La Recherche*, v. 25, p. 436–437.

Gohau, G., 1994b, Lavoisier géologue: *Il y a 200 ans Lavoisier—Colloque International commémorant le bicentenaire de la mort de Lavoisier (Paris du 3 au 6 mai)*, Paris, p. 73–78.

Gould, S.J., 1983(1990), Hutton's Purpose, *in* Gould, S.J., *Hen's Teeth and Horse's Toes*: Penguin, London, p. 79–93.

Gould, S.J., 1987, *Time's Arrow Time's Cycle—Myth and Metaphor in the Discovery of Geological Time*: Harvard University Press, Cambridge, xiii + 222 p.

Graham, L.R., 1993, *Science in Russia and the Soviet Union—A Short History*: Cambridge University Press, Cambridge, x + 321 p. + 17 photographic plates.

Grant, R., 1979, Hutton's theory of the earth, *in* Jordanova, L.J. and Porter, R.S., eds., *Images of the Earth—Essays in the History of the Environmen-*

tal Sciences: The British Society for the History of Science, Chalfont St Giles, p. 23–38.

Gray, A., and Adams, C.B., 1853, *Elements of Geology*: Harper & Brothers, New York, xv + 354 p.

Green, A.H., 1876, *Geology for Students and General Readers Part I. Physical Geology*: Daldy, Isbister, & Co., London, xxvii + 552 p.

Green, A.H., 1877, *Geology for Students and General Readers Part I. Physical Geology* second edition: Daldy, Isbister, & Co., London, xxix + 555 p.

Green, A.H., 1882, *Geology Part I. Physical Geology*, third and enlarged edition: Rivingtons, London, xxiv + 728 p.

Greene, J.C., 1984, *American Science in the Age of Jefferson*: The Iowa State University Press, Ames, xiv + 484 p.

Greene, M.T., 1982, *Geology in the Nineteenth Century. Changing Views of a Changing World*: Cornell University Press, 324 p.

Grierson, J., 1818, Mineralogical Observations in Galloway: *Memoirs of the Wernerian Natural History Society*, v. 2, part 2, p. 373–391.

Grierson, J., 1821, Account of some sandstone petrifactions found near Edinburgh: *Memoirs of the Wernerian Natural History Society*, v. 3, p. 156–166.

Grierson, J., 1826, General observations on geology and geognosy, and the nature of these respective studies: *Memoirs of the Wernerian Natural History Society*, v. 5, part II, p. 401–410.

Grimaux, E., ed., 1892, *Oevures de Lavoisier* publiées par les soins du ministre de l'Instruction Publique: Imprimerie Nationale, Paris, v.5, III + 749 p. + plates (not consecutively numbered).

Grimm, J., 1852, *Grundzüge der Geognosie und Gebirgskunde für Praktische Bergmänner*: k.k. Hofbuchdruckerei von Gottlieb Haase Söhne, Prag, XVI + 71p. + 1 colored plate.

Grimm, J., 1856, *Grundzüge der Geognosie für Bergmänner zunächst für die des Österreichischen Kaiserstaates*, zweite um das doppelte vermehrte und verbesserte Auflage J.G. Calve, Prag, XXX + 360 p. + 1 page of errata + 1 folded table in back pocket.

Guettard, J.-É., 1752, Mémoire dans lequel on compare le Canada à la Suisse par rapport à ses minéraux: *Histoire de l'Académie Royale des Sciences*, year 1752, p. B: 189–220 + plate VII; C: p. 323–360 + plates VIII–XII; (additions) p. 524–538.

Guettard, J.-É., 1770, *Mémoires sur Différentes Parties des Sciences et Arts*, v. 3: Laurent Prault, Paris, [ii] + 544 p. + 71 foldout plates.

Gumprecht, [T. E.], 1850, Einige Beiträge zur Geschichte der Geognosie 1. Ueber den Urheber des Namens Geognosie und die Einführung einiger anderen Namen, Begriffe und Bestimmungen in die Geognosie: *Archiv für Mineralogie, Geognosie, Bergbau und Hüttenkunde*, v. 23, p. 468–484.

Günther, S., 1891, *Lehrbuch der Physikalischen Geographie*: Ferdinand Enke, Stuttgart, XII + 508 p. + 3 plates.

Hack, M., 1832, *Geological Sketches and Glimpses of the Ancient Earth*: London, Harvey and Darton, xix + 393 p. + 3 plates (of which 2 foldouts) + 1 frontispiece.

Haidinger, W., 1826, Account of the method of drawing crystals in true perspective, followed in the Treatise on Mineralogy of Professor Mohs [+ plates XV and XVI.]: *Memoirs of the Wernerian Natural History Society*, v. 5, part II, p. 485–508.

Hall, B., and Playfair, J., 1815, Account of the structure of the Table Mountain, and other parts of the Peninsula of the Cape. Drawn up by Professor Playfair, from observations made by Captain Basil Hall [+ plates XIII–XV]: *Transactions of the Royal Society of Edinburgh*, v. 7, p. 269–278, https://doi.org/10.1017/S0080456800027836.

Hall, J., 1859, Introduction, in Hall, J., *Geological Survey of New York. Palæontology*: volume III. ...Part I: Text., p. 1–96.

Hall, J., 1883, Contributions to the geological history of the American continent: *Proceedings of the American Association for the Advancement of Science, Thirty-first Meeting*, Published by the Permanent Secretary, AAAS, Salem, p. 29–71.

Halley, E., 1724–1725, Some considerations about the cause of the universal Deluge: *Philosophical Transactions of the Royal Society of London*, v. 33, p. 118–125.

Hann, J., von Hochstetter, F. (Ritter) and Pokorny, A., 1881, *Allgemeine Erdkunde. Ein Leitfaden der Astronomischen und Physischen Geographie, Geologie und Biologie*: F. Tempsky, Prag, XII + 646 p. + 16 plates and 1 foldout colored map.

Harrison, E., 1987, Whigs, prigs and historians of science: *Nature*, v. 329, p. 213–214, https://doi.org/10.1038/329213a0.

Hasse, F.C.A., ed., 1841, Quantum geographia novissimis periegesibus et peregrinationibus profecerit—Pars II. Specialis continens. A. Quae ad rationes globi terrestris naturales, maxime ad Gegnosiam et Geologiam, seu Geographiam subterraneam spectant: in *Solemnia Doctorum Philosophiae et Magistrorum Libb. Artium in Ordinis Ampliss. Consessu A.D. XXV, Februarii a. MDCCCXLI Rite Creatorum Renunciatorum*, no publisher, Lipsiae], 24 p.

von Hauer, F., (Ritter), 1875, *Die Geologie und Ihre Anwendung auf die Kenntniss der Bodenbeschaffenheit der Österr.-Ungar. Monarchie*: Alfred Hölder, Wien, VIII + 681 p.

von Hauer, F., (Ritter), 1878, *Die Geologie und Ihre Anwendung auf die Kenntniss der Bodenbeschaffenheit der Österr.-Ungar. Monarchie*, zweite, vermehrte und verbesserte Auflage: Alfred Hölder, Wien, VIII + 764 p.

Hauff, H., 1840, *Skizzen aus dem Leben und der Natur. Vermischte Schriften*, zweiter Band: J.G. Cotta, Stuttgart und Tübingen, [II] + 499 p.

Haüy, R. J. (Abbott), 1784, *Essai d'une Théorie sur la Structure des Crystaux, Appliquée a Plusieurs Genres de Substances Crystallisées*: Gogué & Née de la Rochelle, Paris, [iv] + 236 p. + 8 plates.

Hawking, S., and Mlodinow, L., 2010, *The Grand Design*: Bantam Books, New York, 198 p. + frontispiece.

Heilbron, J.L., 2011a, Historian of nature and mankind, in Heilbron, J.L. and Sigrist, R., eds., *Jean-André Deluc—Historian of Earth and Man*, Slatkine Érudition, Genève, p. 82–103.

Heilbron, J.L., 2011b, Savior of science and society, in Heilbron, J.L. and Sigrist, R., eds., *Jean-André Deluc—Historian of Earth and Man*, Slatkine Érudition, Genève, p. 186–240.

Heilfurth, G., 1954, *Der Bergmannslied—Wesen/Leben/Funktion—Ein Beitrag zur Erhellung von Bestand und Wandlung der sozial-kulturellen Elemente im Aufbau der industriellen Gesellschaft*: Bärenreiter, Kassel und Basel, 789 p. + 32 plates.

Heilfurth, G., 1958, *Glückauf! Geschichte, Bedeutung und Sozialkraft des Bergmannsgrusses*: Verlag Glückauf, Essen, 224 p.

Heilfurth, G., and Greverus, I.-M., 1967, *Bergbau und Bergmann in der Deutschsprachigen Sagenüberlieferung Mitteleuropas*, v. 1, *Quellen*: Veröffentlichungen des Instituts für Mitteleuropäische Volksforschung an der Philipps-Universität Marburg, v. 1, N.G. Elwert, Marburg, 1291 p.

Heim, A., 1878, *Untersuchungen über den Mechanismus der Gebirgsbildung im Anschluss an die Geologische Monographie der Tödi-Windgällen-Gruppe*, v. I (xiv + 346 p.), v. II (246 p.), Atlas (xvii plates), Benno Schwabe, Basel.

Heimann, P.M., 1978, Voluntarism and immanence: conceptions of Nature in eighteenth-century thought: *Journal of the History of Ideas*, v. 39, p. 271–283, https://doi.org/10.2307/2708779.

Hennig, E., 1938, Was ist, kann und will „Planarbeit"?: *Zeitschrift der Deutschen Geologischen Gesellschaft*, v. 90, p. 113–115.

Henry, W.A., III, 1994, *In Defense of Elitism*: Doubleday, New York, viii + [i] + 212 p.

von Herder, S.A.W. (Baron), 1838, *Der Tiefe Meissner Erbstollen—Der Einzige, den Bergbau der Freyberger Refier für die Fernste Zeit Sichernde Betriebsplan*: Brockhaus, F.A., Leipzig, 115 + CXXIV p. + 1 plan, geological map and one profile.

Hiärne, U., 1702, *Den Korta Anledningen Til Åthskillige Malm och Bergarters/ Mineraliers och Jordeslags, &c. Beswarad och Förklarad Jämte Deras Natur Födelse och I Jorden Tilwerckande/ samt Vplösning och Anatomie, I Giörligaste Måtto Beskrifwen*: J.H. Werner, Stockholm, [viii] + 132 p.

Hitchcock, E., 1840, *Elementary Geology*: J.S. & C. Adames, Amherst. Crocker and Srewster, Boston. F.J. Huntington and Co., Gould, Newman and Saxton, New York. Thomas, Cowperthwait and Co. Marshall, and Co., Philadelphia, viii + 329 p. + 1 colored foldout paleontological chart.

Hitchcock, E., 1847, *Elementary Geology*, eighth edition, revised, enlarged and adapted to the present advanced state of the science: William H. Moore & Co., Cincinnati, xii + 361 p. + 1 colored foldout paleontological chart.

Hitchcock, E., 1868, *Elementary Geology*, a new edition [31st], remodeled, mostly rewritten, with several new chapters, and brought up to the present state of the science for use in schools, families, and by individuals: Ivan, Phinney, Blakeman & Co., New York; S.C. Griggs & Co., Chicago, xiv + 430 p.

Hœfer, J.-C.-F., 1872, *Histoire de la Botanique de la Minéralogie et de la Géologie Depuis Les Temps les Plus Reculés Jusqu'à Nos Jours*: Hachette et Cie, Paris, 411 p.

Hoffmann, F., 1838, *Geschichte der Geognosie und Schilderung der Vulkanischen Erscheinungen—Vorlesungen gehalten an der Universität zu Berlin in den Jahren 1834 und 1835*: Hinterlassene Werke von Friedrich Hoffmann, v. 2, Nicolaische Buchhandlung, Berlin, VIII + 596 p.

Hoffmann, R.J., ed., 1987, *Celsus on the True Doctrine—A Discourse Against the Christians*: Oxford University Press, New York, xiii + 146 p.

Hoffmann, R.J. (trans. ed.), 1994, *Porphyry's Against the Christians—The Literary Remains* with an introduction and epilogue by R.J. Hoffmann: Prometheus Books, Amherst, New York, 181 p.

Hoheisel, K., 1979, Gregorius Reisch, *in* Büttner, M., ed., *Wandlungen im Geographischen Denken von Aristoteles bis Kant, Abhandlungen und Quellen zur Geschichte der Geographie und Kosmologie*, v. 1, Ferdinand Schöningh, Paderborn, p. 59–67.

Holland, H., (Sir), 1872, *Recollections of Past Life*: Longmans, Green, and Co. London, viii + 346 p.

Holland, H., (Sir), 1960, *Dagbók í Íslandsferð 1810*, íslenzk Þýding og skýringar eftir Steindór Steindórsson frá Hlöðum: Allmenna Bókafélagið, Reykjavik, 279 p.

Holland, H., (Sir), 1987, *The Iceland Journal of Henry Holland 1810*, edited by Andrew Wawn: The Hakluyt Society, second series, no. 168, London, xvii + 342 p. + 1 portrait.

Hosmann, G.C., 1729, *Des Heiligen Theophili, sechsten Bischofs zu Antiochia, Drey Bücher zu Avtolicvm, einen gelehrten Heiden: Darin Der Aberglaube beschämet, und Die Christliche Religion beschützet wird, Aus dem Griechischen übersetzt, Auch mit einer Vorrede und Anmerkungen vermehret*: Bey sel. Theodor Christoph Felginers Wittwe, Hamburg, [XXXVIII] + 296 p.

Hubbert, M.K., 1967, Critique of the Principle of Uniformity, *in* Albritton, C.C., ed., *Uniformity and Simplicity: A Symposium on the Principle of the Uniformity of Nature*: Geological Society of America Special Paper 89, p. 3–33, https://doi.org/10.1130/SPE89-p3.

Hübner, M., 2010, *Jean-André Deluc (1727–1817)—Protestantische Kultur und Moderne Naturforschung*: Religion, Theologie und Naturwissenschaft, v. 21, Vandenhoeck & Ruprecht, Göttingen, 280 p.

Hübner, M., 2011, Between the Republic of Letters and the Kingdom of God, *in* Heilbron, J.L. and Sigrist, R., eds., *Jean-André Deluc—Historian of Earth and Man*, Slatkine Érudition, Genève, p. 20–43.

von Humboldt, A., (Baron), 1790, *Mineralogische Beobachtungen über einige Basalte am Rhein mit vorangeschickten, zerstreuten Bemerkungen über den Basalt der älteren und neueren Schriftsteller*: Schulbuchhandlung, Braunschweig, 126 p.

von Humboldt, A. (Baron), 1823, *Essai Géognostique sur le Gisement des Roches dans les Deux Hémisphères*: F.G. Levrault, Paris, viij + p.

von Humboldt, A., (Baron), 1845, *Kosmos. Entwurf einer Physischen Weltbeschreibung*, erster Band: J.G. Cotta, Stuttgart und Tübingen, XVI + 493 p.

von Humboldt, A., (Baron), 1850, *Kosmos. Entwurf einer Physischen Weltbeschreibung*, dritter Band: J. G. Cotta, Stuttgart und Tübingen, 644 p.

von Humboldt, A. (Baron), 1858, *Kosmos. Entwurf einer Physischen Weltbeschreibung*, vierter Band: J. G. Cotta, Stuttgart und Tübingen, 649 + [I] p.

Hume, D., 1739[1978], *A Treatise of Human Nature*: Oxford, At the Clarendon Press, xix + 743 p. (Second ed. with text revised and notes by P.H. Nidditch)

Hunt, T.S., 1873, On some points in dynamical geology: *American Journal of Science and Arts*, third series, v. 5, p. 264–270.

Hunt, T.S., 1875, *Chemical and Geological Essays*: James R. Osgood and Company, Boston; Trübner & Co., London, xxii + 489 p.

Hunt, T.S., 1878, *Chemical and Geological Essays*, second edition, revised, with additions: S.E. Cassino, Salem, xlvi + 489 p.

Hunt, T.S., 1886, *Mineral Physiology and Physiography. A Second Series of Essays with a General Introduction*: Samuel E. Cassino, Boston, xvii + 710 p.

Huot, J.-J.-N., 1839, *Cours Élémentaire de Géologie*: Roret, Paris, tome deuxième, VI + 794 + Atlas of 23 plates.

Hutton, J., 1785, Abstract of A Dissertation read in the Royal Society of Edinburgh, upon the Seventh of March, and Fourth of April M, DCC, LXXXV, concerning the System of the Earth, Its Duration and Stability (published anonymously), *in* Craig, G.Y., ed., The *1975 Abstract of James Hutton's Theory of the Earth*: Scottish Academic Press, Edinburgh, xiv + 30 p.

Hutton, C., 1778, XXXIII. An account of the calculations made from the survey and measures taken at Schehallien, in order to ascertain the mean density of the Earth: *Philosophical Transactions of the Royal Society of London*, v. 68, p. 689–788.

Hutton, J., 1788, Theory of the Earth; or An Investigation of the Laws observable in the Composition, Dissolution, and Restoration of Land upon the Globe: *Transactions of the Royal Society of Edinburgh*, v. 1, p. 209–304, https://doi.org/10.1017/S0080456800029227.

Hutton, J., 1793, Extrait d'une dissertation sur le système et durée de la Terre; lue par M. Hutton à la Société Royale d'Edimburg l'an 1785: traduit de l'Anglois, par Iberti, Médecin, pensionnaire de S.M.C. suivie par des observations du traducteur sur le même sujet: *Observations sur la Physique, sur l'Histoire Naturelle et sur les Arts*, v. 43, part 2, p. 3–12.

Hutton, J., 1794, Observations on granite: *Transactions of the Royal Society of Edinburgh*, v. 3, no. 2, p. 77–85, https://doi.org/10.1017/S0080456800020305.

Hutton, J., 1794a[1999a], *An Investigation of the Principles of Knowledge and of Progress of Reason from Sense to Science and Philosophy*, v. 1, with a new introduction by Jean Jones and Peter Jones: Thoemmes Press, Bristol, xi + lxxii + 649 p.

Hutton, J., 1794b[1999b], *An Investigation of the Principles of Knowledge and of Progress of Reason from Sense to Science and Philosophy*, v. 2: Thoemmes Press, Bristol, xxiii + 734 p.

Hutton, J., 1794c[1999c], *An Investigation of the Principles of Knowledge and of Progress of Reason from Sense to Science and Philosophy*, v. 3: Thoemmes Press, Bristol, xvi + 755 p.

Hutton, J., 1795, *Theory of the Earth with Proofs and Illustrations*, v.1: Cadell, Junior and Davies, London, and William Creech, Edinburgh, viii + 620 p. + IV plates.

Hutton, J., 1798, Observations of Granite, &c. Observations sur le Granit, par JAMES HUTTON: *Bibliotheque Britannique; ou Recueil Extrait des Ouvrages Anglais Périodique & Autres, des Mémoires & Transactions des Sociétés & Académies de la Grande-Bretagne, d'Asie, d'Afrique & d'Amerique.*, v. 7 (Sciences et Arts), p. 250–266.

Hutton, J., 1899[1997], *Theory of the Earth with Proofs and Illustrations*, v. 3, edited by Sir Archibald Geikie, The Geological Society of London (reprint of 1899 edition).

Inglehart, A.J., 2018, Filippo Buonanni and the Kircher Museum, *in* Rosenberg, G.D., and Clary, R.M., eds., *Museums at the Forefront of the History and Philosophy of Geology: History Made, History in the Making*: Geological Society of America Special Paper 535, p. 45–58, https://doi.org/10.1130/2018.2535(04).

Inostransev, A.A., 1885, *Geologiya—Obshchiy Kurs Lektsiy. Tom 1. Sovremennyye geologicheskiye yavleniya (dinamicheskaya geologiya)—Petrografiya i Stratigrafiya*: Tipografiya M.M. Stasyulevicha, Sankt-Peterburg, VIII + 495 s.

Jackson, P., and Morgan, D., 1990, *The Mission of Friar William of Rubruck—His journey to the court of the Great Khan Möngke (1253–1255)*: The Hakluyt Society, London, xvi + 312 p. + frontispiece.

Jackson, P.W., 2006, *The Chronologers' Quest—Episodes in the Search for the Age of the Earth*: Cambridge University Press, Cambridge, xvii + 291 p.

Jameson, [R.], 1811, Mineralogical queries, proposed by Professor Jameson: *Memoires of the Wernerian Natural History Society*, v. 1, p. 107–125.

Jameson, R., 1813a, Mineralogical notes, *in* Cuvier, 1813 (see there).

Jameson, R., 1813b, *Mineralogical Travels through the Hebrides, Orkney and Shetland Isles and Mainland of Scotland with Dissertations upon Peat and Kelp*, v. I (xxvi + 243 p. + 7 plates), v. II (iv + 289 + [i] + 6 plates): Archibald Constable and Co., Edinburgh and White, Cochran and Co., London.

Jameson, R., 1822, Notes on the geognosy of the Crif-Fell, Kirkbean, and the Needle's Eye in Galloway: *Memoirs of the Wernerian Natural History Society*, v. 4, part II, p. 541–547.

Jayatilleke, K.N., 1984, Buddhism and the scientific revolution, *in* Kirthisinghe, B.P., ed., *Buddhism and Science*, Motilal Banarsidass, Delhi, p. 8–16.

Jones, J., 1984, The geological collection of James Hutton: *Annals of Science*, v. 41, p. 223–244, https://doi.org/10.1080/00033798400200241.

Jones, J., 1985, James Hutton's agricultural researches and his life as a farmer: Annals of Science, v. 42, p. 573–601, https://doi.org/10.1080/00033798500200371.

Jones, J., Torrens, H.S., and Robinson, E., 1994, The correspondence between James Hutton (1726–1797) and James Watt (1736–1819) with two letters from Hutton to George Clerk-Maxwell (1715–1784): Part I: *Annals of Science*, v. 51, p. 637–653, https://doi.org/10.1080/00033799400200471.

Jones, J., Torrens, H.S., and Robinson, E., 1995, The correspondence between James Hutton (1726–1797) and James Watt (1736–1819) with two letters from Hutton to George Clerk-Maxwell (1715–1784): Part II: *Annals of Science*, v. 52, p. 357–382, https://doi.org/10.1080/00033799500200291.

Kant, I., 1755, *Allgemeine Naturgeschichte und Theorie des Himmels oder Versuch von der Verfassung und dem Mechanischen Ursprunge des Ganzen Weltgebäudes nach Newtonischen Grundsätzen Abgehandelt*: Johann Friedrich Petersen, Königsberg und Leipzig, 146 p.

Kant, I., 1781, *Kritik der Reinen Vernunft*: Johann Friedrich Hartknoch, Riga, xxii + 856 p.

Karpinski, R.W., 1931, Contribution à l'Étude Métallogenique des Vosges Méridionales (Vallées du Rahin, de l'Oignon, du Breuchin, du Raddon): *Mémoires de la Société des Sciences de Nancy*, Anné 1931, p. VI–XI + 1–141 + 4 geological maps *hors-texte*.

Keferstein, C., 1834, *Die Naturgeschichte des Erdkörpers in ihren ersten Grundzügen*, zweiter Theil *Die Geologie und Paläontologe*: Friedrich Fleischer, Leipzig, IV + 896 p.

Keferstein, C., 1840, *Geschichte und Litteratur der Geognosie, ein Versuch*: Johann Friedrich Lippert, Halle, XIV + 281 p.

Kemmerer, A., 1830, Vzglyad na glavnyye geologicheskiye teorii: Vernerovu i Gyuttonovu: *Gornyy Zhurnal*, v. 10, p. 1–18.

Kircher, A.S.I., 1665, *Mundus Subterraneus in XIII Libros Digestus...*v. I: Joannes Janssonius and Elizeus Weyerstrat, Amsteldami, [xxvi] + 220 p.

Kirthisinghe, B.P., 1984, editor, *Buddhism and Science*: Motilal Banarsidass, Delhi, XII + 613 p.

Klaver, J.M.I., 1997, *Geology and Religious Sentiment—The Effect of Geological Discoveries on English Society and Literature Between 1829 and 1859*: Brill's Studies in Intellectual History, v. 80, Brill, Leiden, XVI + 215 p.

Kober, L., 1921, *Der Bau der Erde*: Gebrüder Borntraeger, Berlin, II + 324 p. + 1 foldout map.

Kober, L., 1928, *Der Bau der Erde*, zweite neubearbeitete und vermehrte Auflage: Berlin, Gebrüder Borntraeger, [II] + 500 p. + 2 foldout maps.

Kopelevich, Y.K., Ossipov, V.I., and Shafran, I.A., eds., 1987, *Uchennaya Korrespondentsia Akademii Nauk XVIII veka—Nauchnoe Opisanie 1783–1800*, under the direction of G.K. Mikhailov: Izdatelstvo "Nauka," Moskva, 271 + [I], p.

Kovrigin, N.N., 1836, Zamechaniya o geognosticheskom sostave vostochnoy chasti Sayana i otrogov yego, zaklyuchayushchikhsya v verkhovyakh rek Irkuta i Kitoya, 1835, 1836: *Gornyy Zhurnal*, no. 9, p. 481–533.

Krafft, F., 2009, Goethe zwischen Neptun und Vulkan, *in* Manger, K. and Klöcking, H.-P., eds., *Symbiosen—Wissenschaftliche Wechselwirkungen zu Gegenseitigem Vorteil—Festschrift für Werner Köhler*, Sonderschriften der Akademie Gemeinnütziger Wissenschaften zu Erfurt, v. 39, p. 231–251.

Kristeller, P.O., 1979a, Introduction, *in* Moody, M., ed., *Renaissance Thought and its Sources*: Columbia University Press, New York, p. 1–14.

Kristeller, P.O., 1979b, The humanist movement, *in* Moody, M., ed., *Renaissance Thought and its Sources*: Columbia University Press, New York, p. 21–32.

Krivonogov, S., 2009, Extent of the Aral Sea drop in the Middle Ages: *Doklady Akademi Nauk RF*: Earth Sciences (Paris), v. 428, no. 1, p. 1146–1150.

von Kronstedt, A.F., (Baron), 1780, *Versuch einer Mineralogie*, aufs neue aus dem schwedischen übersetzt und nächst verschiedenen Anmerkungen vorzüglich mit äusern Beschreibungen der Fossilien vermehrt durch Abraham Gottlob Werner, ersten Bandes erster Theil, Siegfried Lebrecht Crusius, Leipzig, 254 p. + two unpaginated Vorreden + unpaginated Mineral-System + 1 p. errata.

Laks, A., 2018, *The Concept of Presocratic Philosophy: Its Origin, Development and Significance*: Princeton University Press, Princeton, x + 137 p.

Lambeck, K., 1996, Shoreline reconstructions for the Persian Gulf since the last glacial maximum: *Earth and Planetary Science Letters*, v. 142, p. 43–57, https://doi.org/10.1016/0012-821X(96)00069-6.

Lamontagne, R., 1963–1964, Lettre de Jean-François Gauthier à Jean-Étienne Guettard (1752): *Revue d'Histoire de l'Amerique Francaise*, v. 17, p. 569–572, https://doi.org/10.7202/302316ar.

Lamontagne, R., 1965, La participation canadienne à l'œuvre de Guettard: *Revue d'Histoire des Sciences*, v. 18, p. 385–388.

Lane Fox, R., 1993, *The Unauthorized Version—Truth and Fiction in the Bible*: Vintage Books, A Division of Random House, New York, 478 p.

Laplace, P.-S., An IV [1796] *Exposition du Système du Monde*: Imprimerie du Cercle-Social, Paris, v. 2, 312 + 1 + [vi] p.

Laplace, [P.-S.] (Count), 1814, *Théorie Analytique des Probabilités*, seconde édition, revue et augmentée par l'auteur: Mme Vve Courcier, Paris, cvj + 506 p.

de Lapparent, A.-A.C., 1883, *Traité de Géologie*: F. Savy, Paris, XVI + 1280 p.

von Lasaulx, A., 1869, *Der Streit über die Entstehung des Basaltes*: C.G. Lüderitz'sche Verlagsbuchhandlung A. Charisius, Berlin, 38 p.

Laudan, R., 1987, *From Mineralogy to Geology—The Foundations of A Science*: The University of Chicago Press, Chicago, xii + 278 p.

Lavoisier, A.-L., 1789, Observations générales sur les couches modernes horizontales, qui ont été déposées par la mer et sur les conséquences qu'on peut tirer de leurs dispositions, relativement à l'ancienneté du globe terrestre: *Mémoires de l'Académie Royale des Sciences*, 1789 (1793), p. 351–371 + 7 foldout plates. Reprinted in v. 5 of the *Oevures de Lavoisier publiées par les soins du ministre de l'Instruction Publique*: Imprimerie Nationale, Paris, v. 1892, p. 186–204.

Le Conte, J., 1873, Formation of the Earth-surface: *American Journal of Science and Arts*, third series, v. 5, p. 448–453.

Le Conte, J., 1895, Critical periods in the history of the earth: *Bulletin of the Department of Geology*, University of California, no. 1, p. 313–336.

Leibniz, G.G., 1749a, *Protogaea sive de prima facie tellvris et antiqvissimae historiae vestigiis in ipsis natvrae monvmentis dissertatio* ex Schedis Manvscriptis in lvcem edita a Christiano Lvdovico Scheidio: Ioh. Gvil. Schmid, XXVIII + 86 + XII plates.

Leibniz, G.W., 1749b, *Protogaea oder Abhandlung Von der Ersten Gestalt der Erde und den Spuren der Historie in den Denkmaalen der Natur aus seinen Papieren herausgegeben von Christian Ludwig Scheid Aus dem lateinischen ins teutsche übersetzt*: Johann Gottlieb Vierling, Leipzig and Hof, 124 p + 2 pages of *Druckfehler*.

Leibniz, G.W., 2008, *Protogaea*, translated and edited by Claudine Cohen and Andre Wakefield: Chicago University Press, Chicago, xlii + 173 p.

[von] Leonhard, G., 1863, *Grundzüge der Geognosie und Geologie*, zweite vermehrte Auflage: J.F. Winter, Leipsig und Heidelberg, XII + 478.

von Leonhard, K.C., 1816, *Bedeutung und Stand der Mineralogie—Eine Abhandlung in der zur Feier des Allerhöchsten Namensfestes Seiner Majestät des Königs am 12. Oktober 1816 gehaltenen öffentlichen Versammlung der Akademie der Wissenschaften zu München*: Joh. Christ. Hermann, Frankfurt am Main, 111 p.

von Leonhard, K.C., 1835, *Lehrbuch der Geognosie und Geologie*: E. Schweitzerbart'sche Verlagsbuchhandlung, Stuttgart, XVI + 869 p. + an Atlas of 8 foldout plates.

von Leonhard, K.C., 1849, *Lehrbuch der Geognosie und Geologie*, zweite vermehrte und verbesserte Auflage: E. Schweitzerbart'sche Verlagsbuchhandlung, Stuttgart, XXII + 1056 p. + 5 colored foldout plates.

Le Strange, G., 1905(1966), *The Lands of the Estern Caliphate—Mosopotamia, Persia, and Central Asia from the Moslem conquest to the time of Timur*, third [amended] impression: Frank Cass & Co., London, xvii+[iii]+536 p.+10 maps hors-texte.

Leuckart, F.S., 1832, *Allgemeine Einleitung in die Naturgeschichte*: E. Schweitzerbart'sche Verlagsbuchhandlung, Stuttgart, [I] + 130 p., https://doi.org/10.5962/bhl.title.100767.

Lichtenberg, G.C., 1994a, *Schriften und Briefe*, erster Band *Sudelbücher II* edited by Wolfgang Promies: Zweitausendeins, Frankfurt-am-Main, 868 + [I] p.

Lichtenberg, G.C., 1994b, *Schriften und Briefe*, erster Band *Sudelbücher I* edited by Wolfgang Promies: Zweitausendeins, Frankfurt-am-Main, 988 + [I] p.

Lichtenberg, G.C., 2013, *Georg Christoph Lichtenberg Vorlesungen zur Naturlehre*, *in* Krayer, A., and Nickol, T., eds., Wallstein, Göttingen, XXVIII + 1060 p.

Linnaeus, C., (Carl von Linné), 1744, *Oratio de Telluris Habitabilis Incremento. Et Andreae Celsii...Oratio de Mutationibus Generalioribus quae in superficie corporum coelestium contingunt*: Cornelium Haak, Lugduni Batavorum, p. 17–84.

Lloyd, J.J., 1986, The innocence of George Hoggart Toulmin: *Earth Sciences History*, v. 5, p. 128–130, https://doi.org/10.17704/eshi.5.2.11vv445500642t40.

Loewinson-Lessing, F.Y., 1954, *A Historical Survey of Petrology*, translated from the Russian by S.I. Tomkeieff: Oliver & Boyd, Edinburgh, x + 105 p.

Lüdemann, K.-F., and Wenzel, W., 1967, Abraham Gottlob Werners Vorlesung über Eisenhüttenkunde, *in* Rosier, H.J., editor-in-chief, *Abraham Gottlob Werner—Gedenkschrift aus Anlaß der Wiederkehr seines Todestages nach 150 Jahren am 30 Juni 1967*: Freiberger Forschungshefte, C223 Mineralogie-Lagerstättenlehre, Leipzig, p. 157–162.

Lupi, B., 1875, *Trattato Storico-Scientifico sull'Origine, su'Progressi, sullo stato della Geologia in Sicilia*: S. Mucumeci, Catania, 74 p.

Lyell, C. (Sir), 1827, Memoir on the Geology of Central France; including the Volcanic Formations of Auvergne, the Velay, and the Vivarais, with a Volume of Maps and Plates. By G.P. Scrope, F.R.S., F.G.S., London, 1827: *The Quarterly Review*, v. 36, p. 437–483.

Lyell, C., (Sir), 1830, *Principles of Geology, being an attempt to explain the former changes of the earth's surface, by reference to causes now in operation*, v. 1: John Murray, London, xv + 511 p.

Lyell, C., (Sir), 1832, *Principles of Geology, being an attempt to explain the former changes of the earth's surface, by reference to causes now in operation*, v. 2: John Murray, London, xii + 330 p.

Lyell, C., (Sir), 1833, *Principles of Geology, being an attempt to explain the former changes of the earth's surface by reference to causes now in operation*, v. 3: John Murray, London, xxxi + [i errata page] + 398 + 109 p. + V plates.

Lyell, C., 1835, On the proofs of a gradual rising of the land in certain parts of Sweden: *Philosophical Transactions of the Royal Society of London* for the year MDCCCXXXV, Part I, p. 1–38.

Lyell, C., (Sir), 1859, On the structure of lavas which have consolidated on steep slopes; with remarks on the mode of origin of Mount Etna, and on the theory of "craters of elevation": *Philosophical Transactions of the Royal Society of London*, part II, v. 148, p. 703–786 + plates 49 and 50.

Lyell, [K. M.], 1881, *Life, Letters and Journals of Sir Charles Lyell, Bart.*: John Murray, London, v.1, xii + 475 p.

von Maack, P., 1844, *Die Revolutionen des Erdballs von Dr. Alexander Bertrand*: Universitäts-Buchhandlung, Kiel, VIII + 314 p. + 5 plates.

MacCulloch, J., 1819, *A Description of the Western Islands of Scotland, including the Isle of Man: Comprising an Account of their Geological Structure; with Remarks on their Agriculture, Scenery, and Antiquities*: Archibald Constable and Co., Edinburgh and Hurst, Robinson, and Co., London, v. I (xv + 587 p.), v. II (vii + 589 p.), v. III (plates, maps and explanations, 91 p. + numerous unnumbered plates).

MacCurdy, E., 1954, *The Notebooks of Leonardo da Vinci*—Arranged, Rendered into English and Introduced…, v. I: The Reprint Society, London, 610 p.

MacGillivray, W., 1840, *A Manual of Geology*: Scott, Webster, and Geary, London, vi + 248 p. + 1 colored foldout map.

MacGillivray, W., 1844, *A Manual of Geology* second edition: Adam Scott, London, xii + 292 p. + 1 colored foldout map.

MacGillivray, W., 1912, *A Memorial Tribute to William MacGillivray M.A., LL.D., Ornithologist; Professor of Natural History, Marischal College and University, Aberdeen*: Printed for Private Circulation, Edinburgh, xiv + 203 p.

McIntyre, D.B., 1997, James Hutton's Edinburgh: the historical, social, and political background: *Earth Sciences History*, v. 16, p. 100–157, https://doi.org/10.17704/eshi.16.2.xr57822t10787lj3.

McIntyre, D.B., and McKirdy, A., 1997, *James Hutton—The Founder of Modern Geology*: The Stationary Office, Edinburgh, xi + 51 p.

Mackenzie, G.S., (Sir), 1812, *Travels in the Island of Iceland, during the Summer of the Year MDCCCX*, second edition: Archibald Constable and Company, Edinburgh; T. Payne; Longman, Hurst, Rees, Orme & Brown; Cadell & Davies; J. Murray; R. Baldwin; C. Law; J. Hatchard; E. Lloyd; W. Lindsell; Cradock & Joy; Gale, Curtis & Fenner; Sharpe & Hailes; and T. Hamilton, London; xv + [i] + 491 + [i] p.

MacKnight, [T.], 1811, On the mineralogy and local scenery of certain districts in the Highlands of Scotland—part I: *Memoirs of the Wernerian Natural History Society*, v. 1, p. 274–369.

Maclure, W., 1809, Observations on the Geology of the United States, Explanatory of a Geological Map: *Transactions of the American Philosophical Society*, v. 6, p. 411–428, https://doi.org/10.2307/1004821.

Maclure, W., 1817, *Observations on the Geology of the United States of America; with some remarks on the effect produced on the nature and fertility of Soils, by the decomposition of the different classes of rocks and an application to the fertility of every state in the Union, in reference to the accompanying Geological Map*: Abraham Small, Philadelphia, 127 p. + 2 plates.

Maclure, W., 1832a, *Essay on the Formation of Rocks, or an Inquiry into the Probable Origin of their Present Form and Structure*: New-Harmony, Indiana, 53 p.

Maclure, W., 1832b, *Observations on the Geology of the West India Islands from Barbados to Santa Cruz, Inclusive*: New Harmony, Indiana, 17 p.

Mansfeld, J., 1999, Sources, in Long, A.A., ed., *The Cambridge Companion to Early Greek Philosophy*, Cambridge University Press, Cambridge, p. 22–44, https://doi.org/10.1017/CCOL0521441226.002.

Mantell, G. A., 1838a, *The Wonders of Geology; or A Familiar Exposition of Geological Phenomena; being the Substance of A Course of Lectures Delivered at Brighton*, v. I: Relfe and Fletcher, London, xvi + 376 p. + unpaginated appendices and glossary.

Mantell, G.A., 1838b, *The Wonders of Geology; or A Familiar Exposition of Geological Phenomena; being the Substance of A Course of Lectures Delivered at Brighton*, v. II: Relfe and Fletcher, London, ix p. + p. 377–689 + 6 plates + unpaginated description of illustrations, index and an errata page.

Mantell, G.A., 1844, *The Medals of Creation; or, First Lessons in Geology and in the Study of Organic Remains*: Bohn, London, v. I (xxvii + 456 p.) + v. II (p. 457–1016) + 6 colored plates distributed to the two volumes).

Mantell, G.A., 1857, *The Wonders of Geology; or A Familiar Exposition of Geological Phenomena*, seventh edition, revised and augmented by T. Rupert Jones, F.G.S. v. I: Henry G. Bohn, London, xxiv + 480 p. + 4 plates.

Marsilli, L.F., Count, 1725, *Histoire Physique de la Mer*: Aux De'pens de la Compagnie, Amsterdam, XI + 173 p. + 12 + 40 levha.

Maskelyne, N., 1775, An Account of Observations Made on the Mountain Schiehallion for Finding Its Attraction: *Philosophical Transactions of the Royal Society of London*, v. 65, p. 500–542, https://doi.org/10.1098/rstl.1775.0050.

Mather, W.W., 1833, *Elements of Geology, for the Use of Schools*: William Lester, Jr., Norwich, [i] + vi + 139 p.

Mather, W.W., 1838, *Elements of Geology, for the Use of Schools and Academies*, second edition: American Common School Union, New York, xii + 286 p.

Mather, K.F., and Mason, S.L., eds., 1939, *A Source Book in Geology*: McGraw-Hill, New York and London, xxii + 702 p.

Mathesius, J., 1578, *Bergpostilla oder Sarepta Darinn von Allerley Bergkwerck vnd Metallen/was jr eygenschafft vnd natur/vnd wie sie zu nutz und gut gemacht/gutter beriche gegeben wird. Mit tröflicher und lehrhafter erklerung aller sprüch, so in heiliger Schrifft von Metall reden/und wie der heilig Geist in Metallen und Bergarbeit die Artickel unsers Christlichen glaubens fürgebildet*. Jetz und mit fleiß widerumb durchsehen, corrigiert vnd gemehret mit einem newen Register vnd kurtzen Summarien /in welchen kürtzlich angezeiget wird/was in einer jeden Predigt gehandelt/ und was für sprüch auß altem und neuem Testament darinnen fürnemlich erkleret warden/zc. Sampt der Jochimsthalischen kurtzen Chroniken, biß auff das 1578. Jahr: (no publisher given) Nürnberg, [26] + ccxlvii p. + 1 p. Register.

Merrill, G.P., 1904, *Contributions to the History of American Geology*: Report of U.S. National Museum, 1904, The Smithsonian Institution, Washington, D.C., p. 191–733.

Merrill, G.P., 1924, *The First One Hundred Years of American Geology*: Yale University Press, New Haven, xxi + 773 p. + 1 folded plate.

Miall, L., 1896, In Memoriam: Professor A.H. Green, F.R.S., F.G.S: *Proceedings of the Yorkshire Geological and Polytechnic Society*, v. 13, p. 232–233, https://doi.org/10.1144/pygs.13.2.232.

Mohs, F., 1842, *Die ersten Begriffe der Mineralogie und Geognosie für junge praktische Bergleute der k. k. österreichischen Staaten*, zweiter Theil. *Geognosie*: Carl Gerold, Wien, XIV + 408 + 18 plates.

Mollâ Lutfî'l Maqtûl, 1940, *La Duplication de l'Autel (Platon et le Problème de Délos)*, texte arabe publié par Şerefettin Yaltkaya, traduction française et introduction par [Adıvar] Adnan, A. und Corbin, H.: Études Orientales publiées par l'Institut d'Archéologie de Stamboul VI, E. de Boccard, Paris, 61 + 23 s.

Mori, G., 1975, Marzari Pencati Giuseppe: in *Österreichisches Biographisches Lexikon 1815–1950 ([Maier] Stefan–Musger August)*: Verlag der Österreichischen Akademie der Wissenschaften, Vienna, v. 6, p. 122.

Moro, A.L., (Abbot), 1740, *De Crostacei e degli Altri Marini Corpi che si Truovano su' Monti*, libri due: Stefano Monti, Venezia, [xii] + 1 + 452 p.

Mühlfriedel, W., and Guntau, M., 1967, Abraham Gottlob Werners Wirken für die Wissenschaft und sein Verhältnis zu den geistigen Strömungen des 18. Jahrhunderts: in *Abraham Gottlob Werner—Gedenkschrift aus Anlaß der Wiederkehr seines Todestages nach 150 Jahren am 30. Juni 1967, Freiberger Forschungshefte*, C223 Mineralogie-Lagerstättenlehre, p. 9–46.

Mühlpfordt, G., 1985, Stahls Grundlegung der neueren Chemie und die Weltwirkung der halleschen Aufklärung, *in* Kaiser, W., and Völker, A., eds., *Georg Ernst Stahl (1659–1734)—Hallesches Symposium 1984*, Wissenschaftliche Beiträge der Martin-Luther-Universität Halle-Wittenberg 1985 66 (E 73), Halle (Saale), p. 117–160.

[Murray, J.], 1802, *A Comparative View of the Huttonian and Neptunian systems of Geology in Answer to Illustrations of the Huttonian Theory of the Earth, by Professor Playfair*: Ross and Blackwood, Edinburgh and T.N. Longman and O. Rees, London, v + 256 p.

Mushketov I.V., 1881[181], *Kurs Geologii, Chitannyy v Gornom Institute*: F. Radlov i N. Koksharov, S. Petersburg, 773 s.

Mushketov, I.V., 1899, *Fizicheskaya Geologiya*, v. 1 *Obshiya Svoistva Sostav Zemli. Tektonicheskie Protsessi (Dislokatsionniya, Vulkanicheskiya i Seismicheskiya Yavleniya)*, second edition: Y.N. Ehrlich, St. Petersburg, VIII + 784 p. + 3 colored foldout maps.

Neill, P., 1814, Biographical account of Mr. Williams, the Mineralogist: *Annals of Philosophy*, v. 4, no. 2, p. 81–83.

Nemitz, R., and Thierse, D., 1995, *St. Barbara—Weg Einer Heiligen Durch die Zeit*: Glückauf, Essen, 553 p.

Neumayr, M., 1887, *Erdgeschichte*: Verlag des Bibliographischen Instituts, Leipzig, v. 1, XII + 653 p.

Newcomb, S., 1990, The ideas of A.G. Werner and J. Hutton in America: *Earth Sciences History*, v. 9, p. 96–107, https://doi.org/10.17704/eshi.9.2.vwv66j323567vv13.

Nicols, A., 1880, *Chapters from the Physical History of the Earth—An Introduction to Geology and Palæontology*: C. Kegan Paul & Co., London, xiv + 281 p.

Niewöhner, F., and Pluta, O., eds., 1999, *Atheismus im Mittelalter und in der Renaissance*: Wolfenbütteler Mittelalter-Studien, v. 12, Harrassowitz, Wiesbaden, VI + 372 p.

Novalis [Friedrich von Hardenberg], 1983, *Schriften*, dritter Band *Das philosophische Werk* II, edited by Richard Samuel in cooperation with H.-J. Mähl and G. Schulz; 3rd revised edition, Kohlhammer, Stuttgart, XV + 1077 p.

Nye, R.B., 1960 (1963), *The Cultural Life of the New Nation 1776–1830*: The New American Nation Series, Harper Torchbooks, New York, xii + 324 p.

Ogilby, J., 1811a, On the Transition Greenstone of Fassney: *Memoirs of the Wernerian Natural History Society*, v. 1, p. 126–130.

Ogilby, J., 1811b, On the veins that occur in the newest flœtz-trap formation of East Lothian: *Memoirs of the Wernerian Natural History Society*, v. 1, p. 469–478.

Oldroyd, D.R., compiler and editor, 1999, *International Commission on the History of Geological Sciences INHIGEO Newsletter* no. 31 for 1998: The University of New South Wales, Australia, 79 p.

Oldroyd, D.R., 2003, Essay Review: A Manichean view of the history of geology: *Annals of Science*, v. 60, p. 423–436, https://doi.org/10.1080/00033790210155220.

Oldroyd, D.R., 2009, Jean-André de Luc (1727–1817): an atheist's comparative view of the historiography, *in* Kölbl-Ebert, M., ed., *Geology and Religion: A History of Harmony and Hostility*, The Geological Society, London, Special Publications, v. 310, p. 7–15, https://doi.org/10.1144/SP310.2.

Oldroyd, D.R., 2010, Essay review: geohistory revisited and expanded: *Annals of Science*, v. 67, p. 249–259, https://doi.org/10.1080/00033790902788048.

Omboni, G., 1894, *Brevi Cenni sulla Storia della Geologia Compilati per i suoi Allievi*: F. Sachetto, Padova, 72 p.

d'Orbigny, A., 1842, *Voyage dans l'Amerique meridionale, (le Brésil, la République orientale de l'Uruguay, la République Argentine, République de Chili, la République de Bolivia, la République de Perou) exécuté pendant les années 1826, 1827, 1828, 1829, 1830, 1831 et 1833*, t. III, 3e partie: Géologie: P. Bertrand, Paris ve V.e Levrault, Strasbourg, 42 (Extrait des Rapports) + 289 + [1] p.

Ortelius, A., 1606, *Theatrum Orbis Terrarum—The Theatre of the Whole World: set forth by that excellent Geographer Abraham Ortelius*: John Norton, London, 114 p. + unnumbered plates of maps.

Ospovat, A.M., 1967, The "Wernerian Era" of American geology: in *Abraham Gottlob Werner—Gedenkschrift aus Anlaß der Wiederkehr seines Todestages nach 150 Jahren am 30. Juni 1967, Freiberger Forschungshefte*, no. C223 Mineralogie-Lagerstättenlehre, p. 237–244.

Pallas, P.S., 1776, *Reise durch Verschieden Provinzen des Rußischen Reichs, dritter Theil Vom Jahr 1772 und 1773*: St. Petersburg, Kaiserliche Académie der Wissenschaften, unpaginated preface + 760 p. + 12 unpaginated indexes + pls. A–M.

Pallas, P.S., 1778 (1986), *Über die Beschaffenheit der Gebirge und die Veränderungen der Erdkugel von Peter Simon Pallas* mit Erläuterungen von Folkwart Wendland: Leipzig, Ostwalds Klassiker der Exakten Wissenschaften, Akademische Verlagsgesellschaft, Geest & Portig K.-G., 112 p.

Patrin, [E.L.M.], 1783, *Relation d'un Voyage aux Monts Altaïce en Sibérie fait en 1781*: Logan Libraire, St Petersburg, 38 p.

Penn, G., 1822, *Comparative Estimate of the Mineral and Mosaical Geologies*: Ogle, Duncan, and Co., London, vii + 460 p.

Penn, G., 1825, *Comparative Estimate of the Mineral and Mosaical Geologies*. Revised, and enlarged with relation to the latest publications in geology: James Duncan, London, v. I (lix + 1 errata page + 353 p.), v II (viii + 426 p.).

Penn, G., 1828, *Conversations on Geology; Comprising a Familiar Explanation of the Huttonian and Wernerian Systems; the Mosaic Geology as Explained by Mr. Granville Penn; and the Late Discoveries of Professor Buckland, Humboldt, Dr. Macculloch and others*: Samuel Maunder, London, xxii + 371 p. + 11 plates.

Pennant, T., 1790, *A Tour in Scotland*, fifth edition: Benj. White, London, vi + 400 p. +1 foldout map + 42 plates.

Phillips, J., 1838, *A Treatise on Geology, forming the article under that head in the seventh edition of the Encyclopædia Britannica*: Adam and Charles Black, Edinburgh, viii + 295 p. + 1 frontispiece.

Phillips, J., 1842, Geology (as part II of the article Mineralogy): *Encyclopædia Britannica or Dictionary of Arts, Sciences, and General Literature*, 7th edition: Adam and Charles Black, Edinburgh, v. 15, p. 173–241.

Phillips, J., 1855, *Manual of Geology: Practical and Theoretical*: Richard Griffin and Company, London and Glasgow, xii + 669 + 1 frontispiece.

Phillips, J., 1860, *Life on Earth its Origin and Succession*: Macmillan and Co., Cambridge and London, X + 224 p. + 1 colored frontispiece.

Phillips, J., 1885a, *Manual of Geology—Theoretical and Practical*, edited by Robert Etheridge and Harry Govier Seeley in two parts Part I *Physical Geology and Palæontology* by H.G. Seeley: Charles Griffin and Company, London, xiii + [i] + 546 p. + 1 colored frontispiece.

Phillips, J., 1885b, *Manual of Geology—Theoretical and Practical*, edited by Robert Etheridge and Harry Govier Seeley in two parts Part II *Stratigraphical Geology and Palæontology* by Robert Etheridge: Charles Griffin and Company, London, xxiv + 712 p. + 1 colored frontispiece.

Phillips, J., and Daubeny, C.G.B., 1845, Geology, *in* Smedley, E., Rose, H.J. and Rose, H.J., eds., *Encyclopædia Metropolitana; or Universal Dictionary of Knowledge*: B. Fellowes and J. Rivington; Duncan and Malcolm; Suttaby and Co.; E. Hodgson; J. Dowding; G. Lawford; J.M. Richardson; J. Bohn; T. Allman; J. Bain; S. Hodgson; p. 529–808.

Philoponus, J., 1987, *Against Aristotle on the Eternity of the World*, translated by Christian Wildberg: Cornell University Press, Ithaca, [i] + 182 p.

Pictet, M.-A., 1801, M. A. Pictet, l'un des rédacteurs de la Bibliothéque Britannique à ses collaborateurs. Glasgow 10 Juillet 1801. Lettre III: *Bibliotheque Britannique; ou Recueil Extrait des Ouvrages Anglais Périodique & Autres, des Mémoires & Transactions des Sociétés & Académies de la Grande-Bretagne, d'Asie, d'Afrique & d'Amerique*, v. 18, p. 64–100.

Pitcher, W.S., 1997, *The Nature and Origin of Granite*, second edition: Chapman & Hall, London, xvi + 387 p.

Planck, M., 1949, *Vorträge und Erinnerungen*, fünfte Auflage der *Wege zur Physikalischen Erkenntnis*: S. Hirzel, Stuttgart, VI + [I] + 380 p.

Playfair, J., 1802, *Illustrations of the Huttonian Theory of the Earth*: Cadell and Davies, London, and William Creech, Edinburgh, xx + 528 p.

Playfair, J., 1805, Biographical account of the late Dr. James Hutton: *Transactions of the Royal Society of Edinburgh*, v. 5, p. 39–99, https://doi.org/10.1017/S0080456800020937.

Pledge, H.T., 1940, *Science Since 1500—A Short History of Mathematics, Physics, Chemistry, Biology*, reprinted with corrections: His Majesty's Stationary Office, London, 357 p.

Polyakhova, E.N., and Bessarabova, N.V., 2011, E.R. Dashkova kak Direktor Sankt-Peterburskoi Akademii Nauk: poddjerjka razvitiya matematicheskoi I astronomicheskoi Nauchnikh Shkol, *in* Besserabova, I.V., ed., *E.R. Dashkova i Yekaterinenskaya Epokha—Kulturnii Fundament Sovremennosti*: Moskovskii Gumanitemii Institut im. E.R. Dashkovoi, Moskva, 318 + [II] p.

Popper, K.R., (Sir), 1935, *Logik der Forschung*: Springer Verlag, Wien, vi + 248 p.

Popper, K.R., (Sir), 1958–59 [1989] Back to the Presocratics, *in* Popper, K.R., *Conjectures and Refutations* 5th edition, Routledge, London, p. 136–165 (first published without the footnotes in *Proceedings of the Aristotelian Society*, New Series v. 59, p. 1–24).

Popper, K.R., (Sir), 1998, *The World of Parmenides—Essays on the Presocratic Enlightenment*, edited by Arne F. Petersen with assistance from Jørgen Mejer: Routledge, London and New York, x + 328 p.

Popper-Lynkeus, J., 1924, *Über Religion*: R. Löwit Verlag, Wien und Leipzig, 230 + [I] p.

Porter, C.M., 1986, *The Eagle's Nest—Natural History and American Ideas, 1812–1942*: History of American Science and Technology Series, The University of Alabama Press, [Tuscaloosa and London,], xii + 251 p.

Porter, R.S., 1977, *The Making of Geology—Earth Science in Britain*: Cambridge University Press, Cambridge, x + [i] + 288 p.

Porter, R.S., 1980, Die Geologie Großbritanniens im Zeitalter der Aufklärung: *Zeitschrift der Geologischen Wissenschaften*, Jg. v. 8, p. 53–61.

Povarennikh, A.S., 1953, Dimitrii Ivanovich Sokolov, *in* Belankin, D.S., and Barsanoe, G.P., eds., *Trudi Mineralogiceskogo Museya*, vipusk 5, Izdatelstvo Akademii Nauk SSSR, Moskva, p. 30–55.

Prestwitch, J., (Sir), 1886, *Geology—Chemical, Physical, and Stratigraphical*, v. I *Chemical and Physical*: At the Clarendon Press, Oxford, xxiv + 477 p. + 3 colored foldout maps + 3 colored plates.

Promies, W., 1994, *Georg Christoph Lichtenberg Schriften und Briefe—Kommentar zu Band I und Band II*: Zweitausendeins, Frankfurt-am-Main, 1500 p.

Proßegger, P., 2017, Eduard Reyer (1849–1914): Jurist, Geologe, Kulturhistoriker, Soziologe und Volksbildner: *Berichte der Geologischen Bundesanstalt*, v. 123, p. 86–94.

Ramsay, W., (Sir), 1918, *The Life and Letters of Joseph Black, M.D.*: Constable and Company, Ltd., London, p. xix + 148 + 7 plates.

Rappaport, R., 1960, G.-F. Rouelle: an eighteenth-century chemist and teacher: Chymia: *Annual Studies in the History of Chemistry*, v. 6, p. 68–101, https://doi.org/10.2307/27757193.

Rappaport, R., 1967, Lavoisier's geologic activities 1763–1792: *Isis*, v. 58, p. 375–384, https://doi.org/10.1086/350270.

Rappaport, R., 1973, Lavoisier's theory of the earth: *British Journal for the History of Science*, v. 6, p. 247–260, https://doi.org/10.1017/S0007087400016253.

Read, H.H., 1943, Meditations on granite: part one: *Proceedings of the Geologists' Association*, v. 54, p. 64–85.
Read, H.H., 1944, Meditations on granite: part two: *Proceedings of the Geologists' Association*, v. 55, p. 45–93 + frontispiece.
Read, H.H., 1951, *Metamorphism and granitisation*: Alex. L. du Toit Memorial Lectures No. 2, The Geological Society of South Africa Annexure to volume LIV, Johannesburg, 27 p.
Read, H.H., 1957, *The Granite Controversy*: Thomas Murby & Co., London, xix + 430 p. + frontispiece.
Rees, A., 1819, *The Cyclopædia; or, Universal Dictionary of Arts, Sciences, and Literature*, v. 36: Longman, Hurst, Rees, Orme & Brown (and numerous others), London, unpaginated.
Reichetzer, F., 1812, *Anleitung zur Geognosie, insbesondere zur Gebirgskunde—nach Werner für die k.k. Berg-Akademie*: Camesianische Buchhandlung, Wien, XII + 292 p.
Reichetzer, F., 1821, *Anleitung zur Geognosie, insbesondere zur Gebirgskunde—nach Werner für die k. k. Berg-Akademie*, zweite umgearbeitete Auflage: Camesianische Buchhandlung, Wien, XVIII + 286 p.
Reingold, N., ed., 1964, *Science in Nineteenth Century America—A Documentary History*: American Century Series, Hill and Wang, New York, xii + 339 p.
Reisch, G., 1517 (2016), *Margarita Philosophica—Perle (Schatz) der Philosophie—Abdruck der vom Verfasser autorisierten verbesserten und vermehrten 4. Auflage Basel 1517*. Deutsche Übersetzung von Otto und Eva Schönberger: Königshausen & Neumann, Würzburg, XX + [I] + 540 p.
Renovantz, H. M., 1788, *Mineralogisch-geographische und andere vermischte Nachrichten von den Altaischen Gebürgen Russisch Kayserlichen Anteils*: privately printed, Reval, XIX + 274 p. + 1 foldout map.
van Rensselaer, J., 1825, *Lectures on Geology; Being Outlines of the Science Delivered to the New York Athanaeum in the year 1825*: E. Bliss & E. White, New York, xiv + 358 p.
Reuß, F.A., 1805, *Lehrbuch der Mineralogie nach des Herrn O. B. R. Karsten Mineralogische Tabellen, dritten Theiles erster Band welcher die Geognosie enthält*: Friedrich Gotthold Jacobäer, Leipzig, IV + 506 p.
Reventlow, H., [L.G.] (Count), 2001, *Epochen der Bibelauslegung*, v. 4, *Von der Aufklärung bis zum 20. Jahrhundert*: C.H. Beck, München, 448 p.
Reventlow, H., [L.G.] (Count), Sparn, W. and Woodbridge, J., editors, 1988, *Historische Kritik und Biblischer Kanon in der Deutschen Aufklärung*: Wolfenbütteler Forschungen, v. 41, Otto Harrassowitz, Wiesbaden, VII + 293 p.
Rey, A., 1933, *La Science dans l'Antiquité—La Jeunesse de la Science Grecque*, in Berr, H., ed., L'Évolution de l'Humanité, La Renaissance du Livre, Paris, XVII + 519 p.
Reyer, E., 1888, *Theoretische Geologie*: E. Schweizerbart'sche Verlagsbuchhandlung (E. Koch), Stuttgart, XIII + 867 p. + 3 maps.
Richardson, G.F., 1842, *Geology for Beginners*: Hippolyte Baillère, London, xx + 530 p. + frontispiece.
Richardson, G.F., 1843, *Geology for Beginners*, second edition: Longman, Brown, Green, and Longmans, London, Wiley and Putnam, New York, xx + 624 p. + frontispiece.
Richardson, G. F., 1851, *An Introduction to Geology, and its Associated Sciences, Mineralogy, Fossil Botany and Conchology, and Palæontology*, a new edition, revised and considerably enlarged by Thomas Wright: H.G. Bohn, London, xvi + 508 p. + frontispiece.
Ritter, C., 1817, *Erdkunde im Verhältnis zur Natur und zur Geschichte des Menschen oder allgemeine, vergleichende Geographie, als sichere Grundlage des Studiums und Unterrichts in physicalischen und historischen Wissenschaften, erster Theil*: G. Reimer, Berlin, XX + 832 p.
Ritter, C., 1822, *Erdkunde im Verhältnis zur Natur und zur Geschichte des Menschen oder allgemeine, vergleichende Geographie, als sichere Grundlage des Studiums und Unterrichts in physicalischen und historischen Wissenschaften, erster Theil, erstes Buch. Afrika*, Zweite stark vermehrte und verbesserte Ausgabe: G. Reimer, Berlin, XXVII+1 page of errata + 1084 p.
Roger, J., 1962, *Buffon Les Epoques de la Nature—Édition Critique*: Mémoires du Muséum National d'Histoire Naturelle, Sciences de la Terre, v. 10, CXLIX + 343 p.
Roger, J., 1989, *Buffon—Un Philosophe au Jardin du Roi*: Fayard, [Paris], 645 p.
Roger, J., 1997, *Buffon—A Life in Natural History*, translated by Sarah Lucille Bonnefoi, edited by L. Pearce Williams: Cornell University Press, Ithaca and London, xvii + 492 p. + frontispiece.
Rogers, H.D., 1859a, *The Geology of Pennsylvania—A Government Survey with a General View of the Geology of the United States—Essays on the Coal-Formation and Its Fossils, and a Description of the Coal-Fields of North America and Great Britain*: J.B. Lippincott & Co., Philadelphia, I, xxvi + 586 p. + unnumbered plates.
Rogers, H.D., 1859b, *The Geology of Pennsylvania—A Government Survey with a General View of the Geology of the United States—Essays on the Coal-Formation and Its Fossils, and a Description of the Coal-Fields of North America and Great Britain*: J.B. Lippincott & Co., Philadelphia, v. II, Part I, xxiv + 666 p.
Rogers, H.D., 1859c, *The Geology of Pennsylvania—A Government Survey with a General View of the Geology of the United States—Essays on the Coal-Formation and Its Fossils, and a Description of the Coal-Fields of North America and Great Britain*: J.B. Lippincott & Co., Philadelphia, v. II, Part II, p. 668–1045 + 23 plates.
Rogers, H.D., 1859d, *The Geology of Pennsylvania—A Government Survey with a General View of the Geology of the United States—Essays on the Coal-Formation and Its Fossils, and a Description of the Coal-Fields of North America and Great Britain*: J.B. Lippincott & Co., Philadelphia, folder with five folded colored maps.
Rogers, W.B., and Rogers, H.D., 1843, On the physical structure of the Appalachian Chain as exemplifying the laws which have regulated the elevation of great mountain chains generally: *Reports of the First, Second and Third Meetings of the Association of American Geologists and Naturalists, at Philadelphia in 1840 and 1841, and at Boston in 1842, embracing its Proceedings and Transactions*, Gould, Kendall & Lincoln, Boston, p. 474–531 (reprinted with few corrections in Rogers, W.B., 1884, *A Reprint of Annual Reports and Other Papers on the Geology of the Virginias*: D. Appleton and Company, New York, p. 601–642).
Rolle, F., [1888], *Zweiter Teil: Geologie und Paläontologie*, in Kenngott, A., and Rolle, F., authors, *Naturgeschichte des Mineralreichs für Schule und Haus*: I.J. Schreiber, München, 74 p. + 24 colored plates + 40 p. + 18 plates + unpaginated index.
Rosenberg, G.D., 2006, Nicholas Steno's Chaos and the shaping of evolutionary thought in the Scientific Revolution: *Geology*, v. 34, p. 793–796, https://doi.org/10.1130/G22655.1.
Rossetter, T., 2018, Realism on the rocks: Novel success and James Hutton's theory of the earth: *Studies in History and Philosophy of Science*, v. 67, p. 1–13.
Roth, J., 1872, Über die Lehre vom Metamorphismus und die Entstehung der kristallinischen Schiefer: *Abhandlungen der königlichen Akademie der Wissenschaften zu Berlin*, for the year 1871, p. 151–232.
Rozet, C. A., 1830, *Cours Élémentaire de Géognosie, fait au Dépot Générale de Guerre*: F.G. Levrault, Paris, viij + 460 p. + 5 plates.
Rozet, C.A., 1837, *Traité Élémentaire de Géologie—première Partie–Géognosie*: Arthus Bertrand, Paris, xiv + 537 p.
Rudwick, M.J.S., 1964, The inference of function from structure in fossils: *The British Journal for the Philosophy of Science*, v. XV, p. 27–40, https://doi.org/10.1093/bjps/XV.57.27.
Rudwick, M.J.S., 2005, *Bursting the Limits of Time The reconstruction of Geohistory in the Age of Revolution*: Chicago University Press, Chicago & London, XXIV + 708 p.
Rudwick, M.J.S., 2009, Author's response to Şengör's review of *Bursting the Limits of Time*: *Earth Sciences History*, v. 28, p. 135–141.
Rudwick, M.J.S., 2011, Geohistory and the historicity of Genesis, in Heilbron, J.L. and Sigrist, R., eds., *Jean-André Deluc—Historian of Earth and Man*, Slatkine Érudition, Genève, p. 242–260.
Rudwick, M.J.S., 2014, *Earth's Deep History—How it was Discovered and Why it Matters*: The University of Chicago Press, Chicago and London, ix + 360 p.
Rupke, N.A., 1983a, *The Great Chain of History. William Buckland and the English School of Geology (1814–1849)*: Clarendon Press, Oxford, xii + [1] + 322 p.
Rupke, N.A., 1983b, The apocalyptic denominator in English culture of the early nineteenth century, in Pollock, M., ed., *Common Denominators in Art and Science*, Aberdeen University Press, Aberdeen, p. 30–41.
Russell, B.[A.W.], (Lord), 1946, *History of Western Philosophy and its Connection with Political and Social Circumstances from the Earliest Times to the Present Day*: George Allen and Unwin, London, 916 p.
Sacchi, G.D., 1980, Un dimenticato pioniere della tettonica su modelli: Edmund [sic] Reyer (1849–1914): *Atti della Accademia delle Scienze di Torino*, v. 114 (Geologia e Paleontologia), p. 87–92.
Sainte-Claire Deville, C.J., 1878, *Coup-d'œil Historique sur la Géologie et sur les Travaux d'Élie de Beaumont—Leçons Professées au Collège de France (mai-juillet 1875)*: G. Masson, Paris, VII + 597 p. + 1 folded table.
Salomon, W., 1918, Tote Landschaften und der Gang der Erdgeschichte: *Sitzungsberichte der Heidelberger Akademie der Wissenschaften*, mathematisch-naturwissenschaftliche Klasse, v. 9A, 1. Abhandlung, p. 1–10.

Sarjeant, W.A.S., 1988, Book review of *From Mineralogy to Geology: The Foundations of a Science*, by Rachel Laudan, University of Chicago Press, 278 p., 1987; $27.50US, cloth: *Geoscience Canada*, v. 15, p. 307–308.

Schlegelberger, G., 1935, *Die Fürstin Daschkowa—Eine Biographische Studie zur Geschichte Katharinas II.*: Neue Deutsche Forschungen, Abteilung Slawische Philologie und Kulturgeschichte, Junker und Dünnhaupt, Berlin, 249 p. + 1 table (pages 100 to 199 were incorrectly paginated as 300 to 399).

Schmeckebier, L.F., 1904, *Catalogue and Index of the Publications of the Hayden, King, Powell, and Wheeler Surveys namely Geological and Geographical Survey of the Territories, Geological Exploration of the Fortieth Parallel, Geographical and Geological Surveys of the Rocky Mountain Region, Geographical Surveys West of the One Hundredth Meridian*: Government Printing Office, Washington, 208 p., https://doi.org/10.5962/bhl.title.55009.

Schmieder, C., 1807, *Theophrasts Abhandlung von den Steinarten*, aus dem Griechischen übersetzt und mit Anmerkungen begleitet: Craz und Gerlach, Freiberg, XII + 84 p.

Schiffner, C., 1935, *Aus dem Leben Alter Freiberger Bergstudenten*: Ernst Mauckisch, Freiberg, XV + 375 p.

Schneer, C.J., ed., 1979, *Two Hundred Years of Geology in America—Proceedings of the New Hampshire Bicentennial Conference on the History of Geology*: University Press of New England, Hanover, [ix] + 485 p.

Schreiber, G., 1962, *Der Bergbau in Geschichte, Ethos und Sakralkultur*: Wissenschaftliche Abhandlungen der Arbeitsgeleinschaft für Forschung des Landes Nordrhein-Westfalen, v. 21, Westdeutscher Verlag, Köln und Opladen, 757 p. + 31 plates.

Scrope, G.P., 1825, *Considerations on Volcanos, the Probable Causes of their Phenomena, the Laws which Determine their March, the Disposition of their Products, and their Connexion with the Present State and Past History of the Globe; Leading to the Establishment of A New Theory of the Earth*: W. Phillips, London, xxxi + 270 p. + 2 foldout plates.

Scrope, G.J.P., 1827, *Memoir on the Geology of Central France; including the Volcanic Formations of Auvergne, the Velay and the Vivarais*: Longman, Rees, Orme, Brown, and Green, London, xiv + 182 p. + 18 plates.

Şengör, A.M.C., 1991, Timing of orogenic events: a persistent geological controversy, *in* Müller, D.W., McKenzie, J.A. and Weissert, H., eds., *Modern Controversies in Geology* (Proceedings of the Hsü Symposium): Academic Press, London, p. 405–473.

Şengör, A.M.C., 1992, The mountain and the bull: The origin of the word "Taurus" as part of the earliest tectonic hypothesis, *in* Başgelen, N., Çelgin, G. and Çelgin, V., eds., *Zafer Taşlıklıoğlu Armağanı (Festschrift für Zafer Taşlıklıoğlu)*: Arkeoloji ve Sanat Yayınları, İstanbul, p. 1–48.

Şengör, A.M.C., and Natal'in, B.A., 1996, Palaeotectonics of Asia: Fragments of a synthesis, *in* Yin, A., and Harrison, M., eds., *The Tectonic Evolution of Asia, Rubey Colloquium*: Cambridge, Cambridge University Press, p. 486–640.

Şengör, A.M.C., 2001, *Is the Present the Key to the Past or the Past the Key to the Present? James Hutton and Adam Smith versus Abraham Gottlob Werner and Karl Marx in Interpreting History*: Geological Society of America Special Paper 355, x + 51 p., https://doi.org/10.1130/SPE355.

Şengör, A.M.C., 2002, On Sir Charles Lyell's alleged distortion of Abraham Gottlob Werner in *Principles of Geology* and its implications for the nature of the scientific enterprise: *The Journal of Geology*, v. 110, p. 355–368, https://doi.org/10.1086/339537.

Şengör, A.M.C., 2003, *The Large Wavelength Deformations of the Lithosphere: Materials for a history of the evolution of thought from the earliest times to plate tectonics*: Geological Society of America Memoir 196, xvii + 347 p. + 3 folded plates in pocket, https://doi.org/10.1130/MEM196.

Şengör, A.M.C., 2009a, Globale *Geologie und ihr Einfluss auf das Denken von Eduard Suess—Der Katastrophismus-Uniformitarianismus-Streit*: Scripta Geo-Historica, v. 2, 181 p.

Şengör, A.M.C., 2009b, Zum Lauf des Oxos, ein Nachtrag, *in* Stückelberger, A., und Mittenhuber, F., eds., *Ptolemaios Handbuch der Geographie Ergänzungsband mit einer Edition des Kanons bedeutender Städte*: Schwabe, Basel, p. 316–318.

Şengör, A.M.C., 2009c, Essay Review: A Rankean view of historical geology and its development: *Earth Sciences History*, v. 28, p. 108–134.

Şengör, A.M.C., 2014a, Eduard Suess and global tectonics: an illustrated "short guide" [Suess special issue]: *Mitteilungen der Österreichischen Geologischen Gesellschaft*, v. 107, p. 6–82.

Şengör, A.M.C., 2014b, *Dahi Diktatör*: Ka Kitap, İstanbul, 150 p.

Şengör, A.M.C., 2016, What is the use of the history of geology to a practicing geologist? The prepaedeutical case of stratigraphy: *The Journal of Geology*, v. 124, p. 643–698, https://doi.org/10.1086/688609.

Silberschlag, J. E., 1780, *Geogenie oder Erklärung der Mosaischen Erderschaffung nach Physikalischen und Mathematischen Grundsätzen*: Verlag der Buchhandlung der Realschule, Berlin, v. I, x + 194 p. + nine plates.

Silliman, B., 1829, *Outline of the Course of Geological Lectures given in Yale College*: Hezekiah Howe, New Haven, 128 p.

Silliman, B., 1839, Appendix to the Third American from the Fifth English Edition of Bakewell's Geology (see under Bakewell), p. 461–579.

[Simond, L.], 1815, *Journal of a Tour and Residence in Great Britain during the Years 1810 and 1811, by a French Traveller: with Remarks on the Country, its Arts, Literature, and Politics, and on the Manners and Customs of its Inhabitants*, volume second: Archibald Constable and company, Edinburgh, and Longman, Hurst, Rees, Orme, and Brown, London, 360 p. + [i] p.

Simonin, L.[L.], 1867, *Histoire de la Terre—Origines et Métamorphoses du Globe*: J. Claye, Paris, 269 p. + 1 errata page.

Sivin, N., 1988, Science and medicine in Imperial China—the state of the field: *The Journal of Asian Studies*, v. 47, p. 41–90, https://doi.org/10.2307/2056359.

Snelders, H.A.M., 1970, Romanticism and Naturphilosophie and the inorganic natural sciences 1797–1840: an introductory survey: *Studies in Romanticism*, v. 9, p. 193–215, https://doi.org/10.2307/25599763.

Snelders, H.A.M., 1993, Johann Joachim Becher und sein Gold-aus-Sand-Projekt, *in* Frühsorge, G., and Strasser, G.F., eds., *Johann Joachim Becher (1635–1682)*, Wolfenbütteler Arbeiten zur Barockforschung, v. 22: Otto Harrassowitz, Wiesbaden, p. 103–114.Sokolov, D., 1825, Uspekhi geognozii. *Gornyy Zhurnal*, no. 1, p. 5–27.

Sokolov, D., 1842, *Rukovodstvo k geognozii, sostavlennoye Korpusa Gornykh Inzhenerov General-Mayorom, S. Peterburgskogo Universiteta Professorom D. Sokolovym* Chast' vtoraya: V tipografiii Eduarda Pratsa, Sanktpeterburg, 343 p.

Spieker, E.M., 1971, Schopf, Maclure, Werner and the earliest work on American geology: *Science*, v. 172, p. 1333–1334, https://doi.org/10.1126/science.172.3990.1333.

Spieker, E.M., 1972, Translator's introduction, *in* Schöpf, Johann David, *Geology of Eastern North America*, translated by Edmund M. Spieker with a foreword by George W. White, Hafner, New York, p. 1–36.

Stahl, G. E., 1700 [25. III.], *Propempticon Inaugurale de Ortu Venarum Metalliferarum*: no publisher, no place of publication, 7 (unnumbered) p.

Stahl, G.E., 1703, *"Specimen Beccherianum" sistens Fundamenta, Documenta, Experimenta, quibus Principia Mixtionis Subterraneae, & Instrumenta Naturali atque Artificialia demonstrantur. Ex Autoris Scriptis, colligendo, corrigendo, connectendo, supplendo, concinnatum, exhibet Georgus Emestus Stahl*: Ex Officina Weidmanniana, Lipsiae, 161 p.

Steffens, H., 1801, *Beyträge zur Innern Naturgeschichte der Erde*, erster Theil: Verlag der Crazischen Buchhandlung, Freyberg, [III] + 317 p. + 1 p. of errata.

Stenonis, N., 1667, *Elementorum Myologiæ Specimen, seu Musculi Descriptio Geometrica. Cui accedunt Canis Carchariæ Dissectum Caput, et Dissectus Piscis ex Canum Genere*: Typographia sub signo Stellæ, Florentiæ, [VI] + 123 p. + 7 foldout plates.

Stenonis, N., 1669, *De Solido intra Solidum Naturaliter Contento Dissertationis Prodromus*: Typographia sub signo Stellæ, Florentiæ, 78 p. 2 foldout plates.

Stickler, F., [C.L.], 1812, Ideen zu einem vulcanischen Erdglobus, oder zu einer Darstellung aller auf der Oberfläche unseres Erdkörpers verbreiteten, ehemaligen und jetzigen Vulcane, nebst den für Naturphilosophie daraus sich ergebenden Resultaten: *Allgemeine Geographische Ephemeriden*, v. 38, erstes Stück, p. 121–190 + a colored foldout map.

Stille, H., 1922, *Die Schrumpfung der Erde*: Gebrüder Borntraeger, Berlin, 37 p.

Stille, H., 1924, *Grundfragen der Vergleichenden Tektonik*: Gebrüder Borntraeger, Berlin, VII + [I] + 443 p.

Stille, H., 1940, *Einführung in den Bau Amerikas*: Gebrüder Borntraeger, Berlin, XX + 717 p.

Stoppani, A., 1873, *Corso di Geologia*, v. II *Geologia Stratigrafica*: G. Bernardoni e G. Brigola, Milano, 868 p.

Strange, P.J., and Shaw, R., 1986, *Geology of Hong Kong Island and Kowloon—1:20 000 Sheets 11 & 15*: Hong Kong Geological Survey Memoir no. 2, Geotechnical Control Office, Civil Engineering Services Department, Hong Kong, 134 p.

Struik, D.J., 1948, *Yankee Science in the Making*: Little, Brown and Company, Boston, xiii + 430 p.

Studer, B., 1844, *Lehrbuch der Physikalischen Geographie und Geologie—erstes Capitel enthaltend: Die Erde im Verhältniss zur Schwere*: J.F.J. Dalp, Bern, Chur and Leipzig, IX + 398 p.

Studer, B., 1847, *Lehrbuch der Physikalischen Geographie und Geologie—zweites Capitel enthaltend: Die Erde im Verhältniss zur Wärme*: J.F.J. Dalp, Bern, Chur and Leipzig, VII + 526 p.

Studer, B., 1863, *Geschichte der Physischen Geographie der Schweiz bis 1815*: Stämpflische Verlagshandlung, Bern and Friedrich Schulthess, Zürich, IX + 696 p.

Suess, E., 1875, *Die Entstehung der Alpen*: W. Braumüller, Wien, IV + 168 p.

Suess, E., 1883, *Das Antlitz der Erde*, v. Ia (Erste Abtheilung): F. Tempsky, Prag and G. Freytag, Leipzig, 310 p. + frontispiece.

Suess, E., 1916, *Erinnerungen*: S. Hirzel, Leipzig IX + 451 p.

Taylor, K.L., 1971, The geology of Déodat de Dolomieu, *in* XIIe Congrès International d'Histoire des Science, Paris 1968, Actes, t. VII, *Histoire des Sciences de la Terre et de l'Océanographie*, Albert Blanchard, Paris, p. 49–53.

Taylor, K.L., 2009, Desmarest's "Determination of some epochs of nature through volcanic products" (1775/1779): *Episodes*, v. 32, p. 114–124, https://doi.org/10.18814/epiiugs/2009/v32i2/004.

Taylor, K.L., 2013, A peculiarly personal encyclopedia: What Desmarest's *Géographie Physique* tells us about his life and work: *Earth Sciences History*, v. 32, p. 39–54, https://doi.org/10.17704/eshi.32.1.w2tq16t77n3585x1.

Teske, R.J., S.J., 1995, William of Auvergne's arguments for the newness of the World: *Mediævalia: Textos e Estudios*, v. 7–8, p. 287–302.

Thomson, J.O., 1948[1965], *History of Ancient Geography*: Cambridge University Press (reprinted in 1965 by Biblo and Tannen, New York), 427 p.

Tomkeieff, S.I., 1946, James Hutton's "Theory of the Earth," 1795: *Proceedings of the Geologists' Association*, v. 57, p. 322–328.

Tomkeieff, S.I., 1948, James Hutton and the Philosophy of Geology: *Transactions of the Edinburgh Geological Society*, v. 14, p. 253–276, https://doi.org/10.1144/transed.14.2.253.

Torrens, H.S., 1998, James Hutton and the historians: *Geological Society of America Abstracts with Programs*, v. 30, no. 7, p. A-101.

Torrens, H.S., 2000, New light on William Maclure's early geology and travels: *Geological Society of America Abstracts with Programs*, v. 32, no. 7, p. 87.

Torrens, H.S., 2004a, Williams, John (1732–1795), *in* Harrison, B., ed., *Oxford Dictionary of National Biography*, https://doi.org/10.1093/ref:odnb/56916 (accessed 26 July 2019).

Torrens, H.S., 2004b, Richardson, George Fleming (1796–1848), *in* Harrison, B., ed., *Oxford Dictionary of National Biography*, https://doi.org/10.1093/ref:odnb/23555 (accessed 2 May 2019).

Torrens, H.S., 2017, Thomas Beddoes and natural history, especially geology, *in* Levere, T., Stewart, L., Torrens, H., and Wachelder, J., eds., *The Enlightenment of Thomas Beddoes—Science, Medicine, and Reform*: Routledge, Abingdon, Science, Technology, Culture, 1700–1945, p. 79–115.

Touret, J.L.R., 2004, Hermann Vogelsang (1838–1874). "European avant la letter," *in* Touret, J.L.R., and Visser, R.P.W., eds., *Dutch Pioneers of the Earth Sciences*, Koninklijke Akademie van Wetenschappen, Amsterdam, p. 87–108.

Tozzetti, G.T., 1752, *Relazioni d'alcuni Viaggi Fatti in Diverse Parti della Toscana, per osservare le Produzioni Naturali, e gli Antichi Monumenti di essa*, v. 3: Stamperia Imperiale, Firenze, VIII + 462 p.

de Tribolet, M., 1883, *La Géologie—Son Objet, Son Développement—Sa Méthode, Ses Applications—Conférence Académique*: James Attinger, Neuchâtel, 49 p.

Uehlinger, C., 2013, Genesis 1–11: Die Urgeschichte, *in* Römer, T., Macchi, J.-D., and Nihan, C., eds., *Einleitung in das Alte Testament*, translated from the French by Christine Henschel, Julia Hillebrand and Wolfgang Hüllstrung[182]: Theologischer Verlag, Zürich, p. 176–195.

Umlauft, F., 1887, *Die Alpen. Handbuch der Gesammten Alpenkunde*: A. Hartleben, Wien, Pest, Leipzig, VIII + 488 p. + 5 colored maps *hors-texte*.

Un Ancien Officier de Marine [Éléonor-Jacques-François de Sales Guyon de Diziers, Count Montlivault], 1821, *Conjectures sur la Réunion de la Lune et la Terre, et des Satellites en Général à leur Planète Principale; à l'aide desquelles on essaie d'expliquer la cause et des effets du Déluge, la disparition totale d'anciennes espèces vivantes et organiques, et la formation soudaine ou apparition d'autres espèces nouvelles, et l'homme lui-même sur le globe terrestre*: Adrien Egron, Paris, 32 p. + 1 foldout plate.

Ure, A., 1829, *A New System of Geology in Which the Great Revolutions of the Earth and Animated Nature are Reconciled at once to Modern Science and Sacred History*: Longman, Rees, Orme, Brown & Green, London, lv + 621 p. + 1 errata page + 7 plates.

Ussher, J. (Archbishop), 1645, *A Body of Divinitie or the Summe and Substance of Christian Religion, Catechistically propounded, and explained by way of Question and Answer: Methodically and familiarly handled, composed long since ... And at the earnest desires of divers godly Christians now Printed and Published. Whereunto is adjoyned a Tract, intituled Immanuel, or The Mystery of the Son of God; Heretofore written and published by the same author*: Tho: Downes and Geo: Badger, London, [i] + 451 + [xi] p.

Vai, G.B., 2003, Aldrovandi's will: introducing the term "geology" in 1603, *in* Vai, G.B., and Cavazza, W., eds., *Four Centuries of the Word "Geology": Ulisse Aldrovandi 1603 in Bologna*, Minerva Edizioni, Bologna, p. 65–110.

Vai, G.B., 2009, The scientific revolution and Steno's twofold conversion, *in* Rosenberg, G.D., ed., *The Revolution in Geology from the Renaissance to the Enlightenment*: Geological Society of America Memoir 203, p. 175–196, https://doi.org/10.1130/2009.1203(14).

Vai, G.B., and Cavazza, W., 2006, Ulisse Aldrovandi and the origin of geology and science, *in* Vai, G.B. and Caldwell, W.G.E., eds., *The Origins of Geology in Italy*: Geological Society of America Special Paper 411, p. 43–63, https://doi.org/10.1130/2006.2411(04).

Vaysey, H.W., 1826a, On some fossil shells found in the Gawilghur Range of Hills, in April 1823 [+ plate VIII.]: *Memoirs of the Wernerian Natural History Society*, v. 5, part II, p. 289–297.

Vaysey, H.W., 1826b, On the geological structure of the Hill of Seetabuldee, Nagpoor, and its immediate vicinity: *Memoirs of the Wernerian Natural History Society*, v. 5, part II, p. 298–302.

Verbrugghe, G.P., and Wickersham, J.M., 1996, *Berossos and Manetho, Introduced and Translated—Native Traditions in Ancient Mesopotamia and Egypt*: The University of Michigan Press, Ann Arbor, ix + [i] + 239 p.

Vernant, J.-P., 1962 (1990), *Les Origines de la Pensée Grecque*: Quadrige/Presses Universitaires de France, Paris, 133 + [I] p.

Vernant, J.-P., 1982, *The Origins of Greek Thought*, translated from the French: Cornell University Press, Ithaca, 144 p.

Verne, J., [1864], *Voyage au Centre de la Terre*: Bibliothèque d'Éducation et de Récreation, J. Hetzel, Paris, 335 p.

Vézian, A., 1862, *Prodrome de Géologie—Introduction—Livre Premier*: F. Savy, Paris, 615 + [ii] p.

Vogelsang, H., 1867, *Philosophie der Geologie und Mikroskopische Gesteinsstudien*: Max Cohen & Sohn, Bonn, [II] + 229 p. + 19 colored plates.

Vogt, C., 1846, *Lehrbuch der Geologie und Petrefactenkunde. Zum Gebrauche bei Vorlesungen und Selbsunterrichte. Theilweise nach L. Élie de Beaumont's Vorlesungen an der Ecole des Mines*: Friedrich Vieweg und Sohn, Braunschweig, v. 1, XIX + 436 p. + 2 unnumbered plates.

Vogt, C., 1847, *Lehrbuch der Geologie und Petrefactenkunde. Zum Gebrauche bei Vorlesungen und Selbsunterrichte. Theilweise nach L. Élie de Beaumont's Vorlesungen an der Ecole des Mines*: Friedrich Vieweg und Sohn, Braunschweig, v. 2, XVI + 476 p. + 13 partly numbered plates.

Voigt, J.C.W., 1789, Ein Wörtchen für die Vulkanität des Basalts: *Magazin für Naturkunde Helvetiens*, v. 4, p. 214–232.

Voigt, J.C.W., 1805, *Erklärendes Verzeichniß seiner Kabinets von Gebirgsarten*, vierte und verbesserte Auflage: Verlag von Industrie-Comptoir, Weimar, v. 47, p. 279–314.

Wagner, A., 1845, *Geschichte der Urwelt mit Besonderer Berücksichtigung der Menschenrassen und des Mosaischen Schöpfungsberichtes*: Leopold Voß, Leipzig, XVIII + 578 p.

Wagner, A., 1857, *Geschichte der Urwelt mit Besonderer Berücksichtigung der Menschenrassen und des Mosaischen Schöpfungsberichtes*, zweite vermehrte Auflage, erster Theil *Die Erdveste nach Ihrem Felsbaue und Ihrer Schöpfungsgeschichte*: Leopold Voss, Leipzig, XVI + 550 p.

Wagner, A., 1858, *Geschichte der Urwelt mit Besonderer Berücksichtigung der Menschenrassen und des Mosaischen Schöpfungsberichtes*, zweite vermehrte Auflage, zweiter Theil *Das Menschengeschlecht und das Their- und Pflanzenreich der Urwelt*: Leopold Voss, Leipzig, VI + 528 p.

Wagner, A., 1860, Betrachtungen über den gegenwärtigen Standpunkt der Theorien der Erdbildung nach ihrer geschichtlichen Entwicklung in den letzten fünfzig Jahren: *Sitzungsberichte der Königlich Bayerischen Akademie der Wissenschaften zu München Jahrgang*, v. 1860, p. 375–424.

Warren, L., 2009, *Maclure of New Harmony—Scientist, Progressive Educator, Radical Philanthropist*: Indiana University Press, Bloomington & Indianapolis, xii + 343 p. + 13 p. of plates.

Watermann, B., and Wrzesinski, O.J., 1989, *Bibliographie zur Geschichte der Deutschen Meeresforschung*: Deutsche Gesellschaft für Meeresforschung, Hamburg, xiii + 248 p.

Watt, G., 1804, Observations on basalt, and on the transition from the vitreous to the stony texture, which occurs in the gradual refrigeration of melted basalt; with some geological remarks [p.]: *Philosophical Transactions of the Royal Society of London*, v. 94.

Wawn, A., 2000, *The Vikings and the Victorians—Inventing the Old North in Nineteenth-Century Britain*: D.S. Brewer, Cambridge, xiii + [iv] + 434 p.

Wegmann, C.E., 1957, Avant-Propos, *in* Raguin, E., ed., *Géologie du Granite*, deuxième édition, revue et mise en jour, Masson et Cie, Paris, p. VII–XVI.

Weinberg, S., 2015, *To Explain the World—The Discovery of Modern Science*: Harper, New York, xiv + 416 p. + [i].

Wellnhofer, P., 1980, The history of the Bavarian State Collection of Palaeontology and Historical Geology in Munich: *Journal of the Society of the Bibliography of Natural History*, v. 9, p. 383–390, https://doi.org/10.3366/jsbnh.1980.9.4.383.

Wells, D.A., 1861, *Wells's First Principles of Geology. A Text-Book for Schools, Academies and Colleges*: Ivison, Phinney, Blakeman & Co., New York, S.C. Griggs, Chicago, viii + 333 p. + 1 frontispiece.

Werner, A.G., 1774, *Von den Äußerlichen Kennzeichen der Fossilien*: Siegfried Lebrecht Crusijus, Leipzig, 302 + [I] p. + 1 page of errata.

Werner, A.G., 1780, Vorrede des Uebersetzers, *in* von Kronstedt, A., ed., *Versuch einer Mineralogie*, aufs neue aus dem schwedischen übersetzt und nächst verschiedenen Anmerkungen vorzüglich mit äusern Beschreibungen der Fossilien vermehrt durch Abraham Gottlob Werner, ersten Bandes erster Theil: Siegfried Lebrecht Crusius, Leipzig, 30 unnumbered pages.

Werner, A.G., 1791, *Neue Theorie von der Entstehung der Gänge, mit Anwendung auf den Bergbau besonders den Freibergischen*: Gerlachische Buchdruckerei, Freiberg, XXXX + 256 p.

Werner, A.G., 1818, Allgemeine Betrachtungen über den festen Erdkörper: *Auswahl aus den Schriften der unter Werner's Mitwirkung gestifteten Gesellschaft für Mineralogie zu Dresden*, v. 1, p. 39–56.

Werner, A.G., 1962, *On the External Characters of Minerals*, translated by Albert V. Carozzi: University of Illinois Press, Urbana, xxix + [ii] + 118 p.

West, A.F., ed., 1889a, *Richard de Bury Philobiblon ex Optimis Codicibus Recensuit*: printed for the Grolier Club, New York, by the De Vinne Press, *Philobiblon*, v. I, 131 p.

West, A.F., ed., 1889b, *The Philobiblon of Richard de Bury*: printed for the Grolier Club, New York, by the De Vinne Press, *Philobiblon*, v. II, 146 p.

West, A.F., 1889c, *The Philobiblon of Richard de Bury* edited from the best Manuscripts and translated into English with an introduction and notes by Andrew Fleming West, part III Introductory matter and Notes: printed for the Grolier Club, New York, by the De Vinne Press, *Philobiblon*, v. III, 173 p.

[Whewell, W.], 1831, *Principles of Geology, being an Attempt to Explain the Former Changes of the Earth's Surface, by Reference to Causes now in Operation. By Charles Lyell, Esq., F.R.S. For. Sec. to the Geol. Soc., &c. In 2 vols. Vol. I with Wood Cuts, Plans, &c. 15s. Murray. 1830.*: *British Critic, Quarterly Theological Review, and Ecclesiastical Record*, v. 9, p. 180–206.

[Whewell, W.], 1832, *Principles of Geology, being an Attempt to Explain the Former Changes of the Earth's Surface, by Reference to Causes Now in Operation. By Charles Lyell, Esq., F.R.S., Professor of Geology in King's College, London, v. II. London. 1832.*: *The Quarterly Review*, v. 47, p. 103–132.

Whiston, W., 1696, *A New Theory of the Earth, From Its Original, to the Consummation of all Things Wherein the Creation of the World in six Days, the Universal deluge, and the General Conflgration As Laid Down in the Holy Scriptures Are Shown to be Perfectly Agreeable to Reason and Philosophy*: Benj. Tooke, London, [II] + 388 + [1] p. + 7 plates.

White, G.W., 1970, William Maclure was a uniformitarian and not a real Wernerian: *Journal of Geological Education*, v. 18, p. 127–128, https://doi.org/10.5408/0022-1368-XVIII.3.127.

Wigley, P., Darrell, J., Clements, D., and Torrens, H., 2018, *William Smith's Fossils Reunited—Strata Identified by Organized Fossils and a Stratigraphical System of Organized Fossils by William Smith with Fossil Photographs from His Collection at the Natural History Museum*: Natural History Museum, Halsgrove, The Geological Society, London, x + 150 p.

Williams, J., 1789, *The Natural History of the Mineral Kingdom*. In three parts: printed for the author by Thomas Ruddiman, Edinburgh, v. I, lxii + 450 p. + 1 errata page.

Williams, J., 1810, *The Natural History of the Mineral Kingdom, Relative to the Strata of Coal, Mineral Veins, and the Prevailing Strata of the Globe*, the second edition. With an appendix, containing a more extended view of mineralogy and geology. Illustrated with engravings by James Millar: Bell & Bradfute, and Laing, Edinburgh; Longman, Hurst, Rees, and Orme; and J. White & Co. London, v. II, vi + 597 p. + 2 plates.

Wilson, L.G., 1980, Geology on the eve of Charles Lyell's first visit to America, 1841: *Proceedings of the American Philosophical Society*, v. 124, p. 168–202.

Wilson, L.G., 1998, Lyell: the man and his times, *in* Blundell, D.J., and Scott, A.C., eds., *Lyell: The Past is the Key to the Present*: Geological Society, London, Special Publication 143, p. 21–37, https://doi.org/10.1144/GSL.SP.1998.143.01.04.

Wilson, L.G., 2009, Book review (Worlds before Adam: the reconstruction of geohistory in the age of reform, by Martin J.S. Rudwick, 2008, Chicago and London: University of Chicago Press, 614 p.) hardcover, US$49.00: *Earth Sciences History*, v. 28, p. 325–328, https://doi.org/10.17704/eshi.28.2.3617653216203282.

Witham, H., 1826, Notice in regard to the trap rocks in the mountain districts of the west and north-west of the counties of York, Durham, Westmoreland, Cumberland, and Northumberland: *Memoirs of the Wernerian Natural History Society*, v. 5, part II, p. 475–480.

Withers, C., 1994, On georgics and geology: Hutton's 'Elements of Agriculture' and agricultural science in eighteenth-century Scotland: *The Agricultural History Review*, v. 42, part I, p. 37–48.

Wittwer, W.C., 1860, Alexander von Humboldt—Sein Wissenschaftliches Leben und Wirken der Freunden der Naturwissenschaften dargestellet: T.O. Weigel, Leipzig, VI + 440 + XIV p. + 1 frontispiece +1 facsimile.

Woolley, C.L. (Sir), 1953, The Flood: *South African Archaeological Bulletin*, v. 8, p. 52–54, https://doi.org/10.2307/3887471.

von Wurzbach, C., 1872, Popowitsch, Johann Sigmund Valentin: in *Biographisches Lexikon des Kaiserthums Oesterreich, enthaltend die Lebensskizzen der denkwürdigen Personen, welche seit 1750 in den österreichischen Kronländern geboren wurden oder darin gelebt und gewirkt haben*, dreiunzwanzigster Theil Podlaha-Prokesch-Osten und Nachträge (Iv. Folge): k.k. Hof- und Staatsdruckerei, Wien, p. 108–111.

Young, G., (Rev.), 1832, On the fossil remains of quadrupeds, &c. discovered in the cavern at Kirkdale, in Yorkshire, and in other cavities or seams in limestone rocks: *Memoirs of the Wernerian Natural History Society*, v. 6, p. 171–183.

Young, D.A., 2003, *Mind over Magma—The Story of Igneous Petrology*: Princeton University Press, Princeton and Oxford, xxii + 686 p.

Yule, (Sir) H., 1914[1966], *Cathay and the Way Thither Being a Collection of Medieval Notices of China. New edition, revised throughout in the light of recent discoveries by Henri Cordier*. Vol III (Missionary Friars—Rashiduddin—Pegelotti—Marignolli): Hakluyt Society Publications, 2nd series, No. 37, Cambridge University Press, Cambridge, xvi + 269 p. (reprinted in 1966 by the Cheng-Wen Publishing Company, Taipei).

Zalasiewicz, J., Sörlin, S., Robin, L., and Grinevald, J., 2018, Introduction: Buffon and the history of the earth, *in* de Buffon, [G.-L.L.], (Comte), *The Epochs of Nature*, translated and edited by Jan Zalasiewicz, Anne-Sophie Milon, and Mateusz Zalasiewicz: The University of Chicago Press, Chicago and London, p. xiii–xxxiv.

Zippe, F.X.M., 1846, *Anleitung zur Gesteins- und Bodenkunde oder das Wichtigste aus der Mineralogie und Geognosie für Gebildete Leser aller Stände insbesondere für Landwirthe, Forstmänner und Bautechniker*: J.G. Calve, Prag, XXVII + 396 p. + 1 errata page.

von Zittel, K.A., 1899, *Geschichte der Geologie und Paläontologie bis Ende des 19. Jahrhunderts*: R. Oldenbourg, München, XI + 868 p.

Manuscript Accepted by the Society 10 March 2020

Notes

[1] Some historians of geology seem to think that writing essays with the title "theory of the earth" is a past fashion. It is not. Even in our own day, such publications, by highly respected scientists, are offered to the scientific world (e.g., Carey, 1988; Anderson, 1989, 2007). A theory of the earth is an all-encompassing synthesis of the totality of the ideas pertaining to the behavior and history of the planet earth. The reason it has become rare in our days to publish essays under such a title is the realization that coming up with an all-inclusive theory is an unattainable goal. We only approach it piecemeal.

[2] This is a sort of scientific equivalent to Lord Russell's interpretation of the main message of Aristotle's ethics: "...the intellectual virtues are the ends, but the practical virtues are only means" (Russell, 1945, p. 179). The view of those historians holding methods as the more important part of science reminds me of Lord Russell's characterization of the Christian ethics:

> Christian moralists hold that, while the consequences of virtuous actions are in general good, they are not *as* good as the virtuous actions themselves, which are to be valued on their own account and not on account of their effects. (ibid.; italics his)

[3] Although science has no fixed rules, it has a definite demarcation from all other sorts of human activity as shown by Popper (1935). That is why, frivolous statements such as "anything goes" in science (e.g., Feyerabend, 1988), are simply not true.

[4] As an outward expression of such joy, let us read what Hutton's friend and first biographer John Playfair (1748–1819) wrote about the discovery of the granitic dykes in Glen Tilt by Hutton: "In the bed of the river, many veins of red granite, (no less, indeed, than six large veins in the course of a mile), were seen traversing the black micaceous schistus, and producing, by the contrast of color, an effect that might be striking even to an unskilful observer. The sight of objects which verified at once so many important conclusions in his system, filled him with delight; and as his feelings, on such occasions, were always strongly expressed, the guides who accompanied him were convinced that it must be nothing less than the discovery of a vein of silver or gold, that could call forth such strong marks of joy and exultation" (Playfair, 1805, p. 68–69). This amusing anecdote became quite famous in Great Britain already during the first half of the nineteenth century. See the quotation from Maria Hack (1832, p. 71–72) below.

[5] Here Tomkeieff overstates his point by ignoring others that contributed to the development of geological theory such as Lavoisier, Cuvier, Lamarck (in his *Hydrogéologie* of 1802), Haüy, among Hutton's contemporaries.

[6] Steno was aware of the Dutch work on the reclaimed land in the Netherlands through his friend Nicolaas Witsen (1641–1717)—the scholarly mayor of Amsterdam, diplomatist, geographer and also a friend of Czar Peter the Great—who had data on drilling (including the 73 m well drilled by Pieter Ente in 1605 near Amsterdam) and worked out a stratigraphy and a local geological history not unlike what Steno developed later. There is little doubt that Steno's conversations with Witsen must have inspired him to his later thoughts (Abrahamse and Feiken, 2019). Gian Battista Vai (2019, personal commun.) recently pointed out that Steno also had first-hand knowledge of parts of the Alps and the Veneto region in Italy (Steno's first job in Italy was in Padua), in addition to his wanderings in Tuscany. He saw these places within a three-year period. Finally, there are the Italian Renaissance paintings, including those by Leonardo da Vinci, which Steno knew and which Vai (2009) thinks have influenced his geological interpretations. Particularly Leonardo's landscape depictions, although always as "background" to another central theme (e.g., Jesus in *Battesimo di Cristo* by Leonardo's teacher Andrea del Verocchio and Leonardo, *Mona Lisa*, the holy family in *Sant'Anna, la Vergine e il Bambino con l'Agnellino*; on Leonardo's landscapes as "background," see especially Gantner, 1958) are so astonishingly detailed and accurate that one can almost "see" the geology in them, as discussed by Gian Battista Vai in a number of his papers.

[7] Theophrastus of Eresos (ca. 371–ca. 287 BCE) and his friend and teacher Aristotle of Stagira (384–322 BCE) had a vision of a dynamic earth much like Hutton's. However, their premises were very different and only the final outcome of their theory can be compared with that of Hutton's theory. See below.

[8] Steno's (1669) discovery of the constancy of the interfacial angles in quartz crystals was actually a distant precursor to Haüy's later studies.

[9] Upon my query about the origin of the title of his *Nature* review of Rudwick's book, *The Worlds Before Adam,* Prof. Victor R. Baker kindly provided the following statement (written communication, 9 August 2019): "The title of the 2008 Rudwick review was not mine; it got imposed by the editors of *Nature* after I had submitted my text. I never saw the title until the article came out in print.... In retrospect I probably should have complained about the title, but there was a rush to get the review into a journal edition, and arguing over the meaning of the title would have been a diversion from the text of my review. Nevertheless, your use of the phrase 'turned upside down' seems to me appropriate for what you are saying about how Hutton has been treated in several recent historical accounts, even though this phrase was never used nor intended in my review of Rudwick's book." Baker's approval of my usage of the title of his review with respect to a number of recent historical accounts of Hutton is my excuse to cite it under his name. Turning a part of the history of science "upside down" in revisions is not always unfortunate. For example, Sivin (1988, p. 50), a collaborator of the great British biochemist and historian of science and technology in China, Joseph Needham (1900–1995), wrote that "Needham proceeded to turn the old issue upside down by asserting Chinese superiority across the board for a millennium and a half, and asking why modern science was not a Chinese invention." Needham's multi-volume *Science and Civilisation in China*, although parts of it are now antiquated, is a monumental milestone in studies of the history of science in China, eclipsing anything that preceded it either in China or outside it.

[10] Mott T. Greene (1945–).

[11] Or should we dethrone Newton from his deservedly lofty position in the history of physics, because his theory of gravitation turned out to be wrong (that it gives good approximations much below the velocity of light does not make it right, because time and space turned out not to be independent constants), as Einstein showed, and his theory of light was only partially true, as Huygens showed. Or are we to belittle Anaximander's invention, together with Thales, of natural science, because his amazing statement that the earth is suspended in space (cited, most likely, after Anaximander's book, even in the *Old Testament*, Book of Job, chapter 26, verse 7), left so many problems unsolved.

[12] The passage cited does not begin with "The question" as appears in Greene's quotation (his p. 31), but with "This question."

[13] There are several such examples of controversies that had started between the Wernerians and Huttonians and then, after the collapse of Wernerism, continued in the Huttonian camp: see, for example, Crook (1933, p. 36–37) on the origin of ores.

[14] My doctoral advisor and one the towering geologists of our times, Professor John F. Dewey, wrote (4 June 2019) the following upon my query about H.H. Read's standing, when Dewey himself was a postgraduate research student at Imperial College in London toward the end of the fifties and the early sixties: "H.H. Read was a distinguished senior and still active figure in Imperial College and Britain in the 1950s–1960s. He attended GSL [The Geological Society, London] meetings and walked around the IC [Imperial College, London] Department on most days chatting to and helping graduate students. I think that he finally realized that granitisation is not grain by grain replacement à la Doris [Livesey] Reynolds [later Doris Holmes: 1899–1985], but essentially, migmatisation as conceived by [Jakob Johannes] Sederholm [1863–1934] where partial melting of meta-sediments yields leucosome melts and mica-rich restite melanosomes. Janet [Watson: 1923–1985] almost certainly had a part in his conversion."

[15] For those not familiar with the twentieth century granite controversy, a modern and refreshingly short paper by W.J. French (2005) is highly recommended. If comprehensive and detailed reviews from most competent hands are required, one should go to Pitcher, 1997, p. 1–18) and Young (2003, chapters 19 and 20).

[16] Unless she and some of her revisionist predecessors were thinking in terms of what Deluc had called *cosmologie* (note, however, not *cosmogonie*) but meant

only the sciences of the earth and not those of the universe (Deluc, 1778, VII, footnote {a}). Use of cosmogony, cosmography and/or cosmology had been earlier common to denote the sciences of the earth as well as the sciences of the universe (let us remember that the first Latin translation of Ptolemy's Γεωγραφικὴ Ὑφήγησις {*Geografike Uefegesis*, i.e., Manual of Geography} by Jacopo Angelis of Scarperia {ca. 1360–1411}, done in 1406 and printed in Vicenza in 1475 by Hermannus Levilapis, carry the title *Cosmographia*. Similarly, such Renaissance geographers as Sebastian Münster {1488–1552} also used the term cosmography for the titles of their geographical books). As late as the eighteenth century, Axel von Kronstedt used the terms *Geographia subterranea* or *Cosmographia specializ* to describe the science of mining (von Kronstedt, 1780, unnumbered p. 14 in the *Vorrede des Verfassers*, i.e., the preface of the author). For an excellent account of cosmogony in the sense Laudan uses it, see Chesnel (1849, columns 105–112). Even Hutton himself used the adjective "cosmological" to label his speculations about historical geology (see the quotation from his third volume below), but by that time, in investigations on the earth that term had been reduced to a manner of speaking inherited from olden times and it soon disappeared from geology entirely. For the later part of the times Laudan employs cosmogony, it was no longer entirely appropriate for geology or geognosy: see Pledge (1940, p. 92–93; see also his table of contents, p. 6: "Geology separates from cosmology"; Pledge puts the separation just after de Buffon) and Porter (1977, p. 89). Joseph Black, for instance, calls those studying geology "natural historians" (see his letter quoted below), not cosmologists or cosmogonists. In 1774, James Watt calls his friend Hutton "fossil philosopher" (Jones et al., 1994, p. 639). Fifteen years later, we see François Dominique de Reynaud, Comte de Montlosier (1755–1838), one of the fathers of modern volcanology, using the word "geology" (Anonymous, 1789, p. 92), exactly in the sense we do today. Lichtenberg also used the word geology between 1796 and 1799, in the notebook L of his aphorisms, as is done in our own day (Lichtenberg, 1994a, p. 513, aphorism 864).

[17] Theophrastus' (ca. 371–287 BCE) model of a dynamic earth is amazingly similar to that of Hutton, despite the fact that most of Theophrastus' writings had long perished by Hutton's time making a complete reconstruction of his earth-behavior model impossible (see Fortenbaugh et al., 1993a, 1993b). We only have a few fragments, the longest—and the most significant—being in *On the Eternity of the Universe* by Philo the Jew (ca. 20 BCE–ca. 50 CE; for a discussion, see Duhem, 1958, p. 241ff.). In this fragment, Theophrastus argues against those who think that the universe cannot have been eternal because, if it were, the irregularities in the topography would long have disappeared (precisely Hutton's initial problem) and the progressive diminution of the sea would have turned the earth into an endless desert. Theophrastus points out that although erosion denudes the land, "…the fiery element that is enclosed in the earth is driven upwards by the natural force of fire, it moves toward its own proper place, and if it finds any short route by which to escape, it drags up with it a great amount of earthly substance, as much as it can. But this, surrounding the fire from outside, is carried (upwards) more slowly; but being compelled to accompany the (fire) for a great distance it is lifted up to a great height, contracts as it reaches a summit and ends up as a sharp peak which imitates the shape of fire" (Fortenbaugh et al., 1993a, p. 349). This model, elements of which had been inherited from the Pythagoreans, Empedocles, Plato, and Aristotle and which allows the topography to be rejuvenated at intervals through earthquakes, volcanoes, and alternating subsidences and uplifts as outlined above, is a perfect precursor to Hutton's theory. Theophrastus therefore does not find it surprising that an eternal universe may accommodate a planet with a dynamic topography. Neither did Hutton, despite the fact that he was not an eternalist, in sharp contrast to what Rudwick (2005, 2014) incomprehensibly claims in the face of Hutton's own (in his 1785 Abstract, where he wrote "so that, with respect to human observation, this world has neither a beginning nor an end") and Playfair's clear testimony to the contrary (Playfair, 1802, paragraphs 117–119; 1805, p. 55).

We know that Hutton read classics in the University of Edinburgh between the ages 14 and 17. Despite that, I very much doubt whether he was made to read Theophrastus as cited by Philo. For the Greek curriculum in Scottish universities during the eighteenth century, see Clarke (1945).

[18] For Buridan's dynamic earth model, see Şengör (2003, ch. 4)

[19] Professor Gian Battista Vai of Bologna informs me (written communication, 24 August 2019) that Steno greatly admired Descartes' appendix entitled *La Géométrie* to his *Discourse* of 1637. Vai ascribes Steno's discoveries in anatomy, especially myology, and then in geology, to his interest in geometry. I think there is much to be said for that view. For a corroboration of Steno's zeal for geometry, see Rosenberg (2006).

[20] Steno's intellectual environment and his religious convictions prevented him from doing what Hutton did, although he had formulated all the necessary agents (including "subterranean fire") and rock relationships that Hutton later employed (minus the igneous rocks). Steno even erected a historical geology for Tuscany in "six facies" (meaning six stages), but tied it to the creation of the earth and Noah's flood.

[21] This passage by Hutton is the best evidence against what Gould wrote: "The very first line of Lyell's *Principles* stakes out his difference from Hutton's ahistorical vision: 'Geology is the science which investigates the successive changes that have taken place in the organic and inorganic kingdoms of nature'" (Gould, 1987, p. 152–153).

[22] On her p. 4, Laudan states "Nor do I intend to criticize geologists for allowing the coexistence of two aims, for such coexistence strikes me as probably the rule (and a good one) rather than the exception in most scientific disciplines." If it is the rule, why bother separating them as distinct aims; moreover, how can non-historical sciences such as chemistry or physics have the same coexistence in their approaches? Thus, even this admission that historical and causal aims in geology often go together is not an indication that Laudan understands *that they can never be separated*.

[23] I cannot understand what Laudan means by Werner's drawing on "Becher-Stahl cosmogonic tradition." Such a "tradition" used to be assumed, even by some good historians of chemistry, but when one reads the publications of the two men with some care, that perception dissolves, because Stahl's views evolved to oppose Becher's (see Chang, 2007). Laudan claims that "Werner was steeped in the tradition of chemical cosmogony" and that he "transformed the Becher-Stahl tradition from which he had taken so much" (her p. 88). I do not think there is any evidence in the historical record to support these statements. Werner had almost nothing to do with chemistry (except when he taught metallurgy in the years 1789, 1791, 1796, 1798, 1803, 1805, 1809, and 1810; the introductory parts of those lessons were devoted to a Stahlian chemistry "like most of his contemporaries"; Lüdemann and Wenzel, 1967, p. 158–161), and his mineralogy and geognosy were entirely what one might term "field geological" and inherited from miners and geologists, not chemists (although, as a student, he had learnt chemistry from a student of Stahl). So, neither of Laudan's statements quoted above are even close to the truth.

Werner indeed mentioned five fundamental building blocks of matter, which however, allegedly could not be isolated by analysis; one is supposed to become aware of them only by theory: they are, flammable, airy, watery, earthy, and metallic elements. Of these, the flammable ("brenliche"[sic]: Werner, 1780, footnote on the unnumbered p. 3 of the *Vorrede des Uebersetzers*, i.e., preface of the translator) is of course Becher's *terra pinguis* ("fatty earth") and later, in improved form, Stahl's phlogiston.

But Stahl's phlogiston theory and his chemistry formed the conventional wisdom of the day. Anybody who had anything to do with chemistry used it, with such few exceptions as the great Dutch physician and chemist, the *Preceptor Europeae*, Herman Boerhave (1668–1738). In the last public, "geognostic" lecture he gave, Werner said:

> Our solid earth consists dominantly of earthy minerals, rock and soil species. In these are mixed further metallic, salty and burnable minerals in much smaller masses distributed here and there and as against the former their occurrence is confined to a very small amount. (Werner, 1818, p. 50)

The fact that he mentions earthy, metallic, salty, and burnable minerals does not make him a chemical mineralogist and a follower of the so-called "Becher-Stahl tradition"; he was simply using the common terminology of his day. Even Jules Verne, in his *Voyage au Centre de la Terre* ([1864], p. 8) used the same terms for mineral classes: "Ce cabinet était un véritable musée. Tous les échantillons du règne minéral s'y trouvaient étiquetés avec l'ordre le plus parfait, suivant les trois grandes divisions des minéraux inflammables, métalliques et lithoïdes" ("This cabinet was a veritable museum. All the specimens from the mineral kingdom were found there labeled in the most perfect order following the three grand divisions of inflammable, metallic and lithic minerals")! Similarly, one's usage of Newton's formula today to calculate the force of gravity does not make one a follower of a

"Newtonian cosmogony" and neither one's employment of some of Hans Stille's (1876–1966) tectonic terms makes one a supporter of his "cosmogony."

When talking about chemistry, which he rarely did, Werner was simply using the popular, if not already outdated in 1818, theory of matter of his times and which he started to question after Lavoisier showed that it was wrong. Moreover, he *never* used Stahl's chemistry in his mineralogical work. In his very first book, *Von den Äußerlichen Kennzeichen der Fossilien* (On the external characteristics of fossils), Werner wrote that in his book "one would find that I regarded mineralogy, in various parts, with completely different eyes from that which had happened until now" (Werner, 1774, p. 8). He then wrote, in the footnote a) on p. 17, that "in mineralogy there can be no place for an assaying book, description of adit constructions, metallurgical work etc." What Werner called the internal characteristics, i.e., the chemical properties, of minerals, he relegated to metallurgical chemistry (his p. 44, footnote f). From all these, it is clear that Werner stood entirely *outside* the Stahlian approach to mineralogy.

Given his interests, I can find nothing that Werner could have learnt from that tradition anyway, as Stahl's closest connection with mining was in metallurgy, except one small work (only seven pages) on the origin of metallic veins *De Ortu Venarum Metalliferarum* (On the origin of metallic veins; 1700), in which Stahl combatted those who thought that metals grew in the earth like plants, i.e., expressly repudiating a central interpretation of Becher (see Snelders, 1993, p. 104, on Becher's medieval idea). Stahl thought that metals had originated in the most ancient terrains in fissures that formed during the creation of the earth. In that seven-page tract one sees Becher's influence only on Stahl's ideas on the filling of the dykes. Stahl repeated what he wrote in 1700 in his 1703 supplement to Becher's *Physica Subterranea*.

Stahl's student Johann Friedrich Henckel (1678–1744) introduced chemical mineralogy in Freiberg, shorn of Becherisms entirely, which was continued by his younger colleague and successor Christlieb Ehregott Gellert (1713–1794), who was, however, in his Freiberger time, mainly a metallurgist. The Swedish chemist-mineralogists Axel Fredrik Baron von Kronstedt (also spelled with a C as Cronstedt; 1722–1765) and Johann Gottschalk Wallerius (1709–1785) were students of Stahl's pupil Henkel [Henckel] in Freiberg (Mühlpfordt, 1985, p. 139) and in his mineralogical publications Werner indeed cites these Swedes, *but only to defend his own ocular mineralogy against their chemical mineralogy*. In his published work, Werner cites Becher and Stahl only once (see below; but he criticizes Baron von Kronstedt for not mentioning Henkel in his historical introduction in the author's preface of his *Versuch einer Mineralogie* {=A mineralogical essay}; see Kronstedt, 1780, 6th unnumbered page of the *Vorrede des Verfassers* {=author's preface}), because he so widely ignored chemistry and in fact was criticized for it by his own pupils.

Steffens (1801, especially p. 22–23, 49–51; on p. 166–167), for example, points out that Werner was interested in the geognostic distribution of metals in the world, whereas Steffens himself tells his readers that the division of metals into two main classes is based on their chemical, oryctognostical, and even in some measure on geognostical characters. Steffens learnt his chemical mineralogy from Lampadius, not from Werner; see also Novalis' criticism, 1983, p. 138ff! The great mineralogist and Werner student Carl Friedrich Christian Mohs (1773–1839) also disapproved of Werner's handling of mineralogy "in an illogical and unmathematical way, without a system and only descriptively" (Schiffner, 1935, p. 20). Another famous Werner student, the mine counsellor, Ludwig August Emmerling (1765–1841), in his much-admired textbook of mineralogy in which he presented a development of Werner's mineralogy entirely on the master's methods and ideas, felt the need to add the chemistry of the minerals because Werner had not done so (see Emmerling, 1799 and 1802). Louis Simond also wrote: "Werner's nomenclature is founded on the mere external appearances of the substances, without any regard to their composition." (For the full quote and citation, see below.)

Did Lichtenberg not say in one of the aphorisms in his *Sudelbücher* that "Den redlichen Mann zu erkennen ist in vielen Fällen leicht, aber nicht in allen, so wie verschiedene Mineralien, bei einigen ist chemische Zerlegung nötig, aber wer gibt sich bei Charakteren damit ab, oder wieviel haben die Fähigkeit dazu? Das schnelle Aburteln [sic] ist größtenteils dem Faulheits-Trieb der Menschen zuzuschreiben, das mühsamere chemische System findet in Praxi wenig Anhänger, wir sind wernerisch gesinnt im Charaktersystem" (Lichtenberg, 1994a, p. 789, aphorism 967; "In many cases it is easy to recognize an honest man, but not in all cases, just like with minerals: with some a chemical analysis is necessary, but who would do this for characters, who indeed can do it? The quick judgement is to be ascribed to human lazyness; the more difficult chemical analysis finds few supporters in practice. We are Wernerians in character system.") One of the foremost experimental physicists among Werner's contemporaries thus knew that Werner did not "do chemistry" and that at the time this was generally well known enough to use in a metaphor. I should perhaps point out here, that in the Commentary volume to Lichtenberg's *Sudelbücher* (Promies, 1994, p. 628, entry 967), "Werner" is identified with Georg Friedrich Werner (1754–1798), who was a military engineer, physicist, and cosmologist and was embroiled in a controversy with Lichtenberg because of his anti-Newtonism (see Promies, 1994, p. 528, entry 64). I think that identification is wrong, as Georg Friedrich had nothing to do with mineral identification, chemical or otherwise.

Indeed, Abraham Gottlob Werner took the "chemists" to task for ignoring the external characteristics in the identification of minerals (e.g., he wrote, in his preface to his translation of Axel von Kronstedt's *Versuch einer Mineralogie*, "I wish to add a third to them [to the two previously mentioned causes of von Kronstedt's neglect of the external characteristics of minerals], that among the chemists it has become the Original Sin to be enthusiastic about, and to declaim in favor of, the identification and recognition of fossils [i.e., minerals] by means of external characteristics": Werner, 1780, 7th unnumbered page of *Vorrede des Uebersetzers*). Moreover, it was Werner himself who recommended appointing Wilhelm August Lampadius (1772–1842), the enthusiastic "antiphlogistonist," i.e., someone who stood outside the Becher-Stahl line in chemistry to replace the "phlogistonist" Gellert in the Bergakademie as teacher of chemistry (Anonymous, 1965, p. 128).

Werner's dislike of chemistry and quantitative methods in general has been emphasized also in the modern historiography of geology. The distinguished Swiss carbonate petrographer and historian of geology, Albert V. Carozzi (1925–2014), wrote that "Several of the handwritten materials reveal…the reasons for his profound dislike of quantitative methods" (Carozzi, 1962, p. xv). At the end of his 1774 book, Werner speaks of chemical, physical and empirical characters in only a few paragraphs. Here Carozzi, in his translation, remarked, in a footnote, "It is surprising that so little attention is given here to chemical tests, two hundred years after Agricola, and in the atmosphere of the Mining Academy of Freiberg, where assaying was such an important tool of applied mineralogy" (Werner, 1962, p. 115, footnote 1).

In geology, Werner's stand vis-à-vis Becher and Stahl was no different from that in mineralogy. In his famous *Neue Theorie von der Entstehung der Gänge mit Anwendung auf den Bergbau besonders den Freibergischen* (New Theory on the origin of Veins with application to Mining, especially that of Freiberg), Werner mentions both Becher and Stahl ("the great German physician, chemist and commentator of Becher" in paragraphs 10 and 11—these paragraphs cover I, I, I, p., 7–8; I, II, I, p. 60; I, II, II, 64; I, II, III, p. 71–72; I, II, IV, p. 87; I, II, V, p. 92–93; I, II, IV, 106–107; II, I, I, p. 129; II, II, II (p. 134; II, II, III, p. 149; II, II, IV, 157; of these only I, II, I, p. 60 I, II, 92–93; II, II, III, p. 149 are about minerals, rocks, and soils: see Stahl, 1703) in his historical review of the ideas on the origin of dykes and veins *only to point out that they were mistaken in their interpretations* as published in Stahl's *De Ortu Venarum Metalliferarum* that appeared in 1700 (Werner, 1791). In that review, Werner says expressly that it was Friedrich Wilhelm von Oppel (1720–1769), one of the two founders of the Bergakademie Freiberg (and mainly a metallurgist and mine surveyor by profession), who most profoundly influenced him in his studies on veins and dykes.

Finally, deposition or crystallization out of water was a popular idea since antiquity and I cannot see why it should be referred to as the Becher-Stahl "chymical cosmology." Is it because Becher's two fundamental "elements" were water and earth? But such ideas go back to the Renaissance during the revival of ancient learning; to Empedocles via Aristotle and reached people such as Paracelsus, from which both Becher and even Stahl derived "knowledge" (but when the two latter lived, the universe had become much better known than before thanks to Kepler's and Newton's efforts, but the "chymists" of the seventeenth and eighteenth centuries were not interested in things outside the earth, as Porter pointed out {except for the alleged influence of the heavenly objects on the origin of metals}; however, Ernst Florens Friedrich Chladni's {1756–1827} *Über den Ursprung der von Pallas gefundenen und anderer ihr ähnlicher Eisenmassen und über einige damit in Verbindung stehende Naturerscheinungen* {On the origin of the iron mass found by Pallas and of similar iron masses and on some natural occurrences related to them} is a notable, epoch-making, exception, encouraged by Georg Christoph Lichtenberg, as it initiated modern research on meteorites: Chladni, 1794). Becher's idea of making rocks out of water by condensing it ("Credendum ergo, lapides oriri ex aqua quidem, sed valde compacta": we believe therefore that stones can arise from water, but much compacted [water]: Becher, 1733, p. 212), is perhaps what Laudan has in mind, but that is clearly not what Werner could have been inspired by in his neptunist theory. Let us see what he wrote in an unpublished note "Our solid earth, as far as we know it, is originally built entirely from water…Its outer surface is built entirely by water; what fire worked does not come into consideration at all against the whole" (Mühlfriedel and Guntau, 1967, p. 26), because Becher makes everything, plants, animals, minerals, etc. by condensing water (he was almost back to Thales), in contrast to Werner, he could hardly have been an inspiration for Werner!

By the time Werner developed his neptunistic theory on the basis of his geological predecessors, the idea that an original menstruum could have had everything in dissolution had long been antiquated among chemists and Hutton indeed pointed this out for the specific case of the silicates: "Siliceous matter, physically speaking, is not soluble in water. ... If, by any art, this substance shall be dissolved in simple water, or made to crystallize from any solution, in that case the assertion which has been made here may be denied" (Hutton, 1788, p. 231–232). Almost half a century later, the great Swedish chemist, one of the founders of modern chemistry, Baron Jöns Jakob Berzelius (1779–1848) said exactly the same thing expressly against neptunism. That is why, Gillispie's surprise that "Objections to [Werner's] conception are so obvious that it is difficult to understand how responsible geologists could ever have accepted it" (see the long quotation from Vogelsang below for precisely the same point), which Laudan cites with disapproval (her p. 94), owing to her quite baseless conviction that Werner's alleged continuation of the chemical tradition had made his theory palatable, is perfectly understandable, because Werner did not continue any chemical tradition whatever! In chemical matters in general, Hutton, just like his intimate friend Joseph Black, was more of a follower of Stahl than Werner ever was (because he was more interested in chemistry than Werner), despite the fact that Black gave up Stahl's phlogiston theory in favor of Lavoisier's oxygen theory sometime in the 1780s. Hutton was still defending Stahl's theory as late as 1788 against his close friend Sir James Hall's support of Lavoisier's discovery in a debate in the Royal Society of Edinburgh (Allchin, 1994; see also Anderson and Jones, 2012, p. 1100–1102; Ramsay, 1918, p. 97–98)! Therefore, Laudan's consideration of Werner in a chemical tradition is entirely unsupported by the available documentation. Werner himself paid lip-service to chemistry indeed, but never let an opportunity slip to emphasize that what he was doing was not what the chemists were doing.

[24] Laudan herself confesses that Werner's inability to explain the sequence of strata he described was a loss for geology (Laudan, 1987, p. 228).

[25] Gehler's book includes not only minerals, but also fossils in the sense we use that term today. That is why I left "fossils" in my English translation of his title as such, although that term for Gehler included both minerals and fossils *sensus praesentis*.

[26] The title *Sarepta* derives from the name of the ancient Phoenician city (present-day Sarafand in Lebanon; צרפת {Zarephath} in Hebrew), which in Hebrew became an eponym for a metalworking shop or a smelter.

[27] Fitton's review was published originally in the *Edinburgh Review*, no. 140 (July 1839), v. 69, p. 406–466). I here cite a reprint published in the same year in book form (Fitton, 1839), because, in the two copies I own of it, Fitton introduced corrections in his own hand. I assume he did it for all the copies of the separately published book. Herein I give the page numbers of the book and then, in curly brackets, those of the journal.

[28] The English translation quoted here is from the fine rendering by Zalasiewicz et al. (de Buffon, 2018). However, I have the greatest difficulty in understanding the unqualified statement in the introduction to that fine translation "He [i.e., Buffon] was aware that the coal strata and the marine shales tended to be tilted (which he thought was due to sedimentary layers accumulating on steep slopes)" (Zalasiewicz et al., 2018, p. XXIX). Moreover, why do they think by "terrein en pente" (i.e., "inclined terrain": de Buffon, 1778, p. 107) de Buffon meant "*steep* slopes"?

[29] Professor Kenneth L. Taylor of Norman, Oklahoma, suggested, in his detailed review of a previous typescript of the present book (written communication, 1 August 2019), that one should interpret Rudwick's statement about Sir Archibald Geikie's role in the Hutton historiography, as not *inaugurating* a remembrance of Hutton, but *establishing* a tradition of considering him the founder of modern geology. However, even then Rudwick's statement would remain incorrect, because considering Hutton the father or founder of modern geology has been expressed so often, by such diverse authors in so many different countries, as in part documented in the present book, that it is clear that it had already been established by a tradition much earlier than anything Sir Archibald wrote on the subject.

[30] In the online catalogues of the Cambridge University libraries, I was unable to find anything under the name Ralph Grant. It seems that he never finished his postgraduate work, at least not in Cambridge.

[31] For my argument that Hutton was actually an atheist, but did not dare say so openly, see Şengör (2001). Heimann (1978) pointed out the entirely non-Christian stand of Hutton and shown that Hutton said god had created the natural laws, but denied him omnipotence, allowing only omniscience. Hutton's god in some ways resembles the god of Descartes and Newton, but Hutton vehemently denied voluntarism, which Newton had admitted, and defended immanence of force and motion within the matter itself (Heimann, 1978). If this is not taking god out of the equation of the functioning of the universe, I cannot imagine what it is. I do not believe that Hutton's thinking was very different from that of de Buffon when de Buffon said to Marie-Jean Hérault de Séchelles (1759–1794), one of the leaders and then a victim of the *terreur*: "I have always named the Creator; but we need only remove this word and, of course, put in its place the power of Nature" (Roger, 1989, p. 566; I took the English version from the beautiful translation by Sarah Lucille Bonnefoi: Roger, 1997, p. 431). In his account, Hutton kept saying that god could have interfered in the workings of the universe, *if he wanted*. If this is not dissimulating, I do not know what it might have been.

[32] Neither did Gould remember in his assessment of Hutton's theory an extremely important point made by one of the greatest geologists of the twentieth century, Marion King Hubbert (1903–1989) in 1967, during a conference that Gould attended as a young student: Hubbert reminded his audience that Hutton's theory only applied to the Phanerozoic, because he had no means of studying, or even recognizing, the Precambrian (Hubbert, 1967).

[33] Gibbon's ire was aroused by Tacitus' ascription of Augustus' testimonial recommendation of not enlarging the boundaries of the Roman Empire to "either fear or jealousy" (*Annales*, I, 11), because the illustrious English historian thought the Emperor's counsel most reasonable, based on careful reflection of the conditions of his time.

[34] In 1966, Victor Eyles wrote that "the treatment of the subject [i.e., history of geology] by professional historians of science as a whole leaves much to be desired. No matter how well qualified authors may be to write a general history of science, it is no doubt too much to expect that they should be expert in all its branches, but that hardly excuses the inadequate and cursory treatment the history of geology sometimes receives at their hands" (Eyles, 1966, p. 77). Eyles, in a way, was echoing Sir Charles Lyell's observation made more than a century earlier, quoted in the third motto to this book and 32 years after Eyles, his point was brought forward again by Hugh Torrens (1998). Although monographic studies on the history of geology have greatly increased in number since then, the examples I just gave above show that the quality has not uniformly improved since Eyles and Torrens wrote, notwithstanding such stellar exceptions as Oldroyd's history of the Highlands Controversy or his book on the history of geological research in the Lakes District, Dean's books and papers on Hutton and Mantell, Baron von Engelhardt's exhaustive studies on Werner and Goethe's geology, Andrea Wulf's superb Humboldt biography, Torrens' own contributions, especially those on William Smith, Edmund Critchley's *Dinosaur Doctor* relating the life and accomplishments of Gideon Mantell (notwithstanding Donovan's scathing review in the Geological Journal, v. 48, pp. 109-110), and a number of others.

[35] The contents of this booklet formed the Inaugural Lecture at the opening of the Class of Geology and Mineralogy in the University of Edinburgh on 6 November 1871. It was reprinted, with slight changes in the wording, in Geikie (1871b) and in Geikie (1882a, p. 286–311).

[36] The contents of this paper were read at the fortieth annual meeting of the Edinburgh Geological Society in its rooms at no. 5, St. Andrew Square in Edinburgh on 6 November 1873.

[37] See footnote 35, above.

[38] It is in this encyclopedia article that Sir Archibald identified Hutton as "one of the great founders of geological science" for the first time (Geikie, 1888, p. 414), essentially repeating what he had said in 1882 in his textbook with different words and greater emphasis on Hutton. This newer statement expresses precisely what most geologists active today (including me) think of Hutton's place in the history of geology.

[39] This address was reprinted in Geikie (1893, p. 3–26 and in 1905, p. 158–197, the latter under the title "The centenary of Hutton's 'Theory of the Earth'").

[40] Princess Dashkova was a very successful director of the Russian Academy and saved it from ruin to which the Academy had been dragged by the singularly irresponsible and unethical administration of her predecessor Sergei Gerassimovich Domashnev (1743–1795; directorship 1775–1783), a friend of the influential Orlov family. Her memoires were published first in London in English after a translation that had been tampered with and misled many into thinking that it was the original version. Currently, the most accessible and reliable edition (after a third manuscript that had been found in Moscow and translated by the Russian socialist and popular science writer Alexander Ivanovich Herzen {1812–1870} in 1857) is the one published by the Winkler Verlag in Munich (Dashkowa, 1970) that has an epilogue by the German historian Irene von Lossow. The Princess' days in Edinburgh (from 1777 to 1779) are recounted in chapter 14 of her memoirs. For a critical biography of the Princess, see the distinguished German historian, philosopher and diplomatist Günther Schlegelberger's (1909–1974) doctoral dissertation in Friedrich-Wilhelms-University in Berlin (1935).

[41] In their splendid two-volume book containing Joseph Black's correspondence, Anderson and Jones point out, in their footnote 3 on p. 903 (in v. 2), that "Princess Dashkova's mineral cabinet, which she presented to Moscow University [since the celebrations during the 185th anniversary of the foundation of the university in 1940, it is called Lomonosov Moscow State University and is the highest-ranked university in the Russian Federation today], contained 15,121 items. Among these were 7,924 stones and ores, which perhaps included the Derbyshire fossils" (Anderson and Jones, 2012). The term "fossils" here does not necessarily refer to fossils in their present meaning of "remains and traces of organisms that lived in the past." A fossil in the late eighteenth century was anything that came out of the ground, including rocks and minerals, in keeping with the original meaning ("dug out from the ground") of the classical Latin adjective *fossilis*.

[42] Carbon dioxide.

[43] As pointed out in endnote 41, by "Fossils" Black here means rocks and minerals and not remains or traces of organisms that lived in the past.

[44] The Xerox copy of the manuscript I have, made by the late Professor Donald B. McIntyre from the Edinburgh University archive, is probably an earlier draft of the letter as suggested also by the many strikes and insertions on it (indeed identified as a draft in Anderson and Jones, 2012, v. 2, p. 901). I have amended and corrected it using the published version in Ramsay (1918, p. 118–125) and Anderson and Jones (2012, v. 2, p. 901–906). A search in the Archive of the Russian Academy of Sciences by Dr. Irena Malakhova failed to find the final clean version of the letter sent to the Princess.

[45] Polyakhova and Bessarabova (2011, p. 44–45) identify Black also as a geologist.

[46] I wish to emphasize that in Black's account there is not a word about Hutton's allegedly Deistic motives in formulating his theory. In Şengör (2001), I argued that Hutton must have been an atheist, but the society in which he lived was not the right environment to express his true conviction. McIntyre described the religious atmosphere in Scotland when Hutton and his friends were working with the following words: "Hume, Smith, Burns, and Hutton had to tread warily to avoid bringing down the wrath of Church and Government on their heads. To judge Hutton while ignoring his historical context, as revisionist historians too often do, is to condemn without hearing the evidence" (McIntyre, 1997, p. 120); taking Hutton at his word on his religious convictions I think also has caused some recent philosophical discussions to lose their justification (e.g., Rossetter, 2018). Although their writings give the impression that the intellectual elite of Edinburgh were living at the time in sublime surroundings of reason and comfort, their reality included the feared Sedition Trials (that made open defense of atheism a punishable crime) and such social upheavals as the Highland Clearances bringing the common people and the government and the upper classes into a serious conflict (the Ross-shire Sheep Riot, for example, occurred during Hutton's lifetime, in 1792). In such a delicate environment, unnecessary radicalism would have been entirely counter-productive and Hutton and his friends certainly knew that. My own reasoning about Hutton's creed is similar to that which Roger (1989) employed in his inference of de Buffon's "latent atheism" (his p. 158). Black's silence on Hutton's "Deism" is a further possible support of my interpretation. However, Adam Ferguson quotes Hutton saying that his favorite specimens of natural history were God's books, and he treated the books of men comparatively with neglect (Ferguson, 1805, p. 115, footnote *). We do not know whether the "God's book" metaphor was Hutton's or whether it is Ferguson's paraphrase; but even if it was Hutton's, let us remember that Albert Einstein also used to employ the God metaphor often, but he always emphasized that he did not believe in a personal one. Perhaps Hutton's and Einstein's god conceptions were similar: they simply saw no necessity for one (like de Buffon and Laplace), but did not think it wise, or even necessary, to say it openly.

[47] In 1984, the Linda Hall Library in Kansas City, Missouri, published a catalogue of an exhibition its staff had organized under the title *Theories of the Earth 1644–1830—The History of A Genre* (Ashworth and Bradley, 1984). It is a very useful listing of books, but, unfortunately, having been written by a historian and a librarian and not by a geologist, the pamphlet brought together geological publications of diverse genres that can under no circumstances be united under the title "Theories of the Earth." Only between its pages 8 and 31 are listed what can be considered books belonging to that peculiar genre of "theories of the earth." What characterizes those books is a dearth of observation and a profusion of speculation on the structure and the history of the earth as a whole, often in as much conformity to the Biblical narrative as the author considers appropriate. After p. 31, Ashworth and Bradley list publications ranging from Hutton's *Theory of the Earth* to such very specific studies as Scheuchzer's *Herbarium Diluvianum* and regional studies as Eaton's *Index to the Geology of the Northern States*, even to such polemical books as Deluc's *An Elementary Treatise on Geology* that can be compared neither with one another nor with the books cited earlier. This catalogue, very useful as a bibliography, is misleading, in the same way as the revisionists' characterization of Hutton's theory is, because of its lumping diverse kinds of geological publications under one "genre."

[48] Ca. 1730–1795; almost an exact contemporary of Hutton, the discoverer of the famous vitrified forts in Scotland and a somewhat disillusioned miner: Torrens, 2004a; about Williams' disappointment as a miner, see also Neill (1814) and the preface by James Millar to the second edition of Williams' book here considered; see Williams (1810, v. I, p. vii–viii).

[49] Dean (1992, p. 296) added to the citation of Williams' 1789 edition in his bibliography the statement that the second edition "was noticeably less vociferous regarding Hutton." Did he not realize that the second edition was not by Williams, but by Millar? Had he been alive at the time Williams might have objected to Millar's change of tone in his book.

[50] I suppose the reference here is to Thomas Thomson's (1773–1852) *System of Chemistry*, published in 1802 and then went through six editions till 1820. The book originally came out in four volumes and then grew to five and dominated chemistry in the UK for more than three decades. The reference here may be to its second or third editions. Thomson was a student of Joseph Black, but became an opponent of Hutton's theory of the earth.

[51] The reference here is to John Murray's four-volume *System of Chemistry* (1806–1807). It is most likely to the first edition.

[52] In the 55th volume of the *Journal de Physique, de Chimie, d'histoire Naturelle et des Arts* (1802) there is a large number of mineralogical and geological articles by such prominent authors as Jean-François d'Aubuisson de Voisins, and Claude Delamétherie.

[53] I here cite the much enlarged and revised second edition, which is usually taken as the definitive text while discussing Kant's ideas expressed in this important book.

[54] This "binary history" is what the Australian Christian fundamentalist, young earth creationist and Christian apologist, Kenneth Alfred Ham (born 1951) defended against the science educator William Sanford Nye (the host of the

PBS science show "Bill Nye the Science Guy" that aired from 1993 to 1998; born 1955) in their famous debate "science versus the *Bible*" on 4 February 2014 at the Creation Museum in Petersburg, Kentucky, USA (for the entire debate, see https://www.youtube.com/watch?v=z6kgvhG3AkI). Such binary histories exist in religions other than Deluc's: In Zarvanism, the most orthodox form of Zoroastrianism during the Sasanid period in Persia, for example, *Zarvan akarana*, i.e., unlimited time, preceded (and will follow) the 12,000 years of limited time created by Ormazd and Ahriman (Eliade, 1965{1974}, p. 125). The idea that the past behavior of the world was fundamentally different from the present (events *in illo tempore*) is a common assumption of religious thinking that always pitted it against scientific considerations. See below Cornford's and Vernant's observations concerning the birth of science among the Ionian Greeks.

[55] See Marita Hübner's (2010) well-documented account concerning the primacy of religion in Deluc's dealings with geology. These sentiments of Deluc closely resemble those expressed by Kenneth Alfred Ham in the debate with Bill Nye referred to above.

[56] It is not generally known nowadays that Lavoisier, the founder of modern chemistry, learnt geology from the chemist and geologist Guillaume-François Rouelle (see Rappaport, 1960, p. 77), started his scientific life as assistant to Jean-Étienne Guettard (1715–1786), i.e., as a geologist and subsequently published many geological papers. However, it is extremely unlikely that Deluc was unaware of Lavoisier's geological work, because it was mainly published in the memoires of the French Academy of Sciences. For Lavoisier's geological work, see Comte (1949), Carozzi (1965), Rappaport (1967, 1973) and Gohau (1994a, 1994b). All of Lavoisier's geological papers are reprinted in Grimaux (1892), *Oevures de Lavoisier* publiées par les soins du ministre de l'Instruction Publique, Imprimerie Nationale, Paris, v. 5.

[57] See Taylor (2009) for the significance of this paper by Desmarest. It was a precursor both to Count Montlosier (Anonymous, 1789) and to Scrope (1827).

[58] Popowitsch's book is almost totally forgotten in our day, but when it was published it was "reviewed with unconditional praise by all the literary journals of the time" (von Wurzbach, 1872, p. 109). One illustration of how forgotten it is, is the ascription of its authorship to C.G. Schwarz in the bibliography by Watermann and Wrzesinski (1989, p. 2).

[59] The German physician and geologist Georg Christian Füchsel (1722–1773) used the retreat of waters into subterranean volumes of looser arrangement of solid matter, not just empty chambers, and the consequent desiccation of the present land surface as the reason of the most recent global geological event, although he regarded this as a part of the ongoing evolution of the globe and a *slow*, irregular event, not a cataclysm (see Anonymous [Füchsel], 1873, especially p. 1–52). Deluc makes no reference to him, because, I suppose, he could not read German.

[60] In the French original, the word "origins" is italicized by Deluc for emphasis (1809b, p. 28). This is an implicit spurn to Hutton's disavowal of all knowledge of the beginnings of things in geology. By "lights" (=*lumière*) Deluc probably refers to the creation of light that, according to him, made the chemical precipitation of granite possible (see the summary of his view in this regard in his nephew Deluc jr.'s 1816 paper, p. 471). Or does he simply mean clarification, illumination?

[61] Klaver points out that this was so, because Whewell also was constrained by his religious convictions: "His religious presuppositions, moreover, conditioned his geological outlook and impeded a more impartial view of uniformity" (Klaver, 1997, p. 134).

[62] There is much resistance to Popper's view of science from the side of the philosophers. David Deutsch very nicely summarized their position in general and lumped them all, with much justification, into what he judged to be "bad philosophy" (see Deutsch, 2011, chapter 12).

[63] Let us ignore the internal inconsistencies within the book of *Genesis* about the Flood for the time being. Although identified by Jean Astruc (1684–1766) in his anonymously published *Conjectures sur la Genèse* (1753) and by Johann Gottfried Eichhorn (1752–1827) in his *Einleitung in das Alte Testament* (1780–1783), they had not yet become widely accepted during the eighteenth century. In fact, Astruc never denied that Moses was the author of the *Torah*, although he did identify two separate sources in it; he thought they represented the sources that Moses had used.

[64] The mountain interpretation of the Paradise is based on *Ezekiel* 28:13–14. In the Jewish *Bible*, in the five books of Moses, i.e., the *Pentateuch* or the *Torah sensu stricto*, and also in the early years of Christianity, there was a large number of Paradise, and hence sacred mountain, images (the Garden of Eden motif was introduced into the Jewish *Bible* by the Yahwist author (J), whose earliest date is ~950 BCE: Lane Fox, 1993, p. 178; more recent research, however, has shown the construction of the *Torah* to have been even more complicated and stretched over a longer time interval, from the seventh century BCE onward, but also began finding indications of pre-Yahwist fragments. For a summary of the recent research about the history of the book of *Genesis*, chapters 1–11, see Uehlinger, 2013): some regarded it on earth, in the east; others thought it had to be in the Promised Land, in fact identical with Jerusalem (Jerusalem indeed sits on an anticline consisting mostly of Upper Cretaceous strata with an elevation of some 760 m asl. with an impressive slope eastward down to the Dead Sea that has the lowest surface of any water body on earth: −430.5 m bsl and dropping). Yet others saw it as a garden, some others as a mountain and some as both. Some made a distinction between the earthly Paradise and the eschatological "Kingdom of Heaven" (this is implied in *Luke*: 23–43, but the term "Kingdom of Heaven" first occurs in *Matthew*, 3:2; the Jewish *Gan 'Eden* also had a terrestrial {"lower Gan"} and a celestial part {"upper Gan"}). The earthly Paradise was viewed as an intermediary residence for the souls awaiting the Resurrection at the end of time (*Olam ha-Neshamot* of the Jews), somewhat reminiscent of the twelfth century Catholic invention of the Purgatory. Let us here remember that the *Tergumim* and the *Peshitta* considered the Garden of Eden a real geographical place on earth (thus justifying St. Ephrem's interpretation), whereas the *Septuagint* and the *Vulgate* treated it as a general concept designating a lush, fertile and high, mythic place (see Anderson, 1988, p. 194–195). The mountain Paradise, a concept taken over from the older Mesopotamian pagan religions (notwithstanding Clifford's, 1972, objection), was thought to be the abode of the Divinity that inhabited its summit (this is where the storm god, the head of the pagan pantheon in Mesopotamia (El), in Palestine (Yahweh) and in Greek pagan (Zeus) religions resided: see Clifford, 1972; Anderson, 1988; Şengör, 1992). The Tree of Knowledge was believed to be located halfway up the mountain and Adam and Eve and their children were made to wander below it. In the Syriac translation of the Bible, the *Peshitta* (the Hebrew original of which closely follows the Masoretic text: Becker, 2015, p. 37), which St. Ephrem used, Cain says to Abel in *Genesis* 4:8 "Let us go to the valley" implying they are on higher ground. St. Ephrem pointed out that the Flood did not reach this mountain, except to "kiss its feet":

b-aynā d-re'yānā
hzītheh l-pardaysā
w-rawmē d-kull tūrīn
sīmīn thēt rawmeh
l-'eqbaw mtā balhud
rīsheh d-māmōlā (this is from *Genesis* 7:19)
l-reglaw nshaq
wa-sged w-ethpnī
d-nessaq ndush rīshā
d-tūrē w-rāmāthā
'eqbaw d-haw nāsheq
w-rīshā d-kull qāphah

The translation is as follows:
With the eye of my mind
I gazed upon Paradise;
the summit of every mountain
is lower than its summit,
the crest of the Flood
reached only its foothills; (*Genesis* 7:19)
these it kissed with reverence
before turning back
to rise above and subdue every peak
of every hill and mountain.
The foothills of Paradise it kisses,
while every summit it buffets.

I must emphasize, however, that St. Ephrem does not replace Noah's ark with the Paradise mountain, as we know, for example, from the first of the Nisibene hymns, stanza 3 (see also Aydin, 2008, p. 17–20). To do so would have been extremely unorthodox, which St. Ephrem tried to avoid like the plague all his life and fought against everything he thought heretical. For the cosmic mountain theme in Canaan and in the *Old Testament*, see the Jesuit father Richard J. Clifford's excellent doctoral dissertation (Clifford, 1972) and the most informative article by Anderson (1988) that specifically deals with St. Ephrem's *Hymns on Paradise* and the anonymous *Cave of Treasures*.

(Much of the discussion above and the Syriac text and translation, are based on Sebastian Brock's superb Introduction in Brock, 1990).

[65] In citing Reisch's famous book, it is important to indicate which edition one refers to. I usually cite the last, i.e., the fourth, edition that Reisch himself prepared before his death in 1525, published in 1517, although the first edition had been published in 1486 in Heidelberg and the second in 1504 in Strasbourg. These two early editions carried the title beginning with *Aepitome omnis Philosophieae...*, which fell out in later editions, including the 1504 Freiburg edition. After his death, various editors expanded the book and continued putting out newer editions until eleven of them had appeared until 1535. Here I cite the newest German translation of the entire fourth edition by Otto and Eva Schönberger (Reisch, 2016 [1517]); the text I refer to occurs on p. 292).

[66] Here I use the English translation published by the Cornell University Press, but always with an eye on the original French text (1962 [1990]).

[67] For the relationships between theory-inventing scientists and the general public in antiquity, see the excellent book by Hans Blumenberg (1987).

[68] Let us remember here that by "meteorological" the ancient Greeks meant all natural phenomena pertaining to the earth. In fact, an earlier term for what Aristotle later called "physicists" was μετεωρολόγοι (*meteorologoi*, i.e., meteorologists), because they thought that all terrestrial phenomena (even including such things as earthquakes and volcanoes) they could observe was governed by "air" or by the stars! The term μετεωρολόγοι is derived from μετέωρος (meteoros): raised from off the ground, on the surface, running along the ground, prominent, shallow; this broad spectrum of meanings enabled the word *meteorologoi* to designate those interested in a primitive group of "sciences of the universe," including geology. See Mansfeld's (1999) p. 41, note 17. There is a very informative discussion in Abel Rey's (1873–1940) *La Jeunesse de la Science Grecque* (The juvenescence of Greek science: 1933, p. 34–36) where he cites with partial approval both the classicist John Burnet (1863–1928) and the historian of philosophy Émile Bréhier (1876–1952) who had argued that the early Ionian "physics" was entirely what we would call meteorology, because Thales and his successors had reduced all natural phenomena in the inanimate world to processes that ultimately originated in the "clouds." Rey himself concludes "their meteorology was thus their physics" (p. 36). See also Laks (2018, p. 10 and 47), where he cites Cicero (106–43 BCE) from his *Tusculanae Disputationes* and Gorgias (ca. 485–380 BCE) from his Ἑλένης Ἐγκώμιον (*Encomium of Helen*) where they equate Aristotle's physicists with meteorologists.

[69] In the Cornell translation this is rendered as positivism, which is incorrect. Vernant wrote "positivité" (Vernant, 1962(1990), p. 101). This word refers to the French meaning of "positive" (I am here quoting from the Littré dictionary): "that on which one can depend, count; which is assured, constant." Positivism, by contrast, is the name of a philosophical school founded by Auguste Comte (1798–1857) in the nineteenth century, very much on the footsteps of Sir Francis Bacon, greatly admired by Deluc. But even the Presocratics, who founded science as Vernant points out, cannot be considered proto-positivists—in fact, quite to the contrary: see Popper (1958–59) [1989]. In the English translation by Glenn W. Most of André Laks' admirable little book, Vernant's "positivité" is also translated as positivity (Laks, 2018, p. 57, 58, 62).

[70] Naturally in its German form, "vorsokratisch."

[71] That is not exactly true: Anaxagoras said that *nous*, a presiding cause present everywhere in the universe, was responsible for making and destroying things. Nothing religious was attached to *nous*, which greatly disappointed Socrates (ca. 470–399) and eventually led Athenians to banish Anaxagoras for irreverence toward religion and allowed Socrates to derail Greek rational thought toward mysticism that reached its climax already with Socrates' pupil Plato. For an excellent history and evaluation of the term and concept "Presocratics" see Laks (2018).

[72] Some historians wish to give the impression that religion vs. science is too nebulous a contrast, because neither science nor religion are concepts that could be reduced to single terms alone and that they not only embody very varied systems, but also interpenetrated many times in history in diverse ways (see Dixon, 2008, for a statement of this view and a rich source of references to the science and religion relationships). While it is a truism that both science and religion embrace a great variety of beliefs, theories and practices, I disagree entirely that they cannot be contrasted *as blanket concepts*, because "science" is an activity that requires free inquiry; "religion" allows any inquiry so long as it (they) do(es) not contradict the fundamental tenets of religion, whatever they are. Moreover, religions always include a supernatural element (even Marxism! See Lord Russell's point that the classless society that Marxism promises cannot really be imagined without a theistic belief: Russell, 1946, p. 816: "Marx professed himself an atheist, but retained a cosmic optimism which only theism could justify."), which science rejects. These are incompatible principles; or, to use Leibniz's term, they are not compossible. This is clearly expressed for the specific case of the Catholic Church in Dixon's summary of the religion-science conflict: "Although the church was certainly not opposed in general to the study of mathematics, astronomy, and the other sciences, there were limits to how far the authority of the *Bible* and of the church could be challenged by an individual layman like Galileo." (Dixon, 2008, p. 23). As Russell (1946, p. 546) pointed out, in the century following the Reformation and the Counter-Reformation, the Jesuit schools gave the best education obtainable "whenever theology did not interfere." In accounting for the great flourishing of the natural sciences in the lands of the Khalifat, it is commonly claimed to be a result of the tolerance of Islam. Then one should ask why the mathematician and astronomer Lutfî, the personal librarian of Mehmed II (1432–1481; ruled 1444–1446 and 1451–1481), the Conqueror, was publicly hanged for godlessness in the Sultan Ahmed Square (the ancient Hippodrome) in İstanbul on 24 December 1494, after his protector died and his pious son Bayezid II (1447–1512; ruled 1481–1512) ascended the throne (see Mollâ Lutfî'l Maqtûl, 1940). Adıvar (1943, p. 47) calls Lutfî "the first martyr of science in Ottoman history." The great mathematician and astronomer, the immortal author of the *Ruba'iat*, Omar Khayyam (1048–1131), narrowly escaped a similar fate by making a pilgrimage to Mecca to prove his orthodoxy to the hostile clergy! It is common knowledge that Muslim pilgrims going from Isfahan to Meshhed used to make a special stop in Nishapur just to spit on Khayyam's grave. Let us recall that Anaxagoras (500–428 BCE) was condemned to death by Athenians, but was able to escape from Athens to Lampsakos (present-day Lâpseki on the Asian side of the Dardanelles) in 434 or 433 BCE, possibly following the advice of Pericles, for contradicting the common religion (his calling the god Helios, i.e., the Sun, a piece of incandescent stone somewhat larger than Peloponnese, was found particularly offensive). In an interesting compendium of articles defending the alleged compatibility of Buddhism with modern science written by Buddhists (Kirthisinghe, 1984), we read such staggering claims as "Rebirth is therefore not a dogma to be accepted on faith but an hypothesis capable of being scientifically verified" (Jayatilleke, 1984, p. 12). If one refuses to believe in such nonsense, the result is the violence of the kind we have recently witnessed in Sri Lanka and in Myanmar and seen earlier in the history of Mongolia, similar to those resulting from any other religious conflict in history anywhere in the world. Thus, the Austrian philosopher, engineer and inventor Joseph Popper-Lynkeus (1838–1921; his birth name was Joseph Popper, Lynkeus being a pen name he adopted in 1899, and he was a relative of the more famous Viennese philosopher Karl Raimund Popper) stated a truism when he wrote:

> One would not really believe that there can be a *tolerant* religion! Until our own day one hears from relatively enlightened, but still somewhat religious people repeatedly that *real religion is tolerant*, it respects the opinions of those who think differently. But this is completely untrue and, in principle, impossible (Popper-Lynkeus, 1924, p. 11; emphasis his).

He further argued that

> One should say it openly: superstition, also religion, are based on a mental halfness, i.e., weakness; in their terms, one also hunts for causalities, as all science also does; but religions rest content just to hunt and to shoot, without paying attention to what one hunts and what one shoots. This is a characteristic of every religion: a shortage in mental functions; they are thus *simply a negativity, an incompleteness, compared with science*. On the basis of this, one realises that religion must be the mortal enemy of all science that can penetrate its haven. How right is the opinion of one of the most truthful of the newer philosophers of religion, L. Feuerbach, that *religion and modernity, i.e., scientific culture are not, and never can be, compatible with one another*. (Popper-Lynkeus, 1924, p. 102–103; italics mine)

That is why most classicists and historians of ancient science regard what happened in Miletus toward the end of the seventh and the beginning of the sixth centuries BCE variously as a revolution or as a miracle that signified a rapid transition from myth, i.e., religion, to science among the Greek-speaking peoples. Laks (2018) presented a good review of the relevant literature and discussed the transition from myth to rationality. Laks also reviews the various rationalities introduced by Max Weber and in that review he mentions Weber's "Jewish rationalism" and "Calvinist rationalism." Although these designations seem oxymorons, they are not. Every system of thought must have some basic axioms to build upon. The fundamental axioms of science are that there is a real Nature outside us (i.e., the World is not my dream) and that we can come into contact with it, however imperfectly. These axioms cannot be defended rationally, because their negation robs us of any possible test for their truth. Once we accept that irrational axiom pair, we can then proceed perfectly rationally to construct our science. The same is true about religions: once one accepts their fundamental axioms, one can then proceed very rationally to develop the further tenets of whatever religion one chooses to believe in. That statements made within the compass of religious thinking are not amenable to testing by observation, but require faith, sounds like surrendering to irrationality forever, but it is indeed rational once one accepts the fundamental axiom of any religion that there is an "extra-natural world" not accessible to us except by revelation. That is why religious people can at the same time be great scientists. They accept more than one set of axioms, but at the expense of being consistent with themselves, i.e., being irrational twice at that initial step. In the past, religious statements about Nature were thought to represent true accounts, because people thinking that, believed that the fundamental axioms of science and religion were not contradictory. In such cases it might seem that there is no boundary between religion and science. That is not true, because religion, by the nature of its fundamental axioms, must delimit freedom of inquiry; science does not. Religion does not permit testing its statements. Science does. (That is why most sane people do not jump off the Eiffel Tower thinking that the axioms of science and axioms of religion are not contradictory and that a prayer in the extra-natural world will save him or her in the natural world. There were people in the past, and even today, who think such a thing possible! Hence, the suicide bombers or the self-incinerating Buddhist monks.) That is why, science seems a better way to cope with the World around us than religion and that is why the two cannot be compatible with one another, however much human irrationality may make individuals believe that they can be. As Sir Karl R. Popper said "In so far as a scientific statement speaks about reality, it must be falsifiable; and in so far as it is not falsifiable, it does not speak about reality." This is true of all statements, not just scientific statements.

In our specific case, to satisfy those who dislike a blanket contrast "religion versus science," one can say, to be pedantically precise, "an attack of Deluc's understanding of his religion on what was then considered Hutton's science."

[73] Whoever criticized Deluc's geological opinions, always met with the same rebuke of not having read his publications (see for example, Schmieder, 1807, p. VII–X).

[74] Some might think the Black Sea prediction by Polybius is farfetched. However, precisely what he predicts for the Black Sea did happen to the North Caspian Depression, which is underlain by oceanic crust. It had formed during the Devonian as a basin similar to the present-day Gulf of Mexico, and was already completely filled by the Permian, i.e., in ~100 million years, so as to start depositing terrestrial redbeds and salt (Burke, 1977; Şengör and Natal'in, 1996).

[75] I have here used Robin Waterfield's 2010 translation of *The Histories* published by the Oxford University Press.

[76] Lyell's criticism of Deluc's "chronometer" in Lake Geneva is quite excellent and very well worth reading to assess the level of knowledge available on this topic about the time when Deluc was making his claims.

[77] *Aral* means island in Turkish. Aral Sea is thus the sea of islands.

[78] Hakluyt used the manuscript then in possession of Lord Lumley, which is now in the British Library under MS Royal 14 C. XIII, ff. 225r–236r. (L) (see Jackson and Morgan, 1990, p. 52). This manuscript is cut off at chapter 26, but for our purposes it is immaterial, because Rubruk passed the area of the Aral Sea already in the narrative of the earlier chapters.

[79] The historian Krivonogov refers to is the Persian scholar Shahabeddin Abdullah ibn Lutfillah ibn Abdurrashid al-Havafi (died 1430), known as Hafiz-i Abru, who was active in the Timurid court; first of Timur (1336–1405) himself and then that of his son and successor Shah Rukh (1377–1447). He wrote a continuation to Rashīd al-Dīn Faḍlullāh Hamadānī's (1247–1318) famous *Jāmiʿ al-Tawārīkh* (Compendium of Chronicles), entitled *Tarih-i Hafiz-i Abru* in 1417 in Persian, in which he described the change of course of the Amu Darya and the disappearance of the Aral Sea (Le Strange, 1905(1966), p. 16, 457). Le Strange says that the text of that book had not been printed and that he had used a manuscript housed in the British Museum under Or. 1577. I am not aware whether it was subsequently printed.

[80] Thomson (1948[1965], p. 128) further underlines, however that there is not one clear description of the Aral Sea inherited from antiquity. I think this is because it did not exist!

[81] For the geological reason of this episodic behavior of the Aral Sea, see Şengör (2009b). Also the reappearance of the Aral Sea after the fifteenth century coincided with the onset of the Little Ice Age in the northern hemisphere.

[82] Jameson does not tell us where he took Deluc's quotation from. It is from Deluc (1810, p. 338–340).

[83] Some may be inclined to think that Sir Charles Lyell had also originally opposed the theory of sea-level drop. He did not. What he was objecting to was Leopold von Buch's claim that the land in Scandinavia was slowly and continuously rising without any earthquakes. After a personal visit to the relevant localities in 1834, he corroborated von Buch's conclusion and retracted his objection (Lyell, 1835; see Şengör, 2003, Chapter 10).

[84] However, I must point out a recurring error in ascribing to Deluc the invention of the term geology as we understand it today (most recently: Hübner, 2010, 2011; Rudwick, 2011). This is such a puzzling mistake, especially among those who ought to know better (it may have stemmed from a misunderstanding of von Zittel's statement in his famous book: von Zittel, 1899, p. 106), that Vai and Cavazza exclaimed (2006, p. 44) "It is mystifying that some researchers and dictionaries still indicate the Swiss naturalists Jean-André Deluc and Horace-Bénédict de Saussure as definers of the modern meaning of the term geology in the years 1778–1779 (e.g., Encyclopedia Britannica, fifteenth edition)." Although the word geology had been invented to denote a kind of science of human laws (although its inventor hastens to add that such a study would not really be a science *proprement dit*) by the English cleric and bibliophile Richard de Bury (1287–1345) in his *Philobiblon* (written 1344; *editio princeps* in Cologne in 1473 made from a very corrupt manuscript and carelessly printed: see West, 1889c, p. 39 and 93; for the occurrence of the word "geology" in the best available Latin edition, see West, 1889a, p. 90: *nec est hæc facultas inter scientias recensenda, quam licet geologiam appropriato vocabulo nominare*; for English translations, see West, 1889b, p. 98: "nor is this faculty to which we give, by an appropriate term, the name *geology*…" and de Bury, 1903, p. 79, for Ernest C. Thomas' translation), the term, in the sense we still use it, was introduced by the Italian polymath Ulisse Aldrovandi (1522–1605) in his testament written on 10 November 1603 (first published in the Bolognese politician and scholar Giovanni Fantuzzi's {1718–1799} anonymously printed memoir on Aldrovandi, 1774, p. 81; see also Vai, 2003). Since then, many authors used it before Deluc, as the following incomplete list of titles shows: Michael Peterson Escholt (?–1666) *Geologica Norvegica* (1657), Erasmus Warren (?–1718) *Geologia: or, a Discourse Concerning the Earth before the Deluge* (1690), Dethlev Cluver (1645–1708) *Geologia Sive Philosophemata De Genesi Ac Structura Globi Terreni. Oder: Natürliche Wissenschafft / Von Erschaffung und Bereitung der Erd-Kugel* (1700).

[85] Deluc's opposition to Lavoisier's new chemistry was also religiously motivated: to keep the influence of the godless French away from Europe as much as possible! See Hübner's (2010, 2011) and Heilbron's (2011b) accounts.

[86] The Yahwist, Elohimist, Deuteronomist and Priestly sources were identified toward the end of the nineteenth century. But as the history of the Biblical exegesis shows, the doubts about the unity of the *Old Testament* had begun surfacing already in the seventeenth century (Reventlow et al., 1988; Reventlow, 2001; Becker, 2015, p. 84). As I said above, in the eighteenth century, Jean Astruc and J.G. Eichhorn documented that something was wrong with the single, inspired,

author hypothesis for the origin of the *Old Testament*, but the details and the general acceptance only followed in the nineteenth century.

[87] Elohim is the plural of El and may refer to the "children of El." The supreme god El, a northwest Semitic word, was variously either the head of the Mesopotamian pantheon or a major deity in it and indicates the Mesopotamian pagan roots of Judaism.

[88] Originally a storm god from northwestern Arabia and thus a part of the early Middle Eastern religious tradition. Yahweh was most likely the god of the Leah tribe related to the Canaanites. The name is possibly cognate with the Arabic *hawwa* (air), with implications of a destroyer (through storms and rain). It is first in the *Book of Isaiah* that the Jewish god reveals himself to have no rivals "I am the Lord and there is no other" (*Isaiah*, 45, 18). Even the Proto-Isaiah is now dated to the eighth century BCE, i.e., some two centuries after the mythical figure of Moses, so monotheism among the Abrahamic religions seems a fairly late invention.

[89] For excellent recent assessments of Deluc's intellectual standing in his time, amply corroborating Gillispie's view of Deluc, see Hübner (2010, 2011). Particularly the latter is not only superb historiography, but almost a literary artwork in itself, where Hübner documented how Deluc ended up as a religious bigot (see especially her subsection 2, entitled "From moderate to bigot": p. 31ff.). See also Heilbron's (2011a, 2011b) humorous, yet incisive analysis of Deluc's motives and aims in his historical (both natural and social) studies and intellectual and political activism. What Rudwick admires as Deluc's "profoundly *historical* perspective" (2011, p. 260; italics Rudwick's), Heilbron (2011a, p. 93) calls (and I think justifiably) "pseudo-historical"! Rudwick strays away from historical objectivity when he says that Deluc's theory was "radically *contingent* and *historical*, without any less emphasis on the natural character of its causes" (Rudwick, 2014, p. 76–77), because he deliberately omits Deluc's *miraculous history* derived from the *Bible*; Deluc refused to take into account any observation that might negate the Biblical mythology as I showed above. (When listing the traits Deluc shared with King George III, Heilbron itemized: "deep religiosity, strong convictions, fear of revolution, liking for natural science, and twinges of madness" (Heilbron, 2011b, p. 186).

[90] Sir Henry Holland kept a detailed journal during their travels and Sir George made extensive use of it while writing his book. Regrettably, this precious journal remained unpublished until an incomplete Icelandic translation appeared in 1960 by Steindór Steindórsson (1902–1997) (Holland, 1960). Later, in 1987, Andrew Wawn finally published an English translation of the entire journal with a fine introduction, in which he noted that Sir Henry was "an empiricist with Huttonian leanings" and that Sir George "was a prominent supporter of the Huttonian position and it was in the hope of finding specimens and other contextual evidence which might confirm the thesis that he was drawn toward Iceland" (Wawn *in* Holland, 1987, p. 20).

[91] Anybody interested in the way in which at least some Huttonians regarded the attitude of the Wernerians toward geology ought to read the contents of that remarkable footnote, which I here reproduce:

> Those to whom bare inspection of individual minerals affords the greatest pleasure, and who conceive the highest enjoyment to be derived from the study of mineralogy, is to be found in that skill which is necessary to enable them to place a specimen in the class assigned it, despise theory, and, on that account, may consider what follows as very unimportant. Fully acknowledging the indispensable utility of Oryctognosy, we cannot but consider Geology as not the least sublime department of mineralogy; and, while we are daily adding to our knowledge of the structure of the globe, we cannot conceive a more noble occupation for the mind, than endeavouring to trace the means which have been employed by the great Creator in the formation and establishment of our terrestrial abode.

[92] From the picture of the island Abel provides on his p. 62, the island seems to be the Green Island separated from the Hong Kong island by the Sulphur Channel. The minimum width of the Sulphur Channel is ~0.46 km. The geology of the island is dominated by late Mesozoic felsic volcanic rocks into which granite dykes of early Cretaceous age were intruded. Abel may have mistaken the rhyolites for granite, or, conversely some of the darker volcanic rocks for basalt. That the area has intrusive rocks showing contacts and that it illustrates nicely the "Huttonian" geology, is, however, certain. For a geological description of the Green island, see Strange and Shaw (1986, especially fig. 8, for a geological map of the island). One should bear in mind, however, that more recent research has further subdivided what previously had been lumped into Tai Mo Shan Formation which still figures in Strange and Shaw's account undivided.

[93] Abel inserts here a footnote to quote a passage from Hall and Playfair (1815).

[94] Here is another footnote quoting a passage from Jameson's *Elements of Geognosy*, 1822, p. 236, supposedly proving the neptunist viewpoint, which Abel shows in his account to be entirely fallacious.

[95] At this point, Abel introduces yet another footnote citing a long passage from Jameson's *Elements of Geognosy*, 1822, p. 237, claiming exactly the opposite of what Abel describes about veins shooting into the overlying rocks of a pluton. This counterpoint nicely underlines the deficient observational basis of neptunism as propounded by Jameson in Edinburgh.

[96] Here is another footnote containing a passage from Jameson's *Elements of Geognosy*, 1822, p. 90, outlining the relevant assertions of the neptunist theory.

[97] The second edition of John Williams' *The Natural History of the Mineral Kingdom* (1810) was not by himself but by James Millar. But since it was still published as the second edition under Williams' authorship, I here cite it simply as Williams (1810), but indicate those passages where Millar is the author.

[98] Dean does not give a complete reference to these quotations. I rectify that deficit here: Anonymous, 1826, *Considerations on Volcanos, the probable Cause of their Phenomena, the Laws which determine their March, the Disposition of their Products, and their Connexion with the present State and past History of the Globe; leading to the Establishment of a New Theory of the Earth*. By G. Poulett Scrope, Esq. Sec. Geol. Soc. 8vo. W. Phillips. London. 1825: *The Monthly Review*, from January to April inclusive, 1826. With an Appendix, v. I New and Improved Series, p. 24–32. Dean's quotes are on p. 29 of the review.

[99] This is the famous author and playwright Henry Mackenzie (1745–1831), an amateur geologist and an ardent Huttonian.

[100] Wawn (2000, p. 52, footnote 119) notes that the manuscript of the play is in the Henry E. Huntington Library, 1151 Oxford Road, San Marino, California, MS Larpent, 1751, 2.iii. 24–34.

[101] Witches were still being executed only 20 years before Hutton's birth in Scotland and witch trials continued until a year after his birth. The famous witchcraft act that had been introduced in 1563 during the reign of Elisabeth I was repealed by the British Parliament when Hutton was already ten years of age!

[102] It was actually his wife Mary Ann (née Woodhouse, 1795–1869) who found the initial fossils of the Iguanodon consisting of some teeth in Surrey while waiting for her husband during his visit to a patient.

[103] Here is inserted a footnote by Mantell: "See Playfair's Illustrations of the Huttonian Theory, vol. i. p. 33, *et seq*. Edin. 1822."

[104] The reference Mantell gives in a footnote is this: "The general reader will find an interesting account of M. Fournet's theory in Jameson's Edinburgh journal No. xlvii. p. 3." The full and correct reference is: Élie de Beaumont, Bequerel, and Alex. Brongniart, 1838, Account of M. Fournet's opinions regarding the metamorphic changes of rocks, and his observations on the Systems of Elevation in the country near Arbresle (Dept. Du Rhône). Contained in a Report made to the Academy of Sciences of Paris.: *The Edinburgh New Philosophical Journal*, v. 24, no. 47, p. 111–121. Mantell may have mistaken 111 for III.

[105] For a brief account of the life and work of William Gillivray, see the small book published by his namesake in 1912. Being primarily an ornithologist, Gillivray, not the author of the biography, but its object, is little-known among geologists today.

[106] In an otherwise undated November 1841 entry in his journal, Mantell wrote of Richardson "this man, to whom I communicated all my scientific plans for my works—knowing that through the misconduct of my publishers, I had been prevented from bringing out my First Lessons [he means his two-volume *Medals of Creation*, published in 1844], so early as I intended, and, that my illness

rendered me incapable of doing so now—has issued a prospectus of 'Geology for beginners' precisely upon the same plan as mine, and without the slightest allusion to my works—am I never to find gratitude or honorable conduct in those I oblige? Of all the villainy I have experienced, this man's is the basest!—)" (Curwen, 1940, p. 150). He vented his anger again in the preface to his *Medals of Creation*, but without naming his target: "In the meanwhile several works professing the same object have issued from the press; among them a volume by a writer, whom a sense of honor, if not of gratitude, should have deterred from interfering, in any manner, with the literary labors of the individual to whom he was mainly indebted for whatever acquaintance with Geology he may possess; and who, in the unrestricted and unsuspecting confidence of personal intercourse, was made fully acquainted with the plan and scope of the intended publication of the author" (Mantell, 1844, p. vii–viii). Immediately after the preface, there is a section called "Address to the reader," the second footnote of which is a direct attack both on Richardson's book and on his person, alluding to his ignorance of the natural world: "'INTELLIGENT READER:' for, as an able Critic in the Athanæum has shrewdly observed, there can be no 'Geology for Beginners'; we should know something of the *natural world as it is*, before we attempt to pry into the mysteries of the world *as it was*" (Mantell, 1844, p. xi; italics Mantell's). With all due respect to Mantell, I have failed to see much of a resemblance between the two books. Mantell's outburst may have been due to his terribly depressed state of mind brought about by calamitous domestic affairs (his wife, his elder daughter, his elder son and even his housemaid Hannah Brooks had left him {Critchley, 2010, p. 179}; his younger daughter had just died following a painful disease, he himself had taken a bad fall from his carriage and hurt his spine that led to recurrent, debilitating pains for the rest of his life) and failing finances. But, when Richardson took his own life, Mantell felt a very sincere remorse (Curwen, 1940, p. 224).

[107] This is really all Élie de Beaumont (1829–1830, 1830, 1831), to which Lyell was vehemently opposed (see his letter to Scrope dated 25 June 1830: Lyell, 1881, p. 274; also Lyell, 1833, p. 341). Élie de Beaumont was a hardcore catastrophist, a stand difficult to reconcile with Hutton's position (although Hutton did have some catastrophist elements in his theory, they were nothing like those of Élie de Beaumont, who was a self-confessed disciple of Cuvier and von Buch: see the introduction to his 1829–1830 paper). Despite all the most justified critique heaped upon it in the second half of the nineteenth century, including by Eduard Suess (see Şengör, 2014a), Élie de Beaumont's view of tectonics was resurrected during the first quarter of the twentieth century by Hans Stille and Leopold Kober based mainly on the earlier publications by James Dwight Dana, Joseph Le Conte, Marcel Bertrand, Émile Haug, Thomas Chrowder Chamberlin, and Ferdinand Löwl and they both defined their positions as "catastrophist" (Stille, 1922, p. 11: "As much as actualism may be generally valid for the majority of geological phenomena, and also for the evolution of organic life, catastrophism must claim some validity as far as the tectonic phenomena are concerned." Stille repeated this in his famous textbook: Stille, 1924, p. 22; in the first edition of his influential textbook, *Der Bau der Erde*, Kober, 1921, p. 3–4—emphasis by Kober himself—wrote: "Geological experience teaches that times of quiet *development* alternate with the so-called *critical times (Le Conte)* [this is a reference to Le Conte, 1895; Kober was notorious about name dropping without giving complete references]. In exaggerating the critical times, especially with respect to life, *Cuvier* came to the *theory of catastrophes (d'Orbigny)* [d'Orbigny published his catastrophist ideas in many places; but the first place where he did so that I am aware of is d'Orbigny, 1842]. This theory of cataclysms was revived in a tempered form in *Salomon*'s theory of paroxysms [the reference here is to Salomon, 1918], in the idea that the evolution of the earth has not had a uniform course, as one did expect for a long time from *Lyell's theory of actualism*, but consisted of larger and smaller paroxysms with intervening periods of quieter development." Kober repeated this conviction in the second edition of his book (Kober, 1928). Stille's and Kober's ideas, almost exact copies of Élie de Beaumont's interpretations and those of his followers in the nineteenth century, very much dominated tectonics even well after the rise of plate tectonics in the twentieth century. It is really quite astonishing how little attention Kober and Stille and their followers paid to Lyell's admonition and Suess' objections to Élie de Beaumont. Even Phillips' convictions were not so rigid as those of Stille and Kober in their "Beaumontism." That is why it is so important for Hutton to be understood properly by practicing geologists even in our own day, lest we lose sight of sound theoretical geology, as, for example, happened in the case of the so-called sequence stratigraphy of the Exxon workers, who repeated many of the errors of Werner and Élie de Beaumont and their twentieth century incarnations: see Şengör, 2016.)

[108] Phillips gives no reference here, but what is meant is the Reverend Conybeare's report to the second meeting of the British Association for the Advancement of Science in 1832 in Oxford (Conybeare, 1835, p. 366–369). Conybeare was evidently much impressed with Leibniz's geology, as he also mentioned it to Lyell in a letter (Wilson, 1980).

[109] Phillips does not give a reference for this sentence. It is from Leibniz's posthumously published book *Protogaea* (1749a, p. 7). Here is a translation by Cohen and Wakefield (Leibniz, 2008, p. 11): "as these conditions subsided and came into equilibrium, there emerged a more settled state of things."

[110] Let us remember that our present global geological time scale is made up of units that are consensual, not natural. There may indeed be natural global stratigraphic units (the ubiquitous, iridium-rich, K-Pg boundary layer is one example), but such may not be assumed à priori, but must be documented by evidence.

[111] Erzgebirge (=Ore Mountains; Krušné Hory in Czech) is the famous region, the silver mines of which formed the backbone of the wealth of the Electorate of Saxony and its predecessor states since the first silver discovery in 1168 in Freiberg, where Werner spent his entire professional career and based his neptunistic theories on the observations made there.

[112] To this day, the word *accident* is used in the French geological parlance for any tectonic discontinuity (fault, shear zone, etc.) the nature of which is not further specified. In most cases it simply means a fault.

[113] Flœtz Period refers to the time of deposition of flœtz, layered sedimentary rocks of gentle dips. This is the time period following what Werner called the Transition Period (comprising rocks from about the Ediacaran to Middle Devonian, but this range changes greatly in different parts of the world) and extends to the base of the Holocene (called Alluvium by Werner).

[114] This is unfair. Hutton was famous for his eloquence and wit when he spoke. The quality of his prose in his book *Theory of the Earth* was because, as Hugh Torrens informed me, "We need to remember that the final version of Hutton's Theory was all written while he was under 'sentence of death.' This explains why his published *Theory* is so disorganized, prolix and, above all, incomplete (cf. the *Lost Drawings* only published at last by Gordon Craig, in 1978)" (written communication, 29 September 2019). At the time Hutton was indeed suffering from a very painful disease of the bladder, which eventually killed him.

[115] Kenneth L. Taylor informed me (written communication, 29 January 2019) that he found in the holdings of the Académie Nationale de Médecine that Iberti was referred to as a *correspondant* and listed as "Iberti, médecin pensionné du roi d'Espagne, résidant à Edimbourg." The dates of these holdings are 1789–1790. Professor Taylor added the following in his communication:

> Iberti's having resided, for a time, in Edinburgh is confirmed as follows: The Scot Alexander Geddes (1737–1802), an unorthodox Catholic priest and intellectual, evidently knew Iberti in Edinburgh in 1790. "In 1790 Geddes played host in Edinburgh to Don Joseph Iberti, physician to the King of Spain, who was promptly elected to the Society of Antiquaries, and whose Benthamite-sounding suggestions regarding the punishment of duelists Geddes included in his little tract [*Reflections on Duelling*]." (Mark Goldie, "The Scottish Catholic Enlightenment," *Journal of British Studies*, 1991, 30: 20–62, at p. 53.) [On Geddes, see Goldie, "Alexander Geddes at the Limits of the Catholic Enlightenment," *The Historical Journal*, 2010, 53: 61–86.]

Professor Taylor points out that Iberti met the intellectual elite of Edinburgh and even became an honorary member of the Society of the Antiquaries of Scotland. He thus thinks it likely that Iberti became aware of Hutton's abstract and of its importance in Edinburgh at the latest by 1790.

[116] Here Desmarest really means British naturalists. Most continental authors at the time and in the nineteenth century subsumed under "English" all inhabitants of Great Britain.

[117] Nowhere in his book does Faujas give the date of his visit to Great Britain. However, Ashworth (2003) proves that it was during 1784.

[118] The following translation is a revised rendering by me from the English translation of Faujas' work (Faujas de Saint-Fond, 1799, p. 230). The translation

curiously had left out two critical passages, where the meticulous observer Faujas emphasizes the importance of the matrices of various stones in illuminating their "natural history" and the fact that Hutton was writing his treatise "in the tranquility of his cabinet."

[119] Here Faujas inserted the following footnote:

> This work, which contains rather general views of the subject than a body of observations on the theory of the earth, appeared in 1785, in the transactions of the Royal Society of Edinburgh, for that year, under the following title: *Theory of the Earth, or an Investigation of the Laws observable in the composition, dissolution, and restoration of land upon the globe by James Hutton, M.D. F.R.S.E. And Member of the Royal Academy of Agriculture in Paris. From the transactions of the royal society at Edinburgh, April 4, 1785.*
>
> In the English translation it was pointed out that that "This work has since been considerably enlarged; in 1795, in was published in 2 vols. 8vo. Under this title, *Theory of the Earth, with Proofs and Illustrations, in four parts.*—Translator." (Faujas de Saint-Fond, 1799, p. 230).

[120] I was unable to find out who this "Hockard" might be and my friends in Edinburgh did not know of such a person either. Upon my request, Professor Kenneth L. Taylor (written communication, 31 January 2019) offered the opinion that he might be William Hogarth (1697–1764), the famous printer and painter, but why this Englishman who lived in London should be listed by Faujas in connection with Scotland he could not tell. The other possibility he mentioned was George Hoggart Toulmin (1754–1817), another Englishman and the famous author of *The Antiquity and Duration of the World*, published in 1780 and *The Eternity of the Universe* published in 1789. Toulmin would make a lot more sense in the present context, because he studied in Edinburgh from 1776 to 1779 and may have plagiarized Hutton on the basis of manuscript material he may have seen as a student (Davies, 1967; however, Lloyd, 1986, disagrees with Davies and thinks Toulmin's ideas original); but why mention him only with his middle name? This may be because Faujas' source(s) may have been oral and something may have been muddled in an across-language transition.

[121] This is another person who is cited in none of the sources available to me and none of the people I consulted could identify. Could it be a mistake for the famous James Watt or his geologist son Gregory Watt? Gregory died of tuberculosis when he was only 26 years of age and James was a much greater celebrity than his unfortunate son. Thus, Faujas may have meant James.

[122] Coleman (1964) cites one letter, from Cuvier to the Reverend Henry de la Fite, written sometime in the mid-1820s, where the great anatomist ventured to offer a timid opinion as to why the inundations occurred:

> Nevertheless, I believe with M. Buckland that this catastrophe is the most recent or most recent but one of those which have affected the totality of or the greatest part of the globe and the one whose souvenir is preserved by nations under the name of the deluge. I also believe that it was sudden, like several of those which preceded it and if I may express myself on [its] physical causes I would say that the most reasonable conjecture was that it was due to several ruptures in the crust of the globe which changed the level and position of the seas as they had already been changed at other periods and by other catastrophes. But I must confine myself to these general terms and I present them only as the expression of a simple conjecture. To desire more precision, to attempt to explain remote causes, and, particularly, to give the how and why of each small fact would be, in my opinion, to go much further than possible given the present state of our knowledge. (Coleman, 1964, p. 135)

Another unpublished opinion by Cuvier as to the cause of these catastrophes was found by Nikolaas Rupke in Mary Lucy Pendered's (1858–1940) 1923 biography of the painter John Martin (1789–1854), who in the first three decades of the 19th century had created great apocalyptic paintings, the grandest of which possibly is *The Deluge* (1826, a mezzotint version in 1828, another oil canvas in 1834; see Rupke, 1983a, fig. 7 for a black-and-white photograph). Cuvier visited Martin's studio in London, reportedly expressed admiration for the conception of *The Deluge* and mentioned the cometary hypothesis as a possible cause of the inundation (Rupke, 1983a, p. 77; 1983b, p. 40). See also Rupke (1983b).

[123] Basset does not specify which of the four illustrious Jussieus he refers to here.

[124] One wonders whether Élie de Beaumont derived his inspiration for his fundamental observation forming the basis of his theory of timing of orogenic events from this statement:

> When examined with some care, it is seen that along almost all mountain chains beds extend in a horizontal position as far as the foot of the mountain, indicating that they were laid down in the sea or in lakes whose shores were partly formed by these mountains. By contrast, other beds, which are upturned and which turn around the flanks of the mountains, reach in some regions as far high as the summit. (Élie de Beaumont 1829–1830, p. 5–9)
>
> Now a distinction, which is always sharp, and which allows no intermediaries, thus results from this observation between the upturned and the horizontal beds. One concludes that the phenomenon of upturning was not continuous and progressive; it operated in a time interval between the periods of deposition of the two consecutive terrains and during which no deposition of regular beds took place. In one word, it was brusque and of short duration. Such a convulsion that upturns the beds in an entire mountain range necessarily interrupts the slow and progressive development of sedimentary terrains, and it is clear that some anomaly must be observed nearly universally at a point in such series that corresponds to the moment at which the upturning of the beds took place. (Élie de Beaumont 1831, p. 242–243)

For the wide-ranging implications of these statements by Élie de Beaumont, see Şengör (1991, 2016).

[125] The references here are to *A Tour in Scotland* published in 1771 by the Welsh naturalist Thomas Pennant (1726–1798) an almost exact contemporary of Hutton, "It is violently agitated by the winds, and at times the waves are quite mountanous [sic]. *November* 1st, 1755, at the same time as the earthquake at *Lisbon*, these waters were affected in a very extraordinary manner: they rose and flowed up the lake from East to West with vast impetuosity, and were carried above 200 yards up the river *Oich*, breaking on its banks in a wave near three feet high; then continued ebbing and flowing for the space of an hour; but at eleven o'clock a wave greater than any of the rest came up the river, broke on the North side, and overflowed the bank for the extent of 30 feet. A boat near *General's Hut*, loaden with brush-wood, was thrice driven ashore, and twice carried back again; but the last time, the rudder was broken, the wood forced out, and the boat filled with water and left on shore. At the same time, a little isle, in a small loch in *Badenoch*, was totally reversed and flung on the beach. But at these parts no agitation was felt on land" (Pennant, 1790, p. 221–222; italics Pennant's) and to the British artist and clergyman William Gilpin's (1724–1804) *Observations Relative Chiefly to Picturesque Beauty Made in the Year 1776 on Several Parts of Great Britain; particularly the High-Lands of Scotland*, where it is stated that "It is remarkable, that at the beginning of November, 1755, when the city of Lisbon was destroyed by an earthquake, this lake was exceedingly agitated. The day was perfectly calm, and its surface still, when its waters arose suddenly many feet in large swells, and overflowed a considerable district. Then in a moment or two retiring, they sank as much below their usual level. Their next flow, and ebb were less than the former; but still very great; and thus they continued rising and sinking for several hours; till the fluctuation gradually subsiding, the waters at length settled within their common bounds. A boat which was thrown upon dry land, was found by mensuration, to be forty yards from its station in the lake…." (Gilpin, 1792, p. 37–38).

[126] See Lyell (1830, p. 477–479) for a discussion and refutation of the ballooning earth ideas.

[127] This is the very first reference that I am aware of in the history of geology mentioning tectonic repetition of a sedimentary section to account for excessive thickness reports, a lesson not learnt by many well into the second half of the twentieth century (see, for example, Burke et al., 1976)!

[128] This whole discussion about the shrinking orange very much reminds me of Élie de Beaumont's theory of thermal contraction of the globe.

[129] The name Sandwich Islands was the one given to the archipelago by Captain James Cook in 18 January 1778, after the First Lord of Admiralty John Montagu, 4th Earl of Sandwich. It was replaced gradually in the middle years of the nineteenth century by the Polynesian designation Hawai'i.

[130] Compare this sketch with fig. 4 and the corresponding facsimile reproduction of the original drawing in the portfolio in Craig et al. (1978) illustrating the very same locality. This figure is the only one of this locality drawn by a supporter

of Hutton's views that I am aware of published before Lyell's *Principles* came out, illustrating what Hutton intended to show in the drawing he had asked his friend John Clerk of Eldin to draw.

[131] See below for the quote from Hutton mentioning the possibility of cutting off the entire Mediterranean from the world ocean to lead to its total evaporation. Both Hutton and Lyell had no problem with immense, continent-sized floods, provided a natural mechanism to cause them was available. Lyell wrote that a sudden discharge of the waters of the Great Lakes in North America may catastrophically inundate much of the continent and that such an event would be perfectly admissible within his uniformitarianism (Lyell, 1830, p. 89). When he wrote that he did not know that such a catastrophe had indeed happened when Lake Missoula in North America catastrophically discharged its waters to the Pacific Ocean when the ice dam holding it broke down at the end of the Ice Age and created the magnificent Channeled Scablands in the U.S. states of Montana, Washington, Idaho, and Oregon (see Baker, 2010).

[132] This is Sir Frederick William Herschel (original: Friedrich Wilhelm Herschel 1738–1822), the German-born musician and the founder of modern astronomy. I cannot understand to what "system of Herschel" d'Aubuisson here refers to. We know, however, that Herschel knew Hutton and read his books (Balderston, 1961). Balderston wrote:

> [Herschel's] second principle was the assumption that the universe is a dynamic one, continually changing and developing. In adopting these views Herschel was not an isolated voice. He was applying in astronomy the principles that contemporaries were applying in geology and biology. James Hutton, the Scottish geologist who is even less appreciated by the ordinary man than Herschel, set out a similar uniformitarian theory of geological change over ages of endless time. It is surely significant that Herschel knew Hutton and read his books. (Balderston, 1961, p. 8)

[133] I.e., UK.

[134] This book has almost no references to anybody, which is characteristic of MacCulloch's writings and which was criticized by Sir Charles Lyell in the otherwise very appreciative obituary he wrote for him.

[135] I have back-translated into English what Huot wrote in French. His rendering is not entirely felicitous and that is why I think the original text of Conybeare and Phillips should also be reproduced here:

> Hutton had the merit of first directing the attention of geologists to the important phenomena of the veins issuing from granitic rocks, and traversing the incumbent strata, and of bringing forward in a striking point of view the circumstances which seem to corroborate the igneous origin of trap rocks: the wildness of many of his theoretical views, however, went far to counterbalance the utility of the additional facts which he collected from observation. He who could perceive in the phenomena of geology nothing but the *ordinary* operation of actual causes, carried on in the same manner through infinite ages, without the trace of a beginning or the prospect of an end, must have surveyed them through the medium of a preconceived hypothesis alone. (Conybeare and Phillips, 1822, p. xliv; italics theirs).

[136] As in the case of Huot's quotation above, here too I translated what de Boucheporn wrote. The corresponding original passage in Playfair's *Illustrations of the Huttonian Theory of the Earth* (1802, p. 45) is worded as follows:

> Though the primary direction of the force which thus elevated them must have been from below upwards, yet it has been so combined with the gravity and resistance of the mass to which it was applied, as to create a lateral and oblique thrust, and to produce those contortions of the strata, which, when on the great scale, are among the most striking and instructive phenomena of geology.

[137] Daubrée cites him as G. Thomson, because in Italy, where he resided toward the end of his short life, he used the Italian version of William, which is Guglielmo. Thomson was the co-discoverer of the Widmanstätten pattern in iron meteorites; the other, independent, discoverer was the Austrian scientist Count Alois von Beckh-Widmanstätten (1753–1849).

[138] We now know that these are the Permo-Carboniferous volcanic, granitic and sedimentary rocks and the Hercynian basement of the Southern Alps, here cropping out in the hanging wall of the southeast-vergent Valsugana Thrust Fault.

[139] There is no book by Kircher that carries a date of 1644. I think this is a misprint, or misquotation, for 1665, during which the first volume of Kircher's *Mundus Subterraneous* was published that treats "fire in the earth" (Kircher's prodromus to *Mundus Subterraneous*, the *Iter Extaticum II*, was much earlier, published in 1657). Kircher was not held in high regard by most natural scientists of his day. Leibniz said of him "The man understands nothing" (see Inglehart, 2018, p. 45). In the *Protogaea*, Leibniz wrote "Qui contra sentiunt, narratiunculis secunduntur, quea apud Kircherum quendam, aut Becherum, aliosque id genus credulos aut vanos scriptores *de miris naturae lusibus & vi formatrice* in magnam specuiem verbis ornantur" (Leibniz, 1749a, p. 44, italics his). Here is the translation by Cohen and Wakefield: "Whoever believes the contrary is seduced by the fairy tales of Kircher or Becher, and of other credulous or *vain writers of this sort, who describe the wonderful games of nature and its formative power*, all embellished with a great display of words." (Leibniz, 2008, p. 73). Such damning verdicts by such a high authority as Leibniz, if Hutton had heard of them or read the *Protogaea*, must not have recommended Kircher to Hutton.

[140] "Cryptoristic" and "troponomic" are André Marie Ampère's (1775–1836) terms defined in his *Essai sur la Philosophie des Sciences, ou Exposition Analytique d'une Classification Naturelle de Toutes les Connaissances Humaines* in 1834, p. 43. Cryptoristic refers to the discovery of the most "hidden" features of an object of study. Troponomic refers to the study of changes affecting the same objects both in time and in space and deducing from such changes the laws governing the changes.

[141] Sainte-Claire Deville cites the passage of Sir Francis Bacon after Playfair (1802, p. 477).

[142] This section was almost entirely researched and translated for me by my scholarly friend and colleague Boris A. Natal'in, professor of geology at ITU.

[143] In Gamauf (2008, p. 953) Dollart (Dollard in Dutch) is rendered as *Doller* and in Lichtenberg (2013, p. 391) as *Dollert*.

[144] The reference here is to the eruption on 31 December 1720 at the submarine Dom João de Castro Bank, which is a submarine volcano with two craters. The eruption formed an island, termed Ilha Nova (=New Island), but it disappeared by 21 July 1722.

[145] Lichtenberg (2013, p.502). Lichtenberg's notes are also completely available in digital form at http://lichtenberg.adw-goe.de/seiten/view/210964/1475768252598#seiten/view/210964.ajax and also published in book form in seven volumes in Göttingen by Wallstein Verlag.

[146] Dean in his wonderful and informative book on Hutton inexplicably says that Leopold von Buch "neglected Hutton completely" (Dean, 1992, p. 247). Exactly the opposite is true, in that the German baron was profoundly influenced by Hutton in many of his physical geological ideas and said so openly as I instanced above. Dean's mistake is possibly a result of his inability to read the original German sources and not having been educated and been active in geological tradition. Cuvier also did not ignore him, as Dean erroneously claims, and neither did Boué as I cite above; they simply disagreed with his actualism. Here again the language barrier, at least in the case of Boué (for Cuvier is available in multiple English translations), may have crippled Dean. But even before von Buch, there were enthusiastic vulcanists in Germany in addition to Voigt. Friedrich Carl Ludwig Stickler (1773–1836) was one of them. This interesting man, a school counsellor (*Schulrat*) and Egyptologist, a friend of Wilhelm von Humboldt, published a remarkable paper in 1812 in which he proposed that volcanoes, ancient and modern, are arranged along nine volcanic lines on the earth's surface, reaching almost from pole to pole (*vulkanische Meridiane*: Stickler, 1812, p. 140; only those on the east and west margins of the Pacific Ocean stood the test of time), in clusters, and that their source was in a layer below the lowest rocks seen on the surface (below Werner's *Urgebirge*!). He illustrated these lines on a colored map attached to his paper. This paper is, to my knowledge, the first that recognized volcanic lines (Stickler called them *vulkanische Linien*, 1812, p. 140), an idea later taken over and refined by Leopold von Buch. However, there is no mention of Hutton in this paper.

[147] Hauff here paraphrases Lichtenberg, as I could not find the statement verbatim in any of Lichtenberg's writings. The original statement is most likely the following:

"There are already 50 theories of the earth, of which certainly 9/10 are more important for the history of the human reason than for the history of the earth." Lichtenberg continues: "It is incredible what revolutions in the human mind the revolutions on earth caused. Just like finding marine animals on top of mountains without a trace of the sea itself, far and wide, so one finds in amazement, in the former, conclusions without a trace of solid premises as far as the eye can see. The Englishman Woodward assumed, in order to explain those revolutions on earth, that some of the laws of Nature had been *ad interim* suspended: should one not assume, in order to explain these revolutions in minds, that in them the laws of thinking were suspended, at least *ad interim*? Who should believe, in view of the existing laws of thought, that the great teeth, found in North America along the Ohio River, are those of fallen angels? Yet, a Frenchman claimed, long before it had become a fashion in France, to suspend Natural laws *ad interim*." (Gamauf, 2008, p. 937–938).

This is the most damning statement about the bibliolatrous geological theories of the seventeenth and eighteenth centuries I have yet encountered anywhere. Its severity reminds me of Leonardo da Vinci's words which he jotted down on leaf 10 recto of the Leicester Codex:

> Of the stupidity and ignorance of those who imagine that these creatures [he is here talking about the fossils] were carried to such places distant from the sea by the Deluge."(MacCurdy, 1954, p. 321)

[148] When German writers write "England" they commonly include the entire UK in this designation, unless they wish to emphasize a particular constituent state such as Scotland or Wales or Ireland (officially until 1949, when Ireland became an entirely independent state under the name Republic of Ireland; after that date only the Protestant northern Ireland remained in the UK.). As we saw above, the French did the same.

[149] This reference is incomplete and its year is wrong: The correct reference is: J.F.B. [Johann Friedrich Blumenbach], 1790, "Dr. Hutton's Theorie der Erde; oder Untersuchung der Gesetze, die bey Entstehung, Auflösung und Wiederherstellung des Landes auf unserm Planeten bemerklich sind. Ein Auszug aus der ausführlichen Abhandlung im 1sten Bande der Transactions of the royal Society of Edinburgh. 1788. Gr. Quart. S. 209 bis 304": *Magazin für das Neueste aus der Physik und Naturgeschichte*, v. 6, part four, p. 17–27. That such a great figure as Blumenbach found it worthwhile to summarize Hutton's paper is in itself an indication of the high esteem his ideas were held in Germany. In fact, Blumenbach introduces his summary by pointing out that Hutton's name would be familiar to his readers through Hutton's theory of rain and the resulting controversy with Deluc. Hutton's 1788 paper was fully translated into German anonymously under the title "Theorie der Erde, oder Erforschung der Gesetze, nach welchen die Bildung, Zerstreuung und Wiederherstellung des festen Landes auf der Erdoberfläche geschieht" and published in 1792 in the *Sammlungen zur Physik und Naturgeschichte von einigen Liebhabern dieser Wissenschaften*, v. 4, 6th part, p. 223–268; 689–725, Dyckischen Buchhandlung, Leipzig. (I could not fathom the weird pagination that skips 421 pages, but without missing a word in the text! In any case, the pagination given by Victor Eyles in his Hutton entry in Hutton, James, *Complete Dictionary of Scientific Biography*, 4 Dec. 2018, https://www.encyclopedia.com seems incorrect.) Also incorrect is the reference given in Gilbert's *Annalen der Physik*, year 1801, 4th piece, p. 385 footnote * to Gren's *Journal der Physik*, v. IV, p.466. In that volume there is an extensive discussion on Hutton's theory of rain (beginning on p. 413, not 466), but not on his theory of the earth.

[150] He reacquired the original nobility of his family in 1858 together with his brothers August and Wilhelm after their father had repeatedly refused to renew it.

[151] Von Fuchs published his neoneptunist geological ideas in his now little-known small book *Ueber die Theorien der Erde, den Amorphismus Fester Körper und den Gegenseitigen Einfluß der Chemie und Mineralogie* (1844; On the theories of the earth, amorphism of solid bodies and the mutual influence of chemistry and mineralogy). This book was severely criticized by Baron Jöns Jacob Berzelius (1840) and had little influence on the development of geology. See Fritscher (1987).

[152] The two poems in the *Zahme Xenien VI* I cite from the Artemis edition of 1949 entitled *Johann Wolfgang Goethe Gedenkausgbe der Werke, Briefe und Gespräche*, v. 1, *Sämtliche Gedichte Erster Teil: Ausgabe letzter Hand*: Artemis-Verlag, Zürich, p. 665–666. The *Zahme Xenien VI* is dated as 1827 (Krafft, 2009, erroneously gives the publication date as 1820 and does not indicate the *Zahme Xenien* number he cites, which is VI) . Both English translations here are mine.

[153] Recall what Louis Simond wrote some 16 years before Suess was born (see the citation above, p. 21: "Mineralogy, in its present state, is really a very barren and uninviting science. We have names of substances; but as to their relative situation, and other facts leading to the true theory of their formation, contrary assertions are brought forward and denied with equal positiveness on either side."

[154] Present-day Banská Štiavnica in southern Slovakia. It is called Selmecbánya in Hungarian.

[155] Julius von Hann's publications were what inspired the great Serbian astronomer and mathematician Milutin Milanković (1879–1958) for his climate studies.

[156] Interestingly, Umlauft calls Werner "the father of German geology" unlike many of his contemporaries who regarded him facilely as one of the founders of geology.

[157] *Professor extraordinarius* in the German speaking countries is equivalent to the academic rank "reader" in the universities of the British Commonwealth. It is also similar to the French *maître de conferences*, particularly to its upper grade and to the *docent* of the Russian system.

[158] Both Breislak's own strong remarks about Werner and his school and the reference he cites to Richard Chenevix's even stronger criticism of Werner's mineralogy show that the charge by some historians of geology that Lyell distorted Werner's image in his *Principles* is thoroughly unjustified as I pointed out on other examples in Şengör (2002).

[159] The reference here is to the long and damning review of Werner's mineralogy by the Irish chemist Richard Chenevix (1808). Chevenix (1774–1830) spent some eighteen months in Freiberg and heard Werner's lectures in person.

[160] Here again is the common mistake of continental Europeans in considering the entire population of Great Britain "English."

[161] In the original, what I translate as "it seems to me that it could not happen if not in the way I indicated above" is "ciò mi pare che non potrebbe succedere se non se nel modo che ho indicato di sopra." Here *se non se* makes little sense; it is simply bad Italian. My learned friend and colleague Professor Gian Battista Vai (written communication, 11 May 2019) thought that the second *se* may be a misprint for *che*. In that case, the translation may be made more elegant: "it seems to me that it could happen *only* in the way I indicated above."

[162] For an easily accessible version of Pallas' memoir, see Pallas (1778 [1986]).

[163] This is a reference to Hutton's paper on granite (Hutton, 1794) translated into French in the Genevan journal *Bibliotheque Britannique*, (Hutton, 1798), originally published in the *Transactions of the Royal Society of Edinburgh*.

[164] This is a reference to Thomas Beddoes' (1760–1808) paper communicated to the Royal Society by Sir Joseph Banks (1743–1820) in 1791. See Beddoes (1791).

[165] This is a reference to the Genevan naturalist, director of the Observatory of Geneva and editor of the *Bibliothéque Britannique*, Marc-Auguste Pictet's report of his visit to Scotland: Pictet (1801, p. 76). Pictet was a friend of Kirwan and mentions Kirwan's introduction of the term "plutonic," because it seemed that the heat required by Hutton lay deeper in the earth than the volcanic hearths, i.e., in the realm of Pluto and not in that of Vulcan!

[166] The reviewer, very knowledgeable about, and partial to, Hutton's theory and aware of the importance of the density arguments against the assumption of vast empty spaces within the earth, may very well have been John Playfair, who had Breislak's book in his personal library and who had carried out a second survey of the Schiehallion in 1811. He also owned Brocchi's book, which was also reviewed in the *Edinburgh Review*, but not, as it seems from the auction catalogue of his library, Breislak's *Voyage Physique et Lithologique dans la Campanie*, which the reviewer says he also reviewed for the same periodical. For the auction catalogue, see Ballantyne (1820).

[167] The publication history of Breislak's book in France is somewhat complicated. After the translation of the 1811 volumes, which the German translator Friedrich Karl Baron von Strombeck (1771–1848) considers defective (Breislak, 1819, p. V–VI: "nicht völlig gelungen"), Breislak wrote a much enlarged version, but preferred to call it *Geological Institutions*, in the introduction of which he had some very nice things to say about Werner, although he had not changed his mind about geological theory. He wrote it in Italian, but never published it in his mother tongue. Instead, the book was translated, "under his own eyes" into French by P.J.L. Campmas and published in 1818 in three text volumes plus an atlas volume.

[168] The German translation is said to be made from the French, but it is not a translation of either of the editions in French, but a mixed text that used both Breislak 1818a, 1818b, 1818c, and Breislak 1811a, 1811b. Baron von Strombeck even checked the references used by Breislak and made some corrections. (Let us remember that the Baron lived near the great library of Wolfenbüttel, where Leibniz was once librarian).

[169] Sarjeant (1988) criticizes Laudan (1987) for giving the credit for the invention of the term Cretaceous to d'Halloy, and says that it was invented by Conybeare and Phillips (1822). That is not true. Conybeare and Phillips indeed use the term cretaceous numerous times in their book, but not as a system name. Moreover, the date of the Conybeare and Phillips' book is 1822, just like d'Halloy's.

[170] Gemmellaro does not say which edition he used, but, as he was appointed professor of Natural History in Catania in 1830, he probably began with the first edition of d'Halloy's book (d'Omalius d'Halloy, 1831). But we know that he used also the second edition (d'Omalius d'Halloy, 1835), because he refers to it in a footnote on p. 104 of his own book. There are no references to Hutton in d'Halloy's book. D'Halloy also published an "Introduction à la Géologie," subtitled "première partie des Éléments d'Histoire Naturelle Inorganique Contenant des Notions d'Astronomie, de Météorologie et de Minéralogie." Except for the mineralogical part, this book, which had three editions between 1833 and 1838, has nothing properly geological, although it was thought (very properly, I think) as an introduction to geology. There is no indication that Gemmellaro made use of it in preparing his own book.

[171] Comprising parts or the whole of the present federal U.S. states of Louisiana, Texas, New Mexico, Arkansas, Oklahoma, Missouri, Kansas, Colorado, Iowa, Nebraska, Wyoming, Minnesota, South Dakota, North Dakota, Montana, and the southernmost parts of the present Canadian provinces of Saskatchewan and Alberta.

[172] Daniels thinks that the United States geologists did not make notable contributions to geology before the twentieth century. That is emphatically not true. Rogers brothers' monumental Appalachian work (Rogers and Rogers, 1843; Rogers, 1859a, 1859b, 1859c, 1859d), James Dwight Dana's three textbooks (*System of Mineralogy*, 1837, *Manual of Mineralogy*, 1848 and *Manual of Geology*, 1863, with their many later editions), Hall's geosyncline interpretation (1859, 1883; although the idea was not original with him) and very especially the work of the four great surveys in the American west (e.g., Schmeckebier, 1904; Bartlett, 1962) had a profound impact on geology in Europe. Work on vertebrate paleontology by Edward Drinker Cope (1840–1897) and Othniel Charles Marsh (1831–1899) completely overshadowed the similar work in Europe. What Leonard Warren wrote about the standing of American geology in the mid-1800s is right on the mark: "…by the midcentury the American geological enterprise, nourished by numerous state geological surveys, was second to none" (Warren, 2009, p. 63; also see his accurate remarks on his p. 69).

[173] The first geological map of North America (east of the Rockies) was published by the French geologist Jean-Étienne Guettard in 1752, drawn, using the data he obtained in part from the great French geographer, architect and geologist Philippe Bouache (1700–1773), the renowned *Premier géographe du Roi* (see Cailleux, 1979; the reproduction of Guettard's map in this paper is unfortunately of very low quality). Spieker (1972, p. 15, footnote 9), the dedicated field geologist, was unwilling to call Guettard's map the "first" geological map of North America simply because Guettard had never been to America and gathered his data from the available French intelligence—conveyed mainly by Roland-Michel Barrin, comte de La Glassonière (1693–1756), interim governor of Nouvelle-France from 1747 to 1749 and a member of the French Academy of Sciences; Jean-François Gaultier, physician and naturalist (1708–1756); Michel-Alain Chartier, Marquis de Lotbinière (1723–1798), French-Canadian administrator; and François Picquet (1708–1781), a French Sulpician priest who lived for some time with the Iroquois Indians and thus became familiar with the area around the Great Lakes—concerning French Canada and "Louisiana" covering the entire middle one-third of the present-day United States (see endnote 173 above). This judgement I find unwarranted; a geological map cannot be evaluated by how and by whom it was made, but by its correspondence with the reality on the ground. When one considers the area covered by Guettard's map, one sees that no one person could have surveyed that area alone! Moreover, Guettard's correspondents sent him not only written information, but actual specimens of minerals and rocks (see Lamontagne, 1963–1964, 1965).

[174] In a footnote Eaton writes "It is much to be regretted, that Bakewell is not yet reprinted in America" (p. vi). It was to happen eight years after Eaton published this footnote. See below.

[175] Here is another footnote by Eaton (p. vii): "M'Clure [sic] says 'Many names I do not use; because I never met with them.'"

[176] Hunt does not give the details of the Bunsen reference. For those, see Bunsen (1851) in the references cited below.

[177] The literature about miners' folklore in central and northern Europe is indeed vast. The best introductions that I am aware of are Schreiber (1962) and Heilfurth and Greverus' immense tome that has the introduction and the sources (Heilfurth and Greverus, 1967; all published). Heilfurth's *Der Bergmannslied* (1954) and his *Glückauf* (Heilfurth, 1958; richly illustrated) respectively recount the history of miners' songs and of the famous miners' greeting, but in the process of doing so, they present a very rich cultural history of mining. Professor Gerhard Heilfurth (1909–2006) himself came from a family of miners in the Erzgebirge and made a sustained effort over the years to collect documents and interview miners in diverse regions of central Europe. For the history of St. Barbara, the patron Saint of miners, geologists, metallurgists, armourers, artillerymen, engineers, and mathematicians (however, doubts about her historicity led to her being removed from the General Roman Calendar in 1969; although the Catholic Church has kept her in its list of saints; her day, 4 December, is still widely celebrated by miners and geologists in Europe) see the excellent, lavishly illustrated volume by Nemitz and Thierse (1995); in that book one also learns about local patron saints and patron hallows of miners such as the Holy Daniel (in Austria and Silesia) and St. Léonard (in the coal mines in Belgium).

[178] "…nous devons généralement conclure que plus un fait est extraordinaire, plus il a besoin d'être appuyé de fortes preuves."

[179] It is often pointed out that Thales' statement that everything is full of gods contradicts his opposition to the idea of deity. But in that statement we have a mockery of the belief in gods, because gods were believed to have independence of action. Thales noticed that a magnet attracted iron objects independent of anything else. So, magnet, he said, must be full of gods. Had he thought it seriously, he would not have attempted to explain earthquakes and storms independently of gods or that he would not have been overjoyed when he discovered that he could know geometry without any help from any divinity.

[180] Planck (1949, p. 282) distinguished between the truth-content of a theory and its "value," whereby by value he meant how much a theory has given rise to fecund ideas that further our understanding of Nature. In certain aspects I already mentioned, the truth content of Hutton's theory was low, but even those parts of it gave rise to useful pursuits by others as exemplified by Count Collegno's argument mentioned above. In fact, the very greatness of Hutton's theory was more in its "value" in Planck's sense, than in its truth content.

[181] The 1881 date is the one on the cover of the book. On the title page inside the date given is 1880.

[182] The second, enlarged edition of the French original is entitled *Introduction à l'Ancien Testament* was published in 2009. The German translation was enlarged by the addition of two new introductions.

Index of Geographical Localities

A

Aboukir Lake, Egypt, 25
Aci Reale, Sicily, Italy, 97
Adirondacks, New York, 104
Aethiopia, 23
Africa, 30, *See also* East Africa
Ailsa, *See* Elza
Aiolia (historic land in Turkey), 24
Albany, New York, 103
Alberta, Canada, 150n171
Allmannagjá, Iceland, 30
Alpine countries, 115
Alps (in Europe), 88, 91, 93, 104, 110, 137
    slopes of, 27
    soils of, 27
    Southern, 63
        Hercynian basement of, 148n138
America, *See* United States of America
America, North, 9, 56, 58, 101, 110, 148n131, 148n147, 150n173
Amsterdam, The Netherlands, 137n6
Amu Lacus (=Amu Lake) (=Sarikamish, swamp region of, Uzbekistan), 26
Andes, 12, 55
    Central, 117
Angleterre (=England), 54, 126
Antioch (present-day Hatay), Turkey, ix, 4
Arabia, 23, 145n88
Arabian Gulf (=Red Sea), 23
Aralo-Caspian lowlands, 88
Aral Sea, Kazakhstan and Uzbekistan, 26, 130, 144n77, 144n78, 144n79, 144n80, 144n81
Arbresle, France, 145n104
Arkansas, United States of America, 150n171
Arran Island, Scotland, 5, 6
Arthur's Seat, Edinburgh, Scotland, 47, 57, 87
Arve valley, France and Switzerland, 23
Asia
    Central, 20, 25, 26, 117
    inner, 16
    Minor, 24
Assos (now called Behramkale), Turkey, xi
Athens, Greece, 120, 143n70
Atlantic coast, 9
Atlantic Ocean, 75, 103, 105, 126
Austria, 88m89, 91, 150n177
Austria-Hungary, *See* Austro-Hungarian Monarchy
Austrian Empire, 88–92, 115
Austro-Hungarian Monarchy, 89, 91, 94
Auvergne, France, 21, 26, 34, 47, 59, 84, 100, 101
Azores, 74

B

Bakırçay River (ancient Kaikos), Turkey, 23, 24
Baltic, 26
Banská Štiavnica, Slovakia, *See* Schemnitz
Barbados, 105

Bavarian Rhine Provinces, 47
Bayreuth, Germany, 58
Berlin, Germany, 2, 3, 8, 77, 78, 141n40
Black Sea, 24, 25, 144n74
Bohemia, 47, 83, 88
Bologna, Italy, x, xi, 138n19
Bonn, Germany, 28, 82
Bozen (now Bolzano), Alta Adige, Italy, 85
Brazil, 117
Britain, Island of, 12
British Empire, 9–49
Burgogne-France-Comté, 68

C

Cairo, Egypt, 23
Caldonazzo Lake, Italy, 67
California, United States of America, 145n100
Calton Hill, near Edinburgh, Scotland, 56
Cambridge, England, 7, 20, 21, 119, 140n30
Campsie Hills, Stirlingshire, Scotland, 45
Cape Colony, 30
Cape of Good Hope, 31
Cape Town, South Africa, 30
Caria (historic land in Turkey), 24
Carpathians, 88
Catania, Sicily, Italy, 97, 100, 150n170
Catskill Mountain, New York, (also known as Catskills), United States of America, 103
Caucasus, 16
Central Asia, *See* Asia, Central
Ceylon (now called Sri Lanka), 21, 143n72
Chemnitz, Germany, 47
Chicago, Illinois, xi
China, 94, 117, 122, 134, 137n9
Cologne, Germany, 144n84
Colony of the Province of Pennsylvania, North America, 33
Colorado, United States of America, 150n171
Cordelieras [*sic*] of South America, *See* Cordillera of South America
Cordillera (American), 57
Cordillera of South America, 57, *See also* The Andes
Cornwall, England, 39, 59, 71, 75, 98
Cremnitz, Germany, 47
Crifell, *See* Crif-Fell
Crif-Fell (=Criffel; in Galloway), 48
Cross Fell, Mountain of, in Cumberland, England, 32, 127
Cumberland, England, 32, 49, 127

D

Danube River, 24, 25
Deccan, India, 48
Delta (of the River Nile), *See* Nile River, Delta of
Denmuir, Scotland, 45
Derbyshire, England, 9, 32, 141n41

Dosenberg near Warburg, Germany, 48
Dunbar, Scotland, 52
Durham, England, xi, 49

E

East Africa, 117
Edinburgh, Scotland, xi, 118
Edko Lake, Egypt, 25
Eger (now Cheb) in the Czech Republic, 47
Egypt, 23, 24, 25, 26
Eisenach, Germany, 47
Elza (Ailsa) rock, Scotland, 5
Ephesus, Turkey, 23, 24
Eresos, Lesbos, Greece, 137n7
Erzgebirge (=Krušné Hory), Germany and The Czech Republic, 46, 146n111, 150n177
Etna, volcano of, 29, 55, 56
Europe, continent of, 1, 6, 8, 24, 26, 37, 50, 56, 57, 63, 68, 73, 74, 84, 94, 95, 96, 101, 107, 109, 110, 116, 118, 121, 144n85, 150n172, 150n177
    rivers of, 24
European rivers, *See* Europe, continent of, rivers of

F

Fassa Valley, Italy, 85
Fassney, Scotland, 45
Feldbach, Styria, Austria, 47
Florence, Italy, 115, 116
France, 49–88, 63, 64, 84, 87, 94, 95, 100, 107, 148n147, 150n167, 150n173
Franconia, Germany, 58
Freiberg, Saxony, Germany, 14, 26, 59, 65, 67, 70, 73, 78, 79, 83, 94, 138n23, 146n111, 149n159
Freiburg i. Br., *See* Freiburg im Breisgau, Germany
Freiburg im Breisgau, Germany, 21
Freyberg, *See* Freiberg, Saxony, Germany

G

Galloway, Scotland, 46, 48
Gelmersdorf, Brandenburg, Germany, 76
General's Hut, Scotland, 147n125
Geneva, Switzerland, 17, 59, 95, 149n165
Germany, 85, 87, 88, 89, 104, 107, 116, 117, 148n146, 149n149
Gibraltar Strait, 123
Glen Tilt, Scotland, 38, 40, 45, 70, 81, 98, 137n4
Göttingen, Germany, 47, 74, 78, 148n145
Grampian Mountains, Scotland, 38, 73, 81
Granville, Ohio, xi
Great Britain, 7, 8, 9, 37, 54, 74, 92, 118, 137n4, 146n116, 146n117, 147n125, 149n160
Great Lakes (in North America), 148n131, 150n173
Greece, 21, 84, 98
Green Island, Hong Kong, China, 145n92

Note: A lowercase "n" indicates the note number, so 143n65 indicates page 143, note 65.

Green Point, South Africa, 31, 41
Gulf of Lyon, 19
Gulf of Mexico, 144n74

H

Halicarnassus (Bodrum in present southwestern Turkey), 23
Hatay, Turkey, ix
Hawaiian Islands, 57, 147n129
Hebrides Islands, Scotland, 59, 89, 119
Hecla volcano, Iceland, 55
Hermus River (now Gediz River, Turkey), 24
Hessen, Germany, 53
Highlands (of Scotland), 45, 140n24
Holy Island (Holy Isle), Scotland, 5
Hong Kong, 145n92
Hong Kong Island, China, 145n92
Hungaria, See Hungary
Hungary, 47, 48, See Austro-Hungarian Monarchy

I

Iceland, 6, 28, 29, 30, 37, 75
Ilion, Turkey, 23
India, 57
Indiana, United States of America, 105
Inner Asia, See Asia, inner
Ionia (historic land in Turkey), 21, 22, 24, 141n54, 143n68
Iowa, United States of America, 150n171
Ireland, 47, 48, 149n148
Isfahan, Iran, 142n63
Isolabela, Italy, 98
Istanbul, Turkey, x, xi, 1, 117, 143n72
Italy, 94–101, 137n6, 148n137

J

Jáchimov, See St. Joachimsthal
Jeikhun River (Amu Darya=Oxus), 26, 144n79
Jerusalem, Israel, 20, 142n64

K

Kansas, United States of America, 150n151
Kansas City, Missouri, 141n47
Karamenderes River (ancient Skamander), Turkey, 23
Keele, England, xi
Khazarian Sea (=Caspian Sea), 26
Khorezmian Lake (=Aral Sea), 26, 130, 144n77, 144n78, 144n79, 144n80, 144n81
Kinneel (=Kinneil), Scotland, 5
Kinneil, See Kinneel
Kirkbean (in Galloway), 48
Koblenz, Germany, 28
Kremnitz (=Kremnica), Slovakia, 47
Küçük Menderes River (ancient Kaistros), 23, 24

L

Lac Léman, See Lake of Geneva
Lake Maeotis (present day Sea of Azov), 24, 25
Lake of Geneva, 23, 26, 144n76
Laki, Iceland, 28, 55

Lamlash, See Lamlass
Lamlass (=Lamlash), Scotland, 5
La Môle mountain, French Alps, 23
Lampsakos (now Lâpseki, Turkey), 143n72
Leipzig, Germany, 149n149
Leri, Italy, 98
Lesser Antilles, 105
Levico Lake, Italy, 67
Lion's Head, South Africa, 30
Lisbon, 12, 52, 56, 147n125
Loch-Ness, Scotland, 56
London, England, 26, 36, 41, 71, 82, 84, 119, 141n40, 147n122
Lothian, Scotland, 45, 49, 97, 125
Louisiana, United States of America, 150n171
Louisiana Purchase (area), United States of America, 101,150n173
Lydia (historic land in Turkey), 24
Lyon, France, 19, 54, 123

M

Maine, United States of America, 102
Meander River (now Büyük Menderes River in Turkey), 24
Médianes Plastiques (in the Préalpes), 23
Mediterranean Sea, 25, 38, 58, 121, 148n131
Meissen-Germany, 116
Meissner, Germany, 47, 53, 78
Mesopotamia (now largely in Iraq, with a small bit in southeastern Turkey), 4, 116, 142n64, 145n87
Meshhed, Iran, 143n72
Mezen group (of volcanoes, France), 47
Miletus, Ionia (now Milet in Turkey), 21, 24, 116, 144n72
Minnesota, United States of America, 150n171
Mississippi River, North America, 56
Missoula Lake (northwestern United States of America; no longer extant), 148n131
Missouri, United States of America, 141n47, 150n171
Mittelgebirge, Germany, 47
Mongolia, 143n72
Montana, United States of America, 148n131, 150n171
Mont Blanc, Franco-Italian Alps, 104
Montdor volcano, France, 76
Monte Nuovo, Italy, 74
Moon (Earth's satellite), 56, 69, 103
Moravia, Czech Republic, 47, 48
Moscow, Russia, xi, 141n40, 141n41
Mount Rosa, 55
Munich, Germany, 74, 117, 118, 141n40
Myanmar (formerly Burma), 141n72
Mysia, (historic land in Turkey), 24

N

Nagpoor, India, 48
Nancy, France, 60
Naples, Italy, 74
Nebraska, United States of America, 150n171
Needle's Eye (in Galloway), 48
Netherlands (see, The Netherlands)

Neuchâtel, Switzerland, 92, 94
New Madrid, United States of America, 56
Newcastle-upon-Tyne, England, 32, 127
New York, New York, 1, 102, 103, 118
Nile River
  Delta of, 23, 24, 25, 26
  five mouths of, 23
Nishapur, Iran, 143n72
Normandy, France, 17
North Caspian Depression, 144n74
North Dakota, United States of America, 150n171
Northern Ocean (north of Europe), 26
Northumberland, England, 49
Nova Scotia, Canada, 63
Nürnberg, Germany, 21

O

Ohio River, United States of America, 148n147
Oich River, Scotland, 147n125
Oklahoma, United States of America, xi, 150n171
Orléans, France, 21
Oxford, England, 13, 45, 146n108

P

Pacific Ocean, 148
Palermo, Sicily, Italy, 100
Palestine, 142n64
Paradise (imaginary mountain in Central Asia), 20, 21m142n62
Paris, France, 34, 37, 50, 54, 94, 98, 118, 147n119
Paris Basin, 27
Pekin (now Beijing), China, 30
Peloponnese, Greece, 143n72
Pergine Valsugana, Italy, 67
Petersburg, Kentucky, 141n54
Pferde Kopf (in Rhonegebirge), 47
Pladda Island, Scotland, 5, 6
Po Plain, 27
Portrush, Ireland, 48
Pozzuoli, Italy, 74, 122
Préalpes, 23
Promised Land, 142n64
Puy de Dôme (in Auvergne), 76
Puy de Graveneire (in Auvergne), 47
Puy de Nugère (in Auvergne), 76
Pyrenees, 63

R

Rhine River, 28
Rhône Delta (in Lake Geneva), 26
Rhône delta (in the Mediterranean), 19, 23, 26
Rhonegebirge, Germany, 47
Riesengebirge (low mountains with the highest elevation being Sniežka, formerly Schneekoppe, with 1,602 m, forming a part of the Sudeten mountain system=the Krkonoše=Karkonosze) now shared by Poland and the Czech Republic), 47
Roman Empire, 140n33
Rome, Italy, 50, 94, 116, 120
Russia, 70–88
Russian Table-Land, 88

Index of Geographical Localities

S

Saleve [sic] (Salève Mountain, France), 93
Salisbury Craigs, Edinburgh, Scotland, 47
Sandwich Islands, *See* Hawaiian Islands
Santa Cruz Island, Lesser Antilles, 105
Sarcou, France, 76
Sarepta (now Sarafand in Lebanon; תפרצ {Zarephath} in Hebrew), 6, 140n26
Sarikamish, swamp region of, Uzbekistan, 26
Saskatchewan, Canada, 150n171
Savoy, France, 93
Saxony, Germany, 14, 25, 51, 66, 70, 79, 94, 106, 116, 146n111
Scandinavian Peninsula, 27
Scheibenberg, Saxony, Germany, 45, 87
Schemnitz, 149n154
Schemnitz (now Banská Štiavnica in southern Slovakia; Selmecbánya in Hungarian), 89, 149n154
Schiehallion, Scotland, 83, 88, 90, 149n166
Schöneberg, Brandenburg, Germany, 76
Scotland, 5, 7, 30, 37, 45, 47, 52, 53, 56, 59, 60, 62, 64, 69, 70, 71, 73, 77, 78, 80, 83, 84, 87, 89, 93, 96, 98, 102, 112, 114, 119, 141m48, 141n46, 145n101, 146n115, 147n120, 147n125, 149n148, 149n165
Sea of Azov, 18, 24, 25
Seetabuldee (now Sitabuldi), India, 48
Selmecbánya, *See* Schemnitz
Siberia, 57, 58, 72
Sicily, Italy, 100
Silesia (former German land; now mostly in Poland), 47, 150n177
Simon's Town, South Africa, 30
Skamander (=Karamenderes), 23
Slighhouses (in Berwickshire, Scotland), 5
Sorèze, France, 53, 59
South Africa, 30
South Dakota, United States of America, 150n171
Southern Tyrol, 85
Soviet Union, 117

Space
 binary, 22
 celestial, 20, 35, 137n11, 149n166
 Minkowski's, 4
 time and, 20, 137n11, 147n125, 148n140
Spain, 50, 98, 146n115
Spanish Empire, 37
Sri Lanka, *See* Ceylon
St. Joachimsthal (present-day Jáchimov in the Czech Republic), 6
St. Petersburg, Russia, 70
St. Thomas, Virgin Islands, 109
St. Thomas Island, Caribbean, 67
Stagira, Greece (ancient city, no longer extant), 137n7
State of Pennsylvania, United States of America, 33
Stirlingshire, Scotland, 45
Strasbourg, France, 143n65
Sub-Lunar "binary space," 22
Sulphur Channel, Hong Kong, China, 145n92
Supra-Lunar "binary space," 22
Sweden, 116
Switzerland, 92–94
Sydney, Australia, xi
Syria, ix, 21, 23

T

Table Mountain, South Africa, 30, 31
Tapi River, India, 48
Tauride (in Crimea), 56
Terceira Island, Azores, 74
Teuthrania (now in Turkey), 23
Texas, United States of America, 150n171
The Netherlands, 50, 87, 137n6
Torino, Italy, 98
Toronto, Canada, 104
Transylvania (now in Romania), 47
Troy, Turkey, 24
Turkey, 1, 23, 24, 117
Turkey, southern, ix
Tuscany, Italy, 3, 17, 115, 137n6, 138n20
Tyre (now Sur in Lebanon), 120

U

United Kingdom, *See* British Empire
United States of America, 101–111
Unkel, Germany, 28
Uzboy Channel (Uzbekistan and Turkmenistan), 26

V

Vallorcine, *See* Valorsine
Valorsine (=Vallorcine), France, 64
Valsugana Thrust Fault, Southern Alps, Italy, 67, 148n138
Velay, France, 34, 59
Veneto, Italy, 137n6
Venice, Italy, 52
Verespatak (=Roşia Montană=Goldbach=Alburnus Maior), Romania, 47
Vermont, United States of America, 108
Vesuvius volcano, 55, 76
Vivarais, France, 34, 47, 59
Vogelsgebirge (=Vogelsberg), Germany, 47
Vosges in Burgogne-France-Comté, France, 68, 98

W

Western Islands (of Scotland), 59, 93
West Indies, islands of, 12, 56
Westmoreland, England, 49
"Wild west" of the United States of America, 101
Wyoming, United States of America, 150n171

Y

York, England, 49

Z

Zarephath (*See* Sarepta), 6, 140n26
Zinnwald, Germany, 47

Index of People

A

Abdallah (or Nurullah) ibn Lutfullah ibn 'Abd-al-Rashid Behdadini known as Hafiz-i Abru, 26, 144n79
Abel, Clarke, 30–1
Abel, Othenio Lothar Franz Anton Louis, 117
Abel (imaginary person in the *Bible;* second born son of Adam and Eve), 142n64
Abraham (imaginary person; allegedly a patriarch of Abrahamic religions), 116, 145n88
Abrahamse, J.E., 137
Adams, Charles Baker, 108, 109, 112
Adıvar, Abdullah Adnan, 143n72
Agassiz, Louis, 92–3
Agricola, Georgius, 6, 86, 116, 139n23
Ahriman (destructive force in Zoroastrianism; prototype of the devil in Abrahamic religions), 142n54
Albert of Saxony, 25
Albertucius, *See* Albert of Saxony
Albertus de Saxonia, *See* Albert of Saxony
von Albrecht (also known as Albertus de Haller), 118
Aldrovandi, Ulisse, 144n84
Allan, Thomas, 57, 87
Allchin, Douglas, 138n23
Ampère, André-Marie, 148n140
An, *See* Anu
Anaxagoras of Clazomenae, 22, 82, 120, 143n71
Anaximander of Miletus, 5–6, 21–2, 24, 116, 137n11
Anderson, Don Lynn, 137n1
Anderson, Gary A., 142n64
Anderson, James, 13
Anderson, Robert Geoffrey William, 9, 13, 140n23
Andrea del Verocchio, *See* Di Michele di Francesco de' Cioni, Andrea
Andrée, Karl Erich, 22
Angelis of Scarperia, Jacopo, 137n16
von Anhalt-Zerbst-Dornburg, Sophie Friederike Auguste (Princess; *See* Catherine II)
Anu (Supreme god and ancestor of all other gods in the Mesopotamian pantheon), 20
Appleby, J.H., 9
Arduino, Giovanni, 63, 67, 100
Aristotle of Stagira, 137n7
Ashworth, William B., Jr., 141n46–141n47, 146n117
Astruc, Jean, 143n63, 144n86
Atatürk, Kemâl, 117
Atra-Hasis (Akkadian Noah), 20
Augustus (first Roman emperor: Imperator Caesar Divi Filius Augustus, *See* Gaius Julius Caesar Octavianus)

B

Babbage, Charles, 99

Bacon, Francis, 1st Viscount St Alban, 143n69
Bailey, Edward Battersby (Sir), 7
Baker, Victor Richard, 4, 115, 121, 137n9, 148n131
Bakewell, Robert, 103, 105–7, 150n174
Baliani, Giovanni Battista, 115
Ballantyne, James, 149n166
Banks, Joseph (Sir), 149n164
Barba, Alvaro Alonso, 116
Barrin, Roland-Michel, comte de La Glassonière, 150n173
Basset, César-Auguste, 53, 77, 93, 113, 147n123
Batista, Giovanni, *See* Fortis, Alberto
Bauer, Georg, *See* Agricola, Georgius
Bayezid II (Ottoman Sultan), 143n72
Beaufoy, Harry ("Little Harry"; fictional character in Maria Hack's *Geological Sketches and Glimpses of the Ancient Earth*), 39
Beaufoy (Mrs.) (fictional character in Maria Hack's *Geological Sketches and Glimpses of the Ancient Earth*), 38, 39
Becher, Johann Joachim, xi, 6, 116, 138n23
Becker, Uwe, 27–8, 142n64, 144n86
von Beckh-Widmanstätten, Alois (Count), 14
Bequerel, Antoine-César, 145n104
Beddoës, *See* Beddoes, Thomas
Beddoes, Thomas, 13, 81, 96, 149n164
Benso, Camillo Paolo Filippo Giulio, Count of Cavour, Isolabella and Leri, 98
Benzenberg, Johann Friedrich, 74
Berger, Jean-François, 59
Bergman, Torbern, 116
Berthollet (Bertholet [*sic*]), Claude-Louis, 53
Berti, Gasparo, 116
Bertrand, Alexandre Jacques François, 70, 110
Bertrand, Marcel Alexandre, 94, 146n107
Berzelius, Jöns Jakob (Baron), 140n23, 149n151
Bessarabova, N.V., 141n45
Beudant, François Sulpice, 61, 70, 79
Beuerlen, Karl, 117
von Beust, Friedrich Constantin (Baron), 79
Bischof, Karl Gustav, 78, 82–84
Bischoff, Gottfried Wilhelm, 77, 78
Bittner, Alexander, 115
Black, Joseph, 9–10, 13, 26, 52, 99, 107, 130n16, 140n23, 141n41, 141n43, 141n45, 141n46, 141n50
Blackadder, Alexander, 49
Blei, Wolfgang, 6, 8
Blum, Johann Reinhard, 77
Blumenbach, Johann Friedrich, 28, 149n140
Blumenberg, Hans, 143n67
Boase, Henry Samuel, 39, 111
Boerhave, Herman, 138n23
Bölsche, Wilhelm, 87, 88
Bonnaire-Mansuy, J.-S., 60
de Bonnard Auguste-Henri, 61
Bonnefoi, Sarah Lucille, 140n31

Bork, Kennard Baker, xi
Bory de Saint-Vincent, Jean-Baptiste-Geneviève-Marcellin, 49
Bouache, Philippe, 150n173
de Boucheporn, René-Charles Félix Bertrand, 62, 148n136
Boué, Ami (Amédée), 46, 48, 59–60, 63, 79, 93, 110, 114, 148n146
de Bourbon, Louis Joseph, Prince de Condé, *See* Louis V Joseph de Bourbon-Condé, 8ᵉ prince de Condé
Bradley, Bruce, 141n47
Bradley, Francis Herbert, ix, 141n47
Brande, William Thomas, 36–7, 111
Bréhier, Émile, 143n68
Breislak, Scipione, 15, 17, 20, 75–6, 82, 94–5, 97, 149n15, 149n166, 150n167, 150n168
Bright, Richard, 28, 37
Brocchi, Giovanni Battista (also as Giambattista), 79, 99, 149n166
Brochant de Villiers, André-Jean-François-Marie, 63
Brock, Sebastian Paul, 21, 143n64
Brongniart, Alexandre, 27, 61, 63, 78, 93, 103, 145n104
Bronn, Heinrich Georg, 77
Brooks, Hannah (Mantell's housemaid), 136
Brucker, Johann Jakob, 22
Bruno, Giardano, 120
Brush, George Jarvis, 102, 147n125
von Buch, Christian Leopold, Baron of Gelmersdorf and Schöneberg, 44, 63–4, 67, 74–80, 84–5, 87, 91, 96, 100, 104, 112, 148n146
Buckland, William (Rev.), 34, 49, 79, 147n122
Buffon, *See* Leclerc, Georges-Louis, Comte de Buffon
Burat, Amédée, 59
Buridan, Jean, 5, 25, 138n18
Burke, Kevin Charles Antony, xi, 144n74, 147n127
Burnet, John, 143n68
Burnet, Thomas, 7
Burns, Robert ("Rabbie"), 141n46
de Bury, Richard, 144n84

C

Cailleux, *See* Cailleux de Senarpont, André
Cailleux de Senarpont, André, 150n173
Cain (imaginary person in the *Bible;* first born son of Adam and Eve), 142n64
Carey, Samuel Warren, 137n1
Carlyle, Alexander (Very Rev.), 13
Carozzi, Albert Victor, 138n23, 142n56
Carus, Titus Lucretius, 120
Casati, Gabrio, 9
Catherine II (Yekaterina Alexeevna; the Great), 9
Cavazza, William, 144n84
Cavendish, Henry, 20, 83

Note: A lowercase "n" indicates the note number, so 143n65 indicates page 143, note 65.

Celsus, 120
Çetinalp, Kutsi Aybars, x, xi
Chamberlin, Thomas Chrowder, 145
de Chanteloup, Jean-Marie Chaptal, *See* Chaptal, Jean-Antoine, Comte de Chanteloup
Chaptal, Jean-Antoine, Comte de Chanteloup, 94
de Charpentier, Jean (see von Charpentier, Jean; also as Johann)
von Charpentier, Jean (also as Johann)
Charpentier, Johann Friedrich Wilhelm, 86, 93
Chartier Michel-Alain, Marquis de Lotbinière, 150n173
de Chassebœuf, Constantin François, comte de Volney, 101
Chenevix, Richard, 95, 149n158, 149n159
de Chesnel, A., 124, 138n16
Chevalier, Casimir (Abbot), 66–7, 112
Chladni, Ernst Florens Friedrich, 138n23
Chorley, Richard John, 17
Christina (one of the fictional children in Granville Penn's *Conversations*), 34
Ciancio, Luca, 100
Cicero, Marcus Tullius, 143n68
Clark, William, 101
Clarke, John Mason, 111
Clarke, Martin Lowther, 138n17
Cleaveland, Parker, 102, 104
Clerk of Eldin, John, 57, 147n130
Clerk-Maxwell of Penicuik, George (Sir), 5, 119
Clifford, Richard J., S.J., 142n64
Cluver, Dethlev, 144n84
Cochon de Lapparent, Albert Auguste, 69–70, 94
Cochrane, Archibald, 9th Earl of Dundonald, 52
Cohen, Claudine, 130, 146n109, 148n139
Coleman, William, 147n122
Di Collegno, *See* Provana, Giacinto Ottavio, Count of Collegno
Comstock, John Lee, 107
Comte, Isidore Marie Auguste François Xavier, 143n69
Conybeare, William Daniel (Very Rev.), 34, 44, 61–2, 79, 87, 146n108, 148n135, 150n169
Cook, James, 147n129
Cope, Edward Drinker, 150n172
Copernicus, Nicolaus (=Niklas Koppernigk), 79, 80, 112, 122
Cordier, Pierre Louis Antoine, 60–1, 63, 77, 100
Cornford, Francis Macdonald, 21–2, 116–17, 141n54
von Cotta, Carl Bernhard, 3–4, 11, 79
Count Cavour, *See* Benso, Camillo Paolo Filippo Giulio, Count of Cavour, Isolabella and Leri
Craig, Gordon Younger, 146n114, 147n130
Critchley, Edmund, 140n34, 145n106
Cronos (Greek Titan, Son of Uranus and Gaia; Father of Zeus), 21
von Cronstedt, Axel Fredrik (Baron), 139n23
Crook, Thomas, 137n13
Cullen, William, 13, 52
Curwen, Eliot Cecil, 145n106
Custis, Peter, 101
Cuvier, Jean Léopold Nicolas Frédéric, called "Georges" (Baron), 21, 26–7, 34, 44, 46, 52–3, 61–2, 70, 77, 79, 94, 103–4, 112, 121, 137n5, 146n107, 147n122, 148n146

D

d'Alembert, Jean-Baptiste le Rond, 19–20, 50
d'Archiac, *See* Desmier de Saint-Simon, Étienne Jules Adolphe, Vicomte d'Archiac
d'Aubuisson de Voisins, Jean-François, 58–9, 61, 79, 114, 118, 141n52
d'Omalius d'Halloy, Jean Baptiste Julien (Baron), 97, 150n170
d'Orbigny, Alcide Dessalines, 21, 146n107
Da Vinci, Leonardo, x, 25, 137n6, 148n147
Dana, Edward Salisbury, 101
Dana, James Dwight, 91, 101, 109–11, 146n107, 150n172
Daniels, G.H., 101, 150n172
Darwin, Charles Robert, ix, 21, 83, 87, 119
Darwin, Frances Crofts, 21
Dashkov, Pavel Michailovich (Prince), 9, 99
Dashkova, Princess, *See* Vorontsova-Dashkova, Yekaterina Romanovna (Princess)
Daubenton, Louis Jean-Marie, 53
Daubeny, Charles Giles Bridle, 43–4, 87, 107
Daubrée, Gabriel-August, 63–4, 66, 88, 112, 114, 148n137
Davies, Gordon Leslie Harries, 7, 27, 93, 118, 147n120
Davy, Humphry (Sir), 102
De la Beche, Henry Thomas (Sir), 79
Dean, Dennis Richard, xi, 6–8, 13–15, 28, 30, 32, 34, 36, 38–9, 104, 111, 140n34, 141n49, 145n98, 148n146
von Dechen, Ernst Heinrich Carl, 78, 79
Delamétherie, Jean-Claude, 48, 50–1, 53–4, 62, 76, 100, 141n52
Deluc, Jean André, 14–17, 19–27, 32–4, 36, 48, 60, 69, 70, 75, 95, 97, 118, 120–1, 129, 137n16, 141n46, 141n47, 142n46, 142n55, 142n59, 142n60, 143n69, 143n72, 143n73, 144n76, 144n82, 144n84, 144n85, 145n89
Deluc, Jean-André (nephew), 28, 142n60
Descartes, René du Perron, 3, 5, 48, 63, 64, 116, 138n19, 140n31
Deshayes, Gérard Paul, 61
Desmarest, Nicolas, 17–18, 49–51, 69, 113, 142n53, 146n116
Desmier de Saint-Simon, Étienne Jules Adolphe, Vicomte d'Archiac, 65–6, 99, 114
Deutsch, David Elieser, v, 20, 22, 147n62
Dewey, John Frederick, xi, 7, 137n14
Diderot, Denis, 50
Dixon, Thomas, 143n72
Doin, Guillaume-Tell, 49
de Dolomieu, Dieudonné Sylvain Guy Tancrède de Gratet, 28, 29, 45, 48, 52–3, 60, 69, 74–5, 77
Domashnev, Sergei Gerassimovich, 141n40
Doskey, John S., 101
Dott, Robert Henry, Jr., 6, 8
Drummond, Henry Home (Esquire of Blair-Drummond), 49
Du Bois-Aymé, Jean-Marie-Joseph (last name also spelled Du Boisaymé), 25–6

Dufrénoy, *See* Petit-Dufrénoy, Ours-Pierre-Armand
Duruy, Jean Victor, 67

E

Eaton, Amos, 102–5, 141n47, 150n174–50n175
Eberhard, Johann August, 22
Edward (one of the fictional children in Granville Penn's *Conversations*), 34
Eichhorn, Johann Gottfried, 142n63, 144n86
Eimer, Theodor, 117
Einstein, Albert, 2–4, 8, 15, 122, 137n11
El (Supreme god of the Mesopotamian pantheon), 145n87
Eliade, Mircea, 22, 141n54
Élie de Beaumont, Jean-Baptiste Armand Louis Léonce, 28, 42, 44, 54, 62–4, 66–7, 77–9, 83, 91, 100, 112
Ellenberger, François, 3, 7
Elohim (imaginary being; plural of El= a Hebrew god, eventually equivalent of Yahweh. Originally imported from the Mesopotamian pantheon), 28, 145n87
El-Sabh, M.I., 116
Emmerling, Ludwig August, 138n23
Emmons, Ebenezer, 104–6
Empedocles of Agrigentum, 5, 19, 82, 138n17, 138n23
Enfield, William, 22
von Engelhardt, Wolf Jürgen (Baron), 74, 116, 140n34
Ente, Pieter, 137n6
Ephrem of Edessa, 21, 142n64
Ercker, Lazarus, 116
Escher von der Linth, Arnold, 92
Escher von der Linth, Hans Conrad, 92
Escholt, Michael Peterson, 144n84
Etheridge, Robert, 42, 44–5
Euler, Johann Albrecht, 70
Euler, Leonhard, 19–20
Evans, Lewis, 101
Ewing, William Maurice ("Doc"), 2, 31
Eyles, Victor Ambrose, ix, 50, 140n34, 149n149

F

Faber, George Stanley (Rev.), 34
Fantuzzi, Giovanni, 144n84
Faujas de Saint-Fond, Barthélemy, 7, 52–3, 100, 146n11, 147n119
Favre, Jean Alphonse, 28, 91
Feiken, R., 137n6
Ferguson of Raith, Adam, 13, 141n46
Ferguson, Adam, 141n46
von Feuerbach, Ludwig Andreas, 143n72
Feyerabend, Paul, 137n3
von Fichtel, Johann Ehrenreich, 116
de la Fite, Henry (Rev.), 147n122
Fitton, William Henry, 6, 9, 90, 111–12, 140n27
Forbes, I., 33
Forster, Westgarth, 15, 32–3, 41
Fortenbaugh, William W., 138n17
Fortis, Alberto (actually Giovanni Batista), 17–18, 67, 94, 100
Fouqué, Ferdinand André, 67

Index of People

de Fourcroy, Antoine-François (Count), 94
Fournet, Joseph Jean Baptiste Xavier, 40, 78, 79
Frängsmyr, Tore Lennart, 21, 127
Franklin, Benjamin, 75, 101
Freeman, Thomas, 101
French, W.J., 137n15
von Fuchs, Johann Nepomuk, 84, 149n151
Füchsel, Georg Christian, 6, 49, 86, 90, 116, 142n59
Fyfe, William Sefton, 4

G

Gaius Julius Caesar Octavianus, 140n33
Galileo, *See* Di Vincenzo Bonaulti de Galilei, Galileo
Gamauf, Gottlieb, 74–5, 148n143, 148n147
Garibaldi, Giuseppe, 98
Gaudant, Jean, 11, 59
Gaultier, Jean-François, 150
Geddes, Alexander, 146n115
Gehler, Johann Carl, 6, 140n25
Geikie, Archibald (Sir), 1–2, 7–9, 38, 40, 49, 61, 72, 77–8, 80, 85, 87, 89, 91–2, 97–9, 105, 111, 113, 115, 118, 120–1, 140n29, 140n35, 141n38, 141n39
Gellert, Christlieb Ehregott, 138n23
Gemmellaro, Carlo, 97–8, 150n170
Genghis Khan (Temuchin; Great Khan of the Mongols), 26
de Gensanne, Antoine, 68
Gessner, Conrad, 92
Gibbon, Edward, vii, 8–9, 113, 119, 124, 140n33
Gilgamesh (=Bilgamesh; legendary Sumerian king of the city of Ur), 20
Gillispie, Charles Coulston, 27–8, 138n23, 145n89
Gilluly, James, 4
Gilpin, William (Rev.), 56, 147n125
Giraud-Soulavie, Jean-Louis (Abbott), 75
God (imaginary being, alleged lord of the entire creation in monotheistic creeds or a member of an imaginary pantheon that supposedly built or fashioned the universe from eternal matter in various religions), 14, 16, 28, 36, 59, 141n46
von Goethe, Johann Wolfgang, 75, 122, 140n34, 149n152
Gohau, Gabriel, 142n56
Goldie, Mark, 146n115
Gorgias, 143n68
Gould, Stephen Jay, 7–8, 119, 138n21, 140n32
Graham, Loren R., 117
de la Grange Tournier, Giuseppe Ludovico, *See* Lagrange, Joseph-Louis
Grant, Ralph, 7, 42, 49, 91, 140n30
de Gratet de Dolomieu, Dieudonné Sylvain Guy Tancrède, *See* De Dolomieu, Dieudonné Sylvain Guy Tancrède de Gratet
Gray, Alonzo, 108, 109, 112
Green, Alexander Henry, 41–2, 79, 113–14, 145n92
Greene, John C., 4–5, 101, 137n12
Greene, Mott Tuthill, 118, 137n10
Greenough, George Bellas, 79

Greverus, Ina-Maria, 150n177
Grierson, James (Rev.), 46, 48, 49
Grimaux, Louis Édouard, 142n56
Grimm, Johann, 89
Guettard, Jean-Étienne, 18, 69, 142n56, 150n173
Guilhelmo Parisiensis, *See* Guillaume d'Auvergne
Guillaume d'Auvergne, 21, 49, 142n56
Guiton-Morveau, *See* Guyton, Louis-Bernard, Baron de Morveau
von Gümbel, Karl Wilhelm (Ritter), 87
Gumprecht, Thaddäus Eduard, 49
Günergun, Feza, xi
Guntau, Martin, 138n23
Günther, Siegmund, 141n40
Guyot, Arnold Henri, 92
Guyton, Louis-Bernard, Baron de Morveau, 81

H

Hack, Maria, 38–9, 137n4
von Haidinger, Wilhelm Karl (Ritter), 49
Hakluyt, Richard, 26, 129, 144n78
Hall, Basil, 30–1
Hall, James (American paleontologist), 29–30, 34, 39, 103, 107, 111–12, 114, 138n23, 145n93, 150n172
Hall of Dunglass, James, (Sir, 4th Baronet), 29, 30, 34, 39, 40, 45, 48, 54, 56, 57, 62–4, 66, 70, 82–4, 88, 91, 96, 103, 106, 107, 140
Halley, Edmond (also as Edmund), 62
Ham, Kenneth Alfred, 141n54, 142n55
von Hann, Julius Ferdinand, 149n155
Hannibal (Carthaginian general and politician), 33
von Hardenberg, Georg Philipp Friedrich (Baron), 138n23
Harrison, Edward Robert, v, 122
Hasse, Friedrich Christian August, 82
von Hauer, Franz (Ritter), 89
Hauff, Hermann, 79–81, 112, 148n147
Haug, Gustave Émile, 75, 146n107
Hausmann, Friedrich, 78, 79
Haüy, René Just (Abbé), 4, 45, 49, 63, 94, 103, 137n5, 137n8
Hawking, Stephen William, 119
Heilbron, John L., 120, 144n85, 145n89
Heilfurth, Gerhard, 150n177
Heim, Albert, 91, 94
Heim, Johann Ludwig, 79
Heimann, P.M., 140n31
Helios (God of the Sun in Greek mythology), 143n72
Henckel, Johann Friedrich, 139n23
Hennig, Edwin, 117
Henry, Joseph, 111, 122
Henslow, John Stevens, 87
Hephaestius, 82, *See also* Hephaistos
Hephaistos (crippled son of Zeus; Greek god of fire, volcanoes, metalworking, smithing, masonry and sculpting), 88
Hephästos, 88, *See also* Hephaistos
Heraclitus of Ephesus, 4, 24
von Herder, Siegmund August Wolfgang (Baron), 78, 79
Herodotus of Halicarnassus, 23–4

Herschel, Frederick William [Sir], 59, 148n132
Herschel, Friedrich Wilhelm, *See* Herschel, Frederick William [Sir]
Herzen, Alexander Ivanovich, 141n40
Hesiod of Boeotia, 21
Hiärne, Urban, 27, 44
Himmler, Heinrich, 117
Hitchcock, Edward, 108, 109, 112
Hobbes, Thomas, 27
von Hochstetter, Christian Ferdinand Friedrich (Ritter), 89
Hockard (unidentified), 52, 147n120
Hœfer, Jean-Chrétien-Ferdinand, 67
Höfer, Johann Christian Ferdinand, *See* Hœfer, Jean-Chrétien-Ferdinand
Hoffmann, Friedrich, 77, 78, 81, 79, 111
Hoffmann, Raymond Joseph, 120
Hoheisel, Karl, 21
von Hohenheim, Philippus Aureolus Theophrastus Bombastus (Paracelsus), 138n23
Holland, Henry (Sir, Baronet), 28, 37, 74, 145n90
Holmes, Doris L., *See* Reynolds, Doris Livesey
Home, John (Rev.), 13
Hope, Thomas Charles, 107
Hörbiger, Hanns, *See* Hörbiger, Johannes Evangelist
Hörbiger, Johannes Evangelist, 117
Hosmann, G.C., 4
Hubbert, Marion King, 140n32
Hübner, Marita, 21, 27–8, 120, 142n55, 144n84, 144n85, 145n89
von Humboldt, Friedrich Wilhelm Heinrich Alexander (Baron), 4, 6, 28, 34, 44, 61, 67, 72, 74–5, 77–9, 83, 85, 87, 91, 93, 100, 112, 121, 140n34, 148n146
Hume, David, 13, 15, 52, 141n46
Hunt, Thomas Sterry, 110–11, 150, 176
Huot, Jean-Jacques-Nicolas, 49, 61–2, 64, 148n135, 148n136
Hutton, Charles, 20, 83, 89, 90
Huygens, Christiaan, 137n11

I

Iberti, José, 146n115
Iberti, Joseph, *See* Iberti, José
Imrie of Denmuir, Ninian, 50–1, 146n115
Inglehart, A.J., 148n139
Inostransev, Alexandr Alexandrovich, 73, 90
Izdubar (Chaldaean Noah), 20

J

Jackson, Peter, 4, 26, 144n78
Jameson, Robert, 26, 40, 45–6, 48, 56–9, 61, 66, 71, 89, 104, 107, 144n82, 145n95, 145n96, 145n104
Jayatilleke, Kulatissa Nanda, 143n72
Jefferson, Thomas, 101
Jenkinson, Anthony, 26
Jesus (Christ, i.e., the Messiah and son of god according to mainstream Christian belief), 137n6
Joannes Philoponos, *See* John Philoponus
Johannes Damascenus (=Yūḥannā ad-Dimashqī), 21

John Damascene, *See* Johannes Damascenus
John Philoponus (=John of Alexandria, John the Grammarian), 26
Jones, Elisabeth Jean, 5, 8–9, 13, 119, 137n16
Jordanova, Ludmilla Jane, xi, 127
Jussieu (uncertain which of the brothers is meant), 53, 100, 147n123

K

Kalm, Pehr, 101
Kant, Immanuel, 3–4, 15, 141n53
Karpinski, Robert Whitcomb, 68
Karsten, Dietrich Ludwig Gustav, 89
Karsten, O.B.R. (=Oberbergrath), *See* Karsten, Dietrich Ludwig Gustav
Kayser, Friedrich Heinrich Emanuel, 87, 126
Keckermann, Bartholomäus, 61
Keferstein, Christian, 81–83, 114
Kemmerer, A., 72
Kennedy, Robert, 29
Kenngott, Gustav Adolph, 87
Kepler, Johannes, 138n23
Kerr, Robert, 26
Khafizi Abru, *See* Abdallah (or Nurullah) ibn Lutfullah ibn 'Abd-al-Rashid Behdadini known as Hafiz-i Abru
Khayyam, Omar, 143n72
Kiöping, Nils Matsson, 21
Kircher, Athanasius, S.J., 19, 68, 148n139
Kirthisinghe, B.P., 143n72
Kirwan, Richard, 27, 34, 69, 89, 97, 103, 149n165
Klaproth, Martin Heinrich, 53
Klaver, Jan M.I., 142n61
Knox, Robert, 52
Kober, Leopold, 146n102, 146n107
Koeberl, Christian, xi
Kopelevich, Y.K., 71
Köprülü, Mehmed Fuad, 117
Kovrigin, N.N., 72
Krafft, Fritz, 149n152
Krayer, Albert, 127, 130
Kristeller, Paul Oskar, 118, 119
Krivonogov, Sergey, 26, 144n70
von Kronstedt, *See* Von Cronstedt, Axel Fredrik [Baron]

L

La Grange, *See* Lagrange, Joseph-Louis [Comte]
La Metherie, *See* Delamétherie, Jean-Claude
Lagrange, Joseph-Louis [Comte], 53, 77
Lagrangia, Giuseppe Luigi, *See* Lagrange, Joseph-Louis
Laks, André, 143n68, 143n69, 143n71, 143n72
Lamanskii, Vladimir Vladimirovich, 90
Lambeck, Kurt, 116
Lamontagne, R., 150n173
Lampadius, Wilhelm August, 138n23
Lane Fox, Robin James, 142n64
Lange, Johann, 116
de Laplace, Pierre-Simon (Marquis), 3, 47, 53, 62, 77, 118, 141
de Lapparent, *See* Cochon de Lapparent, Albert Auguste

von Lasaulx, Arnold Constantin Peter Franz, 86–87
Laudan, Rachel, 5–7, 116, 118, 138n23
Lavoisier, Antoine, *See* De Lavoisier, Antoine-Laurent
de Lavoisier, Antoine-Laurent, 17, 19, 26, 52, 137n5, 138n23, 142n56, 144n85
Le Conte, Joseph, 110–11, 146n107
Leclerc, Georges-Louis, Comte de Buffon, 4
Lehmann, Johann Gottlob, 6, 90, 116
Leibniz, Gottfried Wilhelm, 19, 25, 48, 62, 66–7, 146n109, 148n139, 150n168
Leonardo da Vinci, *See* Di ser Piero da Vinci, Leonardo
Leonhard, Gustav, 70, 77
von Leonhard, Karl Cäsar, 74, 77, 78, 79, 83, 112
Lesueur, Charles Alexandre, 105
Leuckart, Friedrich Andreas Sigismund, 77
Levilapis, Hermannus, 137n16
Lewis, Meriwether, 101
Lichtenberg, Georg Christoph, 74–5, 80, 120, 137n16, 138n23, 148n143, 148n145, 148n147, 149n147
Lieb, Klaus-Peter, 127
Linnaeus, Carolus, *See* Von Linné, Carl
von Linné, Carl, 20
Littré, Paul-Maximilien-Émile, 143n69
Lloyd, J.J., 147n120
Loewinson-Lessing, Franz Yulyevich, 119
von Löhneyß, Georg Engelhardt, 116
Long, Stephen Harriman, 101
Long-Champ, *See* De Longchamp
de Longchamp, 78
Lord Verulam, *See* Bacon, Francis, 1st Viscount St Alban
Lorentz, Hendrik Antoon, 4
Louis V Joseph de Bourbon-Condé, prince de Condé, 59
Löwl, Ferdinand, *See* Löwl von Lenkenthal, Ferdinand
Löwl von Lenkenthal, Ferdinand, 146n107
de Luc, Jean-André, *See* Deluc, Jean-André
Lucretius, *See* Carus, Titus Lucretius
Lulof, Johann, 74
Lumley, John, 1st Baron Lumley, 144n78
Lupi, Benedetto, 100
Lutfî, (called Mevlâna, scholar, perfect, fair or mad; personal librarian of the Ottoman sultan Mehmed II, "The Conqueror"), 142n73
Luther, Martin, 6
Lyell, Charles (Sir), vii, ix, 2, 9, 20, 26, 32, 34, 38–9, 41, 43–4, 49, 60–5, 70, 72, 74, 76–7, 79, 88–9, 91–2, 94, 97, 100–1, 104–5, 107, 111–12, 138n21, 140n34, 144n76, 144n83, 146n107, 146n108, 147n126, 147n130, 148n131, 148n134, 149n158
Lysenko, Trofim Denisovich, 117

M

von Maack, P., 70, 112
MacCulloch, John (also as Macculoch), 34–5, 59, 63, 70, 79, 87, 89, 93, 148n134
MacCurdy, Edward, 148n147

MacGillivray, William (twentieth century biographer of the nineteenth century), 40
Mackenzie, George Steuart (Sir), 28, 37, 75, 87
MacKnight, Thomas (Very Rev.), 45
Maclure, William, 101–2, 105–6, 150n175
de Maillet, Benoît, 6
Malakhova, Irena, xi, 141n44
Malchus, *See* Porphyry of Tyre
Mansfeld, Jaap, 18, 143n68
Mantell, Gideon Algernon, 40, 111, 140n34, 145n103, 145n104, 145n106
Mantell, Mary Ann (*née* Woodhouse), 145n102
de Marignolli, Giovanni, 21
de Marschall, *See* Marschall von Bieberstein
Marschall von Bieberstein, Karl Wilhelm, 48
Marsh, Charles Othniel, 115, 150n172
Marsigli, Luigi Ferdinando (Count), *See* Marsili, Luigi Ferdinando [Count]
Marsili, Luigi Ferdinando [Count], 18, 19, 92
Martin, John, 147n122
Marx, Karl, 143
Marzari-Pencati, Giuseppe (Count), 100
Maskelyne, Nevil (The Rev.), 20, 83
Mason, Shirly Lowell, 17
Mather, Kirtley Fletcher, 17
Mather, William Williams, 106–7
Mathesius, Johannes, 6, 116
Maximilian I Joseph, King of Bavaria, 74
Maxwell, James Clerk, 5, 78, 119
McIntyre, Donald Bertram, xi, 2, 5, 7–8, 141n44, 141n46
McKirdy, Alan, 7, 8
M'Clure, *See* Maclure, William
Mehmed II, "The Conqueror" (Ottoman Sultan), 143n72
Merian, Peter, 63
Merrill, George Perkins, 101
Miall, Andrew Derwent, 41
Mitchell, John (Rev.), 106
Mikhailov, G. K., 130
Milanković, Milutin, 149n155
Millar, James, 14–15, 32–3, 141n48, 141n49, 145n97
Miller, James, *See* Millar, James
Mlodinow, Leonard, 119
Mohs, Carl Friedrich Christian, 49, 89, 138n23
Mojsisovics Edler von Mojsvar, Johann August Georg Edmund (Ödön in Hungarian), 91, 115
Möngke (Great Khan of the Mongols), 26
Monro, Alexander (secundus), 13
Montagu, John, 4th Earl of Sandwich, 147n129
Montesquieu (see de Secondat, Charles-Louis, Baron de La Brède et de Montesquieu)
de Montlosier, *See* De Reynaud, François Dominique, Comte de Montlosier
Morgan, David, 26, 144n78
Mori, G., 100
Moro, Anton Lazzaro, 4–5, 73, 94, 99–100, 121
Morozewicz, Józef, 90
Moschen, Lamberto, 90
Moses (imaginary person; allegedly the greatest of the Hebrew prophets), 19, 27, 36, 103, 142n63, 142n64, 145n88

Index of People

Most, Glenn W., 143n69
Mrs. R. (fictional person in Granville Penn's *Conversations*), 34
Mühlfriedel, Wolfgang, 138n23
Mühlpfordt, Günther, 138n23
Murchison, Roderick Impey (Sir), 77, 79, 87, 91, 112, 115
Murray, John, 53, 107, 122, 141n51
Murty, Tadepalli Satyanarayana, 116
Mushketov, Ivan Vassilievich, 73–4

N

Nall, William (Rev.), 32–3
Natal'in, Boris Alexeich, x, xi, 72, 144n74, 148n142
Nathorst, Alfred Gabriel, 90
Naumann, Carl Friedrich, 78–9
Nechaev, Alexander Pavlovich, 90
Necker-de Saussure, Louis Albert, 63
Necker, Louis Albert, 28, 93
Needham, Joseph, *See* Needham, Noel Joseph Terence Montgomery
Needham, Noel Joseph Terence Montgomery, 137n9
Nemitz, Rolfroderich, 150n177
Neumayr, Melchior, 90–1, 112, 114
Newcomb, Sally, 101
Newton, Isaac (Sir), 7, 85, 137n11, 138n23
Nicols, Arthur, 42
Niewöhner, Friedrich, 27
Ninhursanga (also Ninhursag, Damgalnuna and Ninmah; Sumerian mother goddess of the mountains), 20
Noah (imaginary person; allegedly the tenth and last patriarch of the Abrahamic religions), 21, 138n20, 142n64
Nöggerath, Johann Jacob, 78–9
Nose, Karl Wilhelm, 75–6
Novalis, *See* Von Hardenberg, Georg Philipp Friedrich (Baron)
Nye, Russel Blaine, 101
Nye, William Sanford, 141n54, 142n55

O

von Oeynhausen, Karl (Baron), 79
Ogilby, James, 45
Oldroyd, David Roger, xi, 8, 23, 28, 104, 118, 120–1, 140n34
d'Omalius d'Halloy, Jean Baptiste Julien (Baron), 97, 150n170
Omboni, Giovanni, 100
von Oppel, Friedrich Wilhelm, 138n23
d'Orbigny, Alcide Dessalines, 21, 146n107
Ormazd (Zoroastrian god, Ahura Mazda), 141n54
Ospovat, Alexander Meier, ix, 101, 104, 108
Ossipov, V.I., 130

P

Paine, Thomas, ix, x
Pallas, Peter Simon, 25–6, 50, 56–7, 60, 63, 69–70, 96, 138n23, 149n162
Paracelsus, *See* Von Hohenheim, Philippus Aureolus Theophrastus Bombastus
Parrot, Johann Jacob Friedrich Wilhelm, 78
Patrin, Eugène Louis Melchior, 26
Pawer (Bauer), Georg, *See* Agricola, Georgius
Pendered, Mary Lucy, 147n122
Penn, Granville, 33–4
Penn, William, 33–4
Pennant, Thomas, 56, 147n125
Pericles, 143n72
Peter I, *See* Peter the Great
Peter the Great (Czar), 137n6
Petit-Dufrénoy, Ours-Pierre-Armand, 79
Petrus Albinus, 116
Phillips, John, 23, 26, 34, 42–5, 61, 146n107, 146n109, 148n135, 150n169
Philo the Jew, 138n17
Picquet, François, 150n172
di ser Piero da Vinci, Leonardo, 25, 137n6, 149n147
Pike, Zebulon Montgomery, 101
Pitcher, Wallace Spencer, 4, 137n15
Pitt Amherst, William (First Earl Amherst, Earl of Arrakan), 30
Planck, Max Karl Ernst Ludwig, 3, 78, 150n180
Plato, 19, 138n17, 143n71
Playfair, John, 4–5, 8–9, 13–15, 17–18, 20, 23, 26–8, 30–1, 35, 38–40, 42–3, 45, 48–9, 53–6, 59–63, 66–8, 70–1, 73, 75, 77, 81, 83–4, 87, 89, 91–3, 95, 98, 100, 104, 106–8, 110–12, 114, 119–20, 137n4, 138n17, 145n93, 145n103, 148n136, 148n141, 149n166
Pledge, H.T., 137n16
Pluta, Olaf, 27
Poincaré, Jules Henri, 4
Pokorny, Alois, 89
Polyakhova, Elena N., 141n45
Polybius, 18, 24–5, 144n74
Popowitsch, Johann Siegmund Valentin, 18, 142n58
Popper, Karl Raimund (Sir), 8, 15–16, 20, 22, 86, 137n3, 142n62, 143n69
Popper-Lynkeus, Joseph, 143n72
Porphyry of Tyre, 120
Porter, C. M., 101
Porter, Roy Sydney, 7, 137n16
Prestwich, Joseph (Sir), 4, 45
Prévost, Louis-Constant, 65, 79
Prometheus (Greek Titan), 21
Promies, Wolfgang, 138n23
Proßegger, P., 91
Provana, Giacinto Ottavio, Count of Collegno, 98
Ptolemaios, Claudios, *See* Ptolemy, Claudius
Ptolemy, Claudius, 26, 137n16
Publilius Syrus, ix, x
Pyotr Alekseevich, *See* Peter the Great

R

Raessler, Jon, xi
Raguin, Eugène Paul Antoine Jacques, 4
Ramsay, William (Sir), 13, 138n23, 141n44
Rappaport, Rhoda, 142n56
Rasumovsky, Grigory Kirillovich, 92
Read, Herbert Harold, 4, 9, 118, 137n14

Rees, Abraham, 20, 32
Reichetzer, Franz, 89
Reisch, Gregor, *See* Reisch, Gregorius
Reisch, Gregorius, 21, 143n65
Renovantz, Hans Mikhail, 26
Reuß, Franz Ambros, 89
Reventlow, Henning Lothar Gert (Count), 27–8, 144n86
Rey, Abel, 143n68
Reyer, Eduard Alexander August, 91–2
de Reynaud, François Dominique, Comte de Montlosier, 79, 111, 137n16, 142n57
Reynolds, Doris Livesey, 137n14
Richardson, George Fleming, 40–1, 112, 145n106
von Richthofen, Ferdinand Paul Wilhelm (Baron), 89
Robertson, Alastair Harry Forbes, xi
Robertson, William, 52
Roger, Jacques, 4, 140n31, 141n46, 150n172
Rogers, Henry Darwin, 150n172
Rogers, William Barton, 150n172
Rolle, Friedrich, 87
Rose, Gustavus, 79, 90
Rößler, Balthasar, 116, 118
Roth, Justus Ludwig Adolf, 87
Rouelle, Guillaume-François, 17, 142n56
Rousseau, Jean-Jacques, 54
Rozet, Claude Antoine, 61, 114
Rudwick, Martin John Spencer, ix, 4, 7, 23, 27–8, 118–20, 137n9, 138n17, 140n29, 144n84, 145n89
Rupke, Nikolaas Adrianus, 147n122
Russell, Bertrand Arthur William, 3rd Earl Russell, vii, 3, 26, 28, 137n2, 143n72

S

Sacchi, Rosalino, 91
Sainte-Claire Deville, Charles Joseph, 62, 67–9, 114, 148n141
de Sales Guyon de Diziers, Éléonor-Jacques-François, Comte Montlivault, 62
Salomon, Wilhelm (later Wilhelm Salomon-Calvi), 146n107
Sarjeant, William Antony Swithin, 5, 150n169
de Saussure, Horace Bénédict, 12, 22–3, 28, 45, 52, 60, 62–4, 69–70, 92–3, 144n84
Scheidt, Christian Ludwig, 25
Scherer, Alexander Nicolaus, 70–1
Scheuchzer, Johann Jacob, 92, 116
Scheuchzer, Johannes, 92, 116, 118, 141n47
Schindewolf, Otto Heinrich, 117
Schlegelberger, Günther, 141n40
Schlosser, Ludwig, 54
Schmieder, Carl (also Karl) Christoph, 27, 144n73
Schmieder, Carl Christoph (also as Karl), 27, 144n73
Schneer, Cecil Jack, 101
Schönberger, Eva, 143n65
Schönberger, Otto, 143n65
Schöpf, Johann David, 101
Schreiber, Georg, 150n177
Schwarz, C.G., 142n5
Schweizer, Claudia, 118
Scott, Walter (Sir), 37

Scrope, George Julius Poulett, 34, 42, 58, 79, 87, 107, 142n57, 145n98, 146n107
de Séchelles, Marie-Jean Hérault, 140n31
de Secondat, Charles-Louis, Baron de La Brède et de Montesquieu, 54
Sedgwick, Adam (Rev.), 72, 79, 87, 110, 115
Seeley, Harry Govier, 42, 44
Şengör, Ali Mehmet Celâl, ix, x, 5–6, 8, 14–15, 19, 27, 33, 44, 90, 116–17, 120, 122, 138n18, 140n31, 141n46, 142n64, 144n74, 144n81, 144n83, 146n107, 147n124, 149n158
Shafran, I.A., 130
Shah Rukh (Timur's youngest son and successor), 144n79
Shaw, R., 145n92
Silberschlag, Johann Esaias, 92
Silliman, Benjamin, 101, 105, 107–8
Simon, Richard, 27, 30
Simond, Louis, 31, 54–8, 138n23, 149n153
Simonin, Louis Laurent, 66
Simony, Friedrich, 91
Sivin, Nathan (also known as Xiwen), 117, 137n9
Smith, Adam, 13, 52
Smith, William, 3, 44, 79, 121, 140n34, 141n46
Snelders, Henricus Adrianus Marie (Harry), 116, 139n23, 138n23
Sokolov, Dimitrii Ivanovich, 71–3
Sophia Charlotte of Mecklenburg-Strelitz (Queen of the British Empire), 15
Spieker, Edmund Maute, 101–2, 150n173
Spinoza, Benedict, 27
St. Barbara, 150n177
St. Ephraim, *See* St. Ephrem
St. Ephrem, *See* Ephrem of Edessa
Stahl, Georg Ernst, xi, 6, 116, 138n23
Steffens, Henrik, 138n23
Steindórsson, Steindór, 145
Steno, Nicolas (Niels Stensen), 3, 5–6, 42, 67, 80, 83, 86, 94, 137n6, 137n8, 138n19, 138n20
Stenonis, Nicolaus, *See* Steno, Nicolas
Stensen, Niels, *See* Steno, Nicolas
Stickler, Friedrich Carl Ludwig, 148n146
Stille, Hans Wilhelm, 16, 101, 138n23, 146n107
Stoppani, Antonio, 99, 112
Strange, P.J., 144n79, 145n92
von Strombeck, Friedrich Karl, 76, 124, 150n168
Struik, Dirk Jan, 101
Studer, Bernhard, 63, 79, 92–4
Suess, Eduard Carl Adolph, 6, 16, 73, 88–9, 91, 94, 146n107, 149n153
Suess, Franz Eduard, 91

T

Tacitus, Publius (or Gaius) Cornelius, 113, 140n33
Taylor, Kenneth Lapham, xi, 49–50, 53, 140n29, 142n57, 146n115, 147n120
Thales of Miletus, 21, 82, 116, 137n11, 138n23, 143n68, 150n179
Theophilos of Antioch, 4
Theophrastus of Eresos, 5, 137n7, 138n17
Thierse, Dieter, 150n177

Thomson, B. (most likely a mistransliteration for W. Thomson), 15, 32, 63, 71, 141n50, 144n80, 148n137
Thomson, Guglielmo, *See* Thomson, William
Thomson, Thomas, 141n50
Thomson, William, 63, 147n137, *See also* Thomson, B.
Thucydides, 24
Thurmann, Jules, 79
Timur (the Great; also known as Tamerlane), 144n79
Togan, Zeki Velidi, 117
Tomkeieff, Sergei Ivanovich, 3, 7, 9, 81, 115, 130, 137n5
Torrens, Hugh, xi, 5, 7, 13, 40, 102, 140n34, 141n48, 146n114
Torricelli, Evangelista, 115–16
Toulmin, George Hoggart, 147n120
Touret, Jacques Léon Robert, 85–6
Tozzetti, Giovanni Targioni, 17–18
de Tribolet, Georges, 94
de Tribolet, Maurice, 94

U

Uhlig, Victor, 91
Umlauft, Friedrich, 91, 149n156
Ure, Andrew, 33–6
Ussher, James, 4

V

Vai, Gian Battista, x, xi, 138n19
van Rensselaer, Jeremiah, III, 104
van Rensselaer, Stephen, III, 103–4
Varenius, Bernhard, 61
Vaysey, H.W., 48
Verbrugghe, Gerald P., 4
Vernant, Jean-Pierre, 21, 22, 142n54, 143n69
Verne, Jules Gabriel, 138n23
Vézian, Alexandre, 4, 64–5
Victoria (Alexandrina Victoria: Queen of the United Kingdom of Great Britain and Ireland and the Empress of India), 37
di Vincenzo Bonaulti de Galilei, Galileo, 80, 96, 112, 115–16, 122, 143n72
Vogelsang, Hermann, 85–7, 138n23
Vogt, August Christoph Carl, 83–4, 92, 96
Voigt, Johann Carl Wilhelm (also as Karl, instead of Carl), 6, 45, 76, 81, 86–7, 89, 114, 148n146
Volney, *See* De Chassebœuf, Constantin François, comte de Volney
Voltaire (François-Marie Arouet), vii, 54
Voltz, Philippe-Louis, 61
Vorontsova-Dashkova, Yekaterina Romanovna (Princess), 7, 9–11, 13, 70, 141n40

W

Wagner, Johann Andreas, 84–5, 112
Wakefield, Andre, 146n109, 148n139
Walchner, Friedrich August, 79
Wallerius, Johann Gottschalk, 79

von Waltershausen, Wolfgang Sartorius (Baron), 100
Warren, Erasmus, 144n4
Warren, Leonard, 101, 105, 150n172
Waterfield, Robin, 144n75
Watermann, B., 142n58
Watson, Christopher, 25
Watson, Janet Vida, 137n14
Watt, Benjamin, 52
Watt, Gregory, 81, 84, 98, 102–3, 106, 147n121
Watt, James, 9, 13, 102, 119, 137n16, 147n121
Wawn, Andrew, 30, 37, 140n
Weber, Maximilian Karl Emil, 143n72
Weber, Wilhelm Eduard, 78
Wegmann, Cäsar Eugen, 4
Weinberg, Steven, v, 3
Weiss, Christian Samuel, 78–9, 87
Wells, David Ames, 109, 112
Werner, Abraham Gottlob, 3, 14, 26, 60, 70, 138n23
Werner, Georg Friedrich, 139
Wernicke, Brian, xi
West, A.F., 144n84
Whewell, William (Rev.), 20, 94, 142n61
Whiston, William, 48, 62, 67, 69, 82
White, George Willard, 101
Whitehurst, John, 33, 106
Wickersham, John Moore, 4
William of Rubruck (=Willem van Rubroeck=Gulielmus de Rubruquis), 26
Williams, John, 5, 14–15, 32–3, 74–5, 102, 106, 141n48–141n49, 145n97
Wilson, John Tuzo, 2
Wilson, Leonard Gilchrist, 62, 94, 120, 146n108
Witham, Henry, 49
Withers, Charles William John, 5
Witsen, Nicolaas (also Nicolaes in Old Dutch), 137n6
Woodward, John, 48, 67, 82, 148n147
Woolgar, Stephen William, 122
Woolley, Charles Leonard (Sir), 116
Wright, Thomas, 40–1
Wrzesinski, O.J., 142n58
Wulf, Andrea, 140n34
von Wurzbach, Constantin, Ritter von Tannenberg, 142n58
Wyse Jackson, Patrick N., 4, 26, 144n78

Y

Yahweh (YHWH: imaginary being; originally most likely the god of the Leah tribe related to the Canaanites), 28, 142n64, 145n88
Yahwist author (J), 142n64
Young, Davis A., 137n15
Young, George (Rev.), 49
Yule, Henry (Sir), 21

Z

Zalasiewicz, Jan, 140n28
Zeus (Greek god; head of the Olympian pantheon), 20, 142n64
Zippe, Franz Xaver Maximilian, 89
von Zittel, Karl Alfred, 13–14, 28, 120, 144n84

Index of Subjects

A

Academy of Sciences (Germany), 74, 84
Academy of Sciences (New York), 102
Academy of Sciences of the Russian Federation, 9, 15, 141n44
Academy of Sciences (Paris), 63, 142n56, 145n104, 150n173
Accidents (geological), 47
Actualists, 44
Aetiological, arguments, 68–9
Air, weight of, 115
Alkalis, 29, 70
Alluvial
    land, 23
    plains, 2, 23
Almond-stone, 81
Alumina, 58
American Academy of Arts and Sciences, 108
Amherst College, 108–9
Anatomy, 52, 138n19
Andersonian Institution (University of Strathclyde), 34
Animals
    exuviae of marine, 106
    impression of, 55
    remains of, 6, 57–8
Antediluvian world, 36
Anthracite, 102
Apophyse(s), 84, 91
Application of the experimental method to physical and mechanical phenomena, 64
Archaeopteryx, 84
Aristotle's ethics, 132
Ark (Noah's), 142
Armourers, 150n177
Artillerymen, 150n177
Astronomy, modern, 79, 148n132
Athanaeum (in New York City), 104, 146
Atheism, 14, 120, 141n46
Atmosphere, 5, 12–13, 56, 63, 71, 76, 81, 95, 139, 141n36
Auriferous transition ... of Kremnitz and of Transylvania, 47
Austrian Empire, 88–92

B

Babel-tower, 35
Baconian view of how to do science, 15
Barite, 51
Barometer, 115
Bar(s), 48
Basalt(s)
    fiery origin of, 84
    German, 76
    neptunistic interpretation of, 53
    plutonic origin, 57
    primordial magmatic rock, 52
    volcanic origin, 60, 70, 84, 100, 112
    volcanicity of, 76
Basement, 30, 148
Bavarian Academy of Sciences, 84
Becher-Stahl tradition, 138
Bed(s)
    aqueous, 66
    inclined, 6
    parallel with one another, 6
Bergakademie in Freiberg, 79, *See also* Mining Academy in Freiberg
Bible
    myth in, 15, 17, 20
    scientific authority of, 20
Binary history, 141n54
Biology, 85, 89, 117, 148n132
Biostratigraphy, 44, 61, 101, 104, 108, 121
Bitumen, 51, 102
Bolide impact(s), 62
British Association for the Advancement of Science, 146n108
British Empire, 45–9
British geomorphic thought, 4
Britons, 37
Buddhism, 143n72

C

Calcareous spar, 29–30
Calcareous substance, calcareous, *See* substance(s)
Calcination, 57
Calcite, 51
Caloric, 95, 99, 102, 105
Cambridge University Main Library, 140n30
Canyon(s), 19
Carbon dioxide, 81, 91, 141
Carbonic acid, 29, 64, 106–7
Carse(s), 48
Catastrophe(s), 19, 21, 25, 27, 53–4, 58, 64–5, 80, 146n107, 147n122, 148n121
    Cuvierian, 121
Catastrophists, 44
Catholics, 27
Cause(s)
    almighty, 76
    atmospheric, 40
    creating, 33
    still in action, 22, 24
    that had allegedly ceased to act, 22
Chalk, 12, 38, 47, 64, 107
Chemistry, 5, 13, 15, 34, 37, 52, 82, 85, 95–6, 102, 107–8, 138n22–138n23, 141n50–141n51, 142n56, 144n85, 149n151
Chemists, 95, 138, 138n23
Chinese, 4, 8, 137
Chinese whisper, 8
Christian, *See also* Christianity
    creed, 17, 120
    faith, 22, 39
Christianity, 17, 27–8, 142n64
    Syrian fathers of, 21
Chronology
    Biblical, 4
    Hutton, modern geology, 111–13
    Hutton's theory, 113–14
    Mosaical, 33
    relative, 26, 53
Chronometer, 24–5
    Deluc's natural, 26, 144n76
Church of England, 37
Clastic material, 2, 14
Clay, 12, 35, 48, 57–8, 71, 102, 109
Cleft(s), 55, 58
Clinkstone(s) (also written as clink-stone), 47–8, 57
Coal, 6, 14, 29, 47, 57, 68, 102, 115
Coastal cliffs, 23
Collège de France, 21, 52–3, 62, 67–8
Columnar forms (in volcanic rocks), 46
Comet, impact of, 98
Cometary fly-by (hypothesis of Laplace), 53, 147
Composition, 12, 21–2, 33, 56, 58, 70–2, 89, 138n23, 147n119
    granitic, 39
Confusion between methods and theory in the sciences, 115
Conjecture (in science), 20, 50, 147n122
Continental drift, 7
Continents
    birth of, 19, 27
    production of the mass of (according to Deluc brothers), 22
Contortions of strata, *See* Strata, contortions of
Controversy between the Huttonians and the Wernerians, 32
Convulsions, 55
    internal, 43–4
Copper, 68
Corals, 12, 69
Correlation of strata, *See* Strata, correlation of
Cosmogonic tradition
    Becher-Stahl, 6, 116, 138
Cosmogonical vision (Hutton's), 5
Cosmogonie, 60, 137
Cosmogonists, 109, 138
Cosmogony, 7, 35, 43, 137n16, 138n23
Cosmologie, 137
Cosmologist, 122, 138n23
Cosmos, the forces that produced it, 22
Cratons, 2
Creation
    myth of, in the *Bible,* 36, 39, 80, 138n20, 138n23
    myths, 21
    organized, wrecks of, 58
    processes of, 57
Creator, 60, 140, 145
Cretaceous, 52, 121, 142n64
Crises of extraordinary action, 44

Note: A lowercase "n" indicates the note number, so 143n65 indicates page 143, note 65.

Index of Subjects

Critical Rationalism, 8
Critical times (in earth history), 146n107
Cross-cutting relationships, 5, 23
Crucial instances, 31
Crustal ruptures, See Globe, crustal ruptures of
Crystallisation (crystallization), 12, 51, 58, 60, 62, 90, 108, 138n23
Crystallography, theory of, 4
Currents
    of lava, 96
    river, 24–5
    in a sea, 17
    subterranean, 54
Cycles, 59, 119

D

Dale(s), 17, 49
Day, See Yom
Days (Biblical), 66, 103
Debris, 14, 17, 25, 63, 78, 83, 94
Deccan traps, 48
Deduction(s), 3
    logical, 16, 68
    scientific, 26
    temporal, 5
Deformation, 14, 19, 42, 62, 64, 68, 83, 92
Deistic motives (of Hutton), 141n46
Delta(s), 23, 25, 48, 74, See also specific delta
Deluge (Biblical), 16, 20, 26, 28, 32, 49, 104, 147n122, 148n147
Deluge (Biblical), See also Flood, Biblical
Demonstrative arguments, 15
Denudation, 5, 42, 114, 118, See also Theory, fluvial theory of denudation
Deposition (of sediment), 14, 17, 30, 38, 44–5, 67–8, 75, 90, 114, 138n23, 146n113, 147n124
Deposits
    basaltic, 47
    delta, 23
    sedimentary, 40–1, 98
Depressions, 54
Desert, 42, 138n17
Desiccation of land surface, 117, 142
Destruction, 5–6, 29, 57–9, 63, 65, 69, 71–2, 76, 78, 95
Deucalion story, 20
Devonian (System, Period), 45, 144n74, 146n113
Diluvialists, 109
Discordant relationship (in stratigraphy), 68
Doctrine, of alternating periods of convulsion and repose, 44
Dogmatism, 97, 118
Dolerite, 45, 85
Doller, See Dollart
Doller, See Dollart
Dollert, See Dollart
Dolomite, 85
Dykes
    basaltic, 30, 84
    directional characteristics of, 62
    granitic, 59, 75, 137n4
    in Scotland, 77

E

Earth
    age of, 17, 25–6
    central fire in, 5, 34, 48–9, 56, 60, 66, 77, 95, 100, 104, 106
    construction of, 77, 85
    corruptible, 22
    density (of), 20, 32, 83, 90, 149n166
    dynamic, models of, 5, 138n18
    formation, history of, 13, 44, 61, 72, 81, 94, 103, 114
    former, mountains of, 35
    history of, 13, 16, 33, 42–3, 51, 62–3, 65, 70, 79–82, 137n1, 141n47
    inhabitants (of), 34, 58
    interior of, 43, 57–8, 60–1, 66, 71, 82, 95
    as a large organism, 84
    molten initial condition of, 70, 112
    origin of, 34
    originally a hot mass, 72
    physical events of, 16
    present appearance of, 46
    solid pavement of, 57
    surface, inequalities of, 10, 42, 56
    temperature of the interior of, 60
Earthquake(s), 10, 56, 68
    of Lisbon, 12, 56
    of New Madrid, 56
Earths, 29, 54, 88
Earth's rocky crust, 35, 65
École des Mines de Saint-Etienne, 66
École Normale, 53
École Spéciale d'Architecture, 66
Edinburgh, 2, 4, 7, 9–11, 13, 32, 37, 40, 44, 46–7, 49, 52, 54, 56–7, 59, 61, 64, 70–1, 81–2, 87, 97, 107, 114, 118, 140n36, 141n40, 141n46, 145n104, 146n115, 147n120, 149n95, 149n140
Edinburgh University, 13, 113, 141n44
Egyptians, 4, 23
Electricity, 29, 41, 85
Elements
    airy, 138n23
    earthy, 138n23
    flammable, 138n23
    metallic, 138n23
    watery, 138n23
Engineers, 25, 83, 150n177
Enlightenment, 7, 9, 22, 50, 54, 94, 96, 120
Eocene, 52
Epochs of disturbance, 43–4
Erosion
    fluvial, 7, 17, 93
    headward, 23
    subaerial, 18, 23, 41
Erratic blocks, 27, 92–3
Eruptive processes, See Processes, eruptive
European ideas of the Enlightenment, 7, 9, 50, 54, 120, 146n115
Evidence
    Neptunian, 17, 96, 111
    Plutonian, 37, 111
Expeditions (in the United States), 101
Experiment(s), 4, 28–9, 32, 34, 37, 40, 45, 48, 54–5, 57–9, 61–4, 66, 68, 70–1, 74–5, 81–2, 84, 87–9, 91, 95–6, 100, 103, 106–8, 114, 119, 138n23
    Sir James Hall's, 34, 39–40, 48, 64, 66, 70, 82, 84, 91, 96, 111
Explanation(s) (in science)
    bad, 20, 22
    good, 20, 117
    hard to replace by any other explanation, 20
    hard to vary, 20
External characters of fossils, 6
Extinctions
    animal, 21
    universal, 23, 121

F

Fault, 92, 146n112, 148n138
Fauna(s)
    changing, of the planet, 6, 62
    succession of, 53
Feldspar, 45, 51, 57, 73, 75
Feltspat [sic] (See Feldspar)
Fire(s)
    activity of, 77
    agency of, 10, 38
    central
        eternally burning, 104
    internal, 35, 55, 102
    natural force of, 138n17
    subterranean, 10, 12, 30, 35, 55, 74–5, 85, 104, 138n20
First geological map of North America, 150n173
Fixed air, 12, 51
Flame, 68
Flint(s), 57–8
Flood(s),
    Biblical, 20–1, 142n63–142n64
    continent-sized, 148
    myth, 20, 23
    universal, 20–1, 97, 116
Flora (changing, of the planet), 62
Flœtz
    layers, 81, 146n113
    Period, 47, 146n113
Flœtz-trap formation(s), 45, 48, 59, 102
Fluid, expandible, 96
Fluorine, 51
Fluor(s), 51
Folding, 62, 64, 66, 70, 88, 111
Foliation, 72
Force(s)
    acting *from beneath* the strata, 38
    expansive, 76, 96
    lateral, 62
Formation(s), flœtz-trap, 45, 78, 89, 102
Fossil Botany, 40
Fossilists, 109
Fossils, See also Petrifactions
    calcareous, 81
    characteristic, 44
    external characteristics of, 139 (See also Minerals, external characteristics of)

plays of Nature (as), 80
siliceous, 50
France, 49–70
Freedom of inquiry, 144
French Revolution, 17, 38, 54, 74
Frost, 10, 54, 57

G

Galvanic power, 100
Garden of Eden, 142n64
General archive of land and sea war and of geography (of France), 61
General geography, *See* Geography, general
Genesis, Mosaic account of, 34, 84
Geogeny, 59, 86
Geognosy, 46, 49, 58–9, 70–2, 77, 80–2, 86, 89–90, 100, 138
　treatises on, 48, 56, 58, 71–2, 81, 83, 89, 145n94–145n96
Geographia generalis, 61
Geography
　general, 61, 89
　history of, 82
　physical, 9, 50, 74–5, 77, 88, 93, 99
Geological
　appearances, 58
　claims of Hutton, 17
　community, 2, 8, 42, 84, 98
　evolution, 51
　history, 8, 16, 137n6
　literature, 1, 5–8, 34, 74, 76, 115, 120
　　German, 74, 76
　　primary, 118, 121
　nomenclature, 94
　phenomena (Hutton's interpretation of), 3, 39, 49, 64–6, 95
　practice, 71
　question, 115
　record, 53
　　religious reading of, 103
　surveys, 115, 150n172
　system (of Hutton), 15, 63, 65, 71, 97, 120
　system (of Werner), 32, 67, 71
　systems, 16–17, 102, 105
　theory
　　aim of, 20
　　prevailing, 39, 111
　　structure of, 118
　　thinking (German), 84, 117, 148n149, 149n156
Geological ideas (Hutton's), 9, 63, 148n146
Geological Museum (London), 119
Geological phenomena, 16, 31, 39–40, 62, 64–6, 95, 97, 104, 146n107
Geological Society of America Annual Meetings, 104, 122
Geological system (of Werner), 102, 105
Geologist(s), mineral, 64
Geology
　aetiological school of, 68, 114
　causal aim of, 6
　experimental, 28, 62–4, 84
　fundamental issues in, 62
　goal of, 2

historical, 6, 45, 87, 137m16, 138n20
historical aim of, 6
history of, 1, 5, 7–8, 37, 41–2, 59, 61, 66–7, 69, 77, 80, 83, 87, 90, 99–100, 106–7, 115, 119, 121, 140n34, 141n38, 147n127
Huttonian, 17, 39, 84, 88, 100, 107, 145n92
methods of, 45
mineral, 34
modern, 1–2, 7, 9, 13, 32, 41, 62–3, 66, 77, 79–80, 85, 87, 89, 94, 99, 111–15, 118, 121, 140n29
Mosaical, 33–4
naming of, 30, 55, 84, 145n106
philosophy of, 3, 44, 85, 122
physical, 44, 87
progress made by, 62
Russian, 90
scientific, 9, 70, 74, 112
state of, 7, 29, 63, 89, 100
structural, 79
theoretical, 44, 53, 90–1, 94, 98, 146n107
theory of (*See also* Geological theory)
　Hutton's, 93
　Steno's, 3
　Werner, 5, 79, 93
Geology, aetiological school of, 114
Geology, history of
　in the Austrian Empire, 88–92
　in the British Empire, 45–9
　in France, 49–70
　in Italy, 94–101
　in the Russia Empire, 70–88
　in Switzerland, 92–4
　in the United States, 101–11
Geology, traditions, 6–7, 12, 21, 33, 110, 113, 115–18, 138n23, 140n29, 145n88, *See also under* Physiographers' tradition and Miners' tradition
Geometric relationships, 5
Geomorphology, 5, 37, 64, 118
Geostrophic cycle, 2, 4, 84
Geosyncline, 150
Geosyncline controversy, 111
Gilgamesh Epic, 20
Glaciers, 27, 92–3
Glass, 57, 81–2
Globe, crustal ruptures of, 53
Globe (terrestrial)
　construction of ... parts of, 63, 77, 85
　contraction of (thermal), 2, 147n128
　crust of, 2, 32, 35–6, 39, 43–4, 54
　　exterior, 14, 49, 54–5
　　ruptures in, 53
　　sinking, 35, 55, 57–8, 147n125
　degradation of one part of, 63
　interior of
　　gradual cooling of, 71
　lines of fracture of, 64
　surface of
　　irregularities of, 15, 99
Gneiss, 34–6, 40–1, 47–8, 51, 65, 71, 73, 88, 110
Gondwana-Land, 30
Good and bad principles of the Persians, 57
Gornyy Zhurnal, 71–2

Granilite, 89
Granite
　anatectic origin of, 65
　blocks of (erratic), 27, 97
　dykes as protuberances of, 59
　endogenic nature of, 4
　fluidity of, 40
　general appearance, 46
　genesis of (Hutton's theory of), 91
　graphic, 52, 89, 96
　igneous origin of, 4, 31, 96
　metamorphic interpretation of, 4
　metamorphosis of, 76
　mountains of, 12
　origin of, 4, 23, 31, 38, 65, 96, 110
　plasticity of, 65
　problem, 4
　red, 38, 81, 137n4
　relations to the neighboring rocks, 46
　research, history of, 4
Granite, basement, 30, 148
Granitelle, 89
Granitic magma, *See* Magma, granitic
Granitization, by grain by grain replacement, 137
Graphic granite, *See* Granite, graphic
Gravel, 102, 108–9
Great Geographical Discoveries, 80
Greek(s)
　ancient, 24, 120, 143n72
　curriculum, 138n17
　miracle, 22
　mythology of, 116
　philosophy, history of, 22
　ports, 24
　reason, 21
　science, 12, 24, 143n68, 143n71
　thought, 21
　vulcanist theories, 80
Greenstone, 20, 29, 44–5, 76–7, 80
Greywacke, 47, 89
Griphosaurus, 84
Gypsum, 51, 88, 101

H

Harvard (University), 92
Heat
　central source of, 29–30
　expansive, 96, 104
　induration, 19, 35, 39, 77, 111, 121
　internal source of, 29
　underground, 71–2
Heaven, divine, 21–2, 28, 80, 139, 142
Heroic Age (of Hesiod), 21
Highland Clearances (in Scotland), 141
Hindus, 4
Historia, 22
Historiography
　of geology, ix, 1, 6, 118, 121, 139
　of science, 2–4
History
　of botany, 67
　of geology (*See* Geology, history of)
　natural, 5, 7, 13–14, 26, 40, 45, 48–9, 51–3, 66–7, 77, 81, 87, 99, 102, 108, 113, 145n97, 146n118, 150n170, 141n46

Holm(s), 48
Hornblende, 45, 51, 102
Hospitals, 30
Huttonian framework, 4
Huttonian hypothesis, *See* Hypothesis, Huttonian
Huttonian theory, ix, x, 30, 38–9, 45, 47–8, 53, 56–8, 64–5, 96, 103–4, 106, 148n136, 107145n103
Huttonism, 45, 49, 76, 87, 97, 102
Hutton's legacy, 111, 123
Hypotheses, framing of, 3
Hypothesis
 of Dr. Hutton, 40
 igneous, 47
 Huttonian, 41
 Wernerian, 110
Hypothesis, Wernerian, 110

I

Ice, 93, 148n131
Idols of the cave, 16
Igneous
 action, 40, 107
 energy, 108
Imperial Academy of Arts and Sciences in Russia, 9
Imperial and Royal Geological Survey (in Vienna), 89
Independent formations, 14
Inductive method, *See* Method, inductive
Instantiæ crucis, 31
Interfacial angles (in quartz), 137n8
Internal inconsistencies in the book of *Genesis*, 142n63
Intrusive body(ies), 5
Island(s) thrown up from the sea, 102
Isothermal surfaces, 99
Italy, 94–101
Izdubar Epic, 20

L

Landslide, 74
Language
 German, 18, 22
 mystic, 37
 symbolical, 37
Lateral thrust (in mountain-making), 62
Lava, subterraneous, 81
Law of relativistic velocity addition, 4
Laws
 of the physical world, 46
 suspended natural, 148n147
Layers,
 consolidated, 96
 sedimentary, 51, 83, 121, 140n28
Legitimate conclusions, 15
Library of Wolfenbüttel, 150m168
Life, succession of, 62
Lignite, 89
Lime, 55–6, 64, 91, 106
Limestone(s), 12, 38, 43, 45, 52, 64, 67, 71–3, 81, 84–5, 90, 94, 102, 106–7, 110
 semi-crystalline, 35
Linnean Society (London), 119

Literature, 1, 4–5, 7–8, 27, 29, 31, 34, 52–3, 74–6, 81, 90, 102, 113, 115–16, 118, 120–1, 143n73, 150n77
Lithoclases, 64
Logical consistency, 20
Logos, 22
Lorentz's transformations, 4
Lyceum of Natural History in New York City, 102
Lysenkoism, 117

M

Machines
 electromagnetic, 78
 galvanic, 108
Madrepore(s), 55
Magma, granitic, 4, 65
Maps
 by Abraham Ortelius, 26
 by Ptolemy, 26, 138
Marble(s)
 origin of, 66, 82–3, 90, 106
Mathematicians, 150n177
Matter
 five fundamental building blocks of, 138 (*See also under* Elements)
 stratified, 10, 12, 99
 vegetable inflammable, 12
Medical instruction (in universities), 50
Meghalayan-age, 116
Mercury, 47
Mesopotamians, 4
Messinian Salinity Crisis, 19, 121
Metallurgists, 150n177
Metallurgy, 116, 138n23
Metals, 29, 50, 116, 138n23
Metamorphic rocks, 31, 40, 73, 91, 109–10
Metamorphism
 contact, 5, 63
 Hutton's theory of, 62, 87
 initial ideas on, 64
Meteorites, 138n23, 148n137
μετεωρολόγοι, *See* Meteorologoi
Meteorologoi, 143n68
μετέωρος, *See* Meteoros
Meteoros, 143n68
Method
 experimental, 64
 inductive, 34, 48–9
 scientific, 119, 122
Method, experimental, 64
Method, inductive, 34, 48–9
Mica, 51, 67, 73, 81
Mica-slate, 35
Middle Ages (European), 25–6, 112
Middlebury College, 108
Migmatisation, 137
Mineralogical-geognostical tradition, 116
Mineralogical research, 29
Mineralogy
 ocular, 138n23
 study of, 28, 145n91
 system of, 90, 150n172
Mineral(s)

 burnable, 138n23
 cabinet, 141n41
 characteristics, 138n23
 chemical
  character of, 31, 81, 138n23
  properties of, 138n23
 earthy, 138n23
 empirical character of, 138n23
 external characteristics of, 139
 fluid agent in the formation of, 102
 formation of ... in an aqueous fluid, 102
 identification of, 6, 138n23
  guides to, 6
 inflammable, 138n23
 internal characteristics of, 138n23
 kingdom, 29, 46, 69, 87, 102, 105, 138n23
 lithic, 138n23
 metallic, 138n53
 salty, 138n23
 substance(s), 12, 14, 51–2, 65, 69
 taxonomy, 6
Miners' tradition, 33
Mines
 coal, 150n177
 copper, 68
 of Hungary, 47
 silver, 68, 146n111
 of Transylvania, 47
Mining
 folklore, 116
 tradition, 116
Mining Academy in Freiberg, 14
Mining Cadet Corps in St. Petersburg, 71
Minkowski's space, 4
Miracles, 20, 22
Modernity, 17, 143
Moisture, 35, 54
Moon, 56, 69, 103
Morphology, 64, *See also* Geomorphology
Mosaic account, 34, 62, 84
Mountain-building, 6, 28, 51, 54, 90–1, 93, 111
 by horizontally-acting forces, 91
Mountain ranges, 2, 16, 71, 81
Mountainous mass(es), uplifted by subterranean explosions, 51, 78
Mountain(s)
 greatest height of, 56
 primitive, 35–6, 119
 ranges, 2, 16, 71, 81, 147n124
 secondary, 12
Mud, 78, 108–9
Muscovite, 51
Museum National d'Histoire Naturelle (Paris), 63
Myology, mystic, 138n19
Myth
 in the *Bible* (*See* Bible, myth in)
 of *Genesis*, 22–3, 26–8
Mythological thought, 21–2
Mythology
 Hebrew, 24
 unscientific, 22, 28, 86
Mythology
 Greek (*See* Greek(s), mythology of)

N

Napoleonic blockade, 9
Nappe(s), 23, 47, 94
Natural history, *See* History, natural
Natural law, 20, 24, 65, 86, 97, 140n23, 148n147
Natural philosophers of Ionia, 22
Natural theology, 4, 108
Nature
    comprehensive view of *(theoria)*, 21
    detached and systematic investigation of *(historia)*, 21
    wisdom and order in, 7
Nazi, 117
Neoneptunist position, 82
Nepheline, 45
Neptunian (person)
Neptunian school, 37
Neptunism, Wernerian, 80, 102
Neptunist (person), 31, 33, 38–9, 44, 55, 59–60, 65–6, 69–71, 73–6, 80, 82, 84, 90, 94, 100, 109, 113, 138n23, 145n94, 145n96, 146n111, 149n151, 8788
Neptunist system, 30
Neptunist teaching, 59
Neptunist theory, 138n23, 145n96
Neptunist-Vulcanist Debate, *See also* Vulcanist-Neptunist debate
Neptunist-Vulcanist Debate, 76
New York Academy of Sciences, 102
Newest Flœtz Trap, 45, 48, 102
Noah's ark, 142
Nodule(s), septarian, 53, 97
Nous, 143n71, 150n178
NSDAP, 117

O

Observatory of Geneva (Switzerland), 149
Ocean
    basins, 2
    diminution of the waters of, 26
    illimitable, 35
    universal, 6, 14
Oceanography, 18
Old Red Sandstone, 47, 101
Oligocene, 52
Olympians, 21
Ore(s), 52, 78, 116, 118, 146n111
Orogeny, globally synchronous, 44
Orthodoxy (Christian), 9, 143n72
Orthogenetic evolution theory, *See* Theory of orthogenetic evolution
Oryctognosy, 141n91
Oryctognosy, *See also* Mineralogy
Outcrop(s), 38, 48, 97
Oxford University, 144n75

P

Paleontology, 53, 74, 80, 83–4, 99, 115
    vertebrate, 150n172
Pan-African orogenic belt, 30
Paradise, mountain interpretation of, 142n64
Periods of ordinary action, 44
Permian, 144n74
Perpetual frost, 57
Petrifactions, 46
Petrified wood, 50
Phanerozoic, 140n32
Philosophical Society (in Edinburgh), 10
Philosophy
    historical accounts of, 21–2, 32, 44, 52, 69, 85–6, 94, 102, 119, 122, 142n62, 143n68
    moral, 21
    natural, 21
    of science, 3, 5, 16, 20
    scientific, 22
Physics, history of, 137n11
Physiographers' tradition, 33
Physis, 22
Plane of invariable temperature below the land, 99
Planet(s)
    density of, 20, 32, 83, 90, 149n166
    evolution of our, 94, 104, 112, 117, 121, 142n59, 146n107
    functioning of our, 38
    history of our, 137n1
    interior of, 43, 57–8, 60–1, 66, 71, 82, 95, 150
Plants, 57–8, 138n23
    impression of, 55
    marine, 55
    remains of, 14, 54, 57
Plutonism, 39, 76, 82, 97
Plutonist(s), 38–9, 44, 66, 69, 72–3, 87, 90–1, 94, 100, 109, 113
Polymath(s), 1–2, 18, 20, 26, 88, 92, 101, 144n84
Pontifex maximus, 37
Porphyry, porphyries, 47, 67, 71–2, 76, 81–2, 85, 120
    auriferous transition ... of Kremnitz and of Transylvania, 47
    granite, 23, 51
    masses, 5, 47, 51
    metalliferous, 47
    ores, 47
    quartz, 85
Poseidon's Empire, 88
Positivism, 143n69
Positivity, 22, 143n69
Post-Diluvial world, 18
Precambrian, 140
Presocratics, 143n69, 143n71
    opinions of, 22
Pressure, 12, 39–40, 45, 48, 56, 58, 61, 63–4, 68, 71, 81–2, 91, 99, 106–7, 114, 117
Primary rocks, 40–1, 73
Primordial powers (of Greek mythology), 21
Princeton (University), 92
Prismatic ore (of Dunbar), 52
Processes
    eruptive, 65
    internal, 51, 80
    neptunian, 76
    volcanic, 76
Puddingstones (in Valorsine), 64
Pumice, 46–7
Pyrite, 102

Q

Quadersandstein, 47
Quartz, 12, 30, 45, 51, 57, 73, 85, 137n8
Quaternary vertebrates, 17

R

Rabbinic tradition, 21
Rains, 9–10, 54, 93, 145n88, 149n140
Reclaimed land (in The Netherlands), 137
Regeneration, 57
Regression, 6, 24–5
Religion(s)
    Abrahamic, 116, 145n88
    attack on, 38
    fundamental axiom of, 143n72
    irreverence toward, 143n71
    revealed, 16–17
    vs. science, 143n72
Religious
    fervor, 15, 38
    sentiment, 37
    thinking, post-Socratic, 22
Renaissance, 21, 25, 27, 67, 92, 115, 119, 137n16, 138n23
Renaissance paintings (Italian), 137
Research
    cryptoristic, 68, 148m140
    natural, 86
    rules of, 37
    scientific, 3, 22
    troponomic, 69, 148n140
Réseau pentagonal, 67
Restite, 132
Revisionist
    accounts (of Hutton's place in the history of geology), 4–9
    histories, ix, 2, 113, 115, 118, 141n46
    literature, 4, 6, 8, 13, 113, 122
Revolution (in Nature), 19, 33, 46, 54, 57–8, 62, 73–4, 103, 140, 148n147
Rex sacrorum, 37
Rhinoceros, carcass of, found in frozen earth, 57–8
Rift valleys, 2
Rocks
    abnormal, 77, 80
    alluvial, 101
    azoic, 110
    crystalline, 34, 63, 67, 73, 81–4, 91, 98, 110, 114
    crystalline igneous, 47
    effects of high temperature on, 39
    eruptive, 65, 91
    of Etna, 29
    exoplutonic, 111
    exotic, 111
    formation of, 30, 64
    granitic, 62, 81, 110, 148n135
    igneous
        fabric of, 82
        field relationships of, 31
        flattening structures in, 92
    intrusive, 89, 98, 145n92
    magmatic, 4, 52, 72, 76

mechanical, 73
metamorphic, 31, 40, 73, 91, 109–10
molten, 63
plutonic, succession of formation of, 72
primary, 40–1, 73
primitive
    plutonic nature of, 97
primordial, 81–2, 114
pyric, 82, 114
Secondary, 14, 101, 106, 110
sedimentary
    of mechanical origin, 73
    transformation of, 64, 114
slaty, 41
stratified, 43–5, 54–5, 78, 80, 82, 107, 109–10, 114
Transition, 14, 101
trap, 48–9, 62–3, 73, 81, 84, 93, 102, 105–7, 109, 148n135
unstratified, 39, 43, 45, 56, 80, 108–9
vitreous, 47
volcanic, 46, 67, 73, 77, 100–1, 145n92
Wernerian arrangement of, 102
Ross-shire Sheep Riot (in Scotland), 141n46
Royal Geographical Society (London), 42, 119
Royal Prussian Academy of Sciences, 2, 8
Royal Society of
    Edinburgh, 30
    London, 27, 34
Rule of original horizontality (Steno), 8
Runic characters, 52
Russian Academy of Sciences, 71
Russian Empire, 70–88
Russians, 37

S

Sand, 12, 81–2, 102, 108–9
Sandbank (at the mouth of the Danube), 24–5
Sanidine, 75
Schiehallion Experiment, 83, 89
Schists
    crystalline, 73, 87, 90
    general, 65, 83, 89
    micaceous, 67
Schistus, 12, 31, 137
Schlieren, 97
School of Mines (in Paris), 52, 119
Schorl, 12, 51
Science(s)
    agricultural, 117
    demarcation of (from non-science), 137n3
    earliest infancy of, 117
    has no fixed rules, 137n3
    history of, ix, 8, 116, 117, 118, 122, 137, 140
    natural, 17, 40, 48, 62–3, 80, 86–7, 94, 109, 115, 116, 119, 137n11, 143n72, 145n89
    nature of, 22
    non-historical, 138n22
    of the earth, 2, 88–9
    pantheon of, 7, 20, 142n64
    philosophy of, 3, 5, 16, 20
    physical, 16, 18, 59, 75,
    Popper's view of, 86, 142n62
    theoretical, 3, 24, 29

    of the universe, 143n68
Scientific revolution, 22
Scoriae, 47
Scottish
    geological milieu, 7
    school (of geology), 64, 66, 68, 77, 114
Sea
    floor, 25, 74, 78, 93, 95
        temperature of, 95
    greatest depth of, 56
    land invaded by, 54
    level of, 14, 17, 26–7, 68, 72, 116, 118, 121, 144n83
    precipitation at the bottom of, 23
    primeval, 35, 110
    progressive diminution of, 138n17
    receding from land, 54
Second-hand information (in doing historiography), 7
Secondary rocks, 14, 101, 106, 110
Sedition Trials (in Scotland), 141n46
Seven Years' War, 116
Shear zone, 92, 146n112
Shelf (continental), 19
Shell limestone (Muschelkalk), 47–8
Shells, 5, 12, 33, 35, 47, 54–5, 57, 67, 106–7
Shock
    lateral, 62
    oblique, 62
Siderite, 51, 88
Silica, 58, 140
Sill(s), 53, 62
Silting up (of the Black Sea and the Sea of Azov), 19, 24–5
Silver, crystallized native, 12
Silver Age (of Hesiod), 21
Smelter, 140n26
Snow White and the seven dwarfs (Grimm brother's German folk tale), 20, 120
Social considerations in history of science, x, 1
Sociological analysis, 1
Soil
    agricultural, 27
    erosion, by means of rain, 9
    formation, 23, 27
Solar System
    Laplace's improvement of Kant's theory of, 3, 53, 118–19
Solar System
    Kant's theory of the origin of, 3, 141n53
Solubility, 51
Sources of the *Old Testament*
    Deuteronomist, 144n86
    Elohimist, 144n86
    Prisetly, 144n86
    Yahwist, 142, 144n76
Space, Aristotelian binary (supra- and sub-Lunar), 22
Spaniards, 37
Spheroid, *See also* Globe
    primordial, 15
St. Petersburg University, 71
Steam, 96, 103
Stones, calcareous, 51, 55
Storm surge, 74, 116

Strata, contortions of, 62, 148n136
Strata, correlation of, 116
Stratified matter, *See* Matter, stratified
Stratigraphic units, global, 146n10
Stratigraphy, 6, 27, 34, 45, 53, 90, 99, 115, 121, 137n6, 146n147
Stratum (plural: strata)
    bendings, fractures, erections and dislocations of, 12
    of cenozoic [sic] times, 100
    consolidation of, 50
    contortions of, 148n136
    correlation of, 116
    formation of, 54–6, 95
    Hutton's definition of, 68
    indurated, 12, 19, 57, 74, 102
    induration of, 19, 77
    of paleozoic [sic] times, 100
    pyric, 81
    shifting and sliding of, 56
    slaty, 45
    steepening of, 78
    succession of, 6, 102
    twisting of, 62
    uplifted, 62
    upturned, 68
Structure
    geological theory, 118
    of our planet, 78, 89, 102, 145n91
Substance(s)
    calcareous, 54–5, 57
    elastic, 62
    glassy, 84
    vegetable, 54
Succession of life, *See* Life, succession of
Sulphur, 102, 145n92
Sumerian pantheon, 20
Sun, 55, 64, 69, 98, 103, 120–1, 143
Supernatural, 21–2, 76, 104, 143
Superposition (of strata), 97
Supreme Governor of the Church of England, 37
Switzerland, 92–4
Syriac text, 142n64
Syrian fathers (of Christianity), *See* Christianity, Syrian fathers of
System of Werner, *See* Geological system (of Werner)
Systems of Elevation (Élie de Beaumont's), 145n104

T

Tectonic phenomena, 82, 146n107
Tectonic repetition (of a sedimentary section), 147n127
Terrain(s)
    consecutive, 147n124
    metamorphic, 65, 92
    sedimentary, 62, 147n124
Terrestrial structure, 94
Terreur, 50, 140n31
Text-book(s), 9, 76, 104, 109
The Geological Society (London), 8, 41, 43, 71, 73, 82, 119, 137n14
Theologian(s), 7, 16, 22, 28, 33, 74, 83
Theological, 4, 28, 33

Theology, 4, 94, 108, 143n72
Theoria, See Nature, comprehensive view of *(theoria)*
Theory(ies)
   bibliolatrous, 7, 80, 116, 120
   of coral reefs (Darwin's), 119
   cosmogonical, 7
   crystallization (of Delamétherie), 62
   Deluc's, 28
   descension, 78
   of the Earth (genre), 1, 7, 13, 15, 36, 46, 48–9, 53–4, 80, 107, 113, 120–1, 141n47, 149n147, 149n151
   explanatory, 3, 116
   fluvial theory of denudation, 42
   geognostic, 6
   geological
      neptunian, 31, 49, 60, 96, 102, 104–5, 110
      volcanic, 60
         Greek, 80
   glacial, 93
   Huttonian, 30, 38–9, 45, 47–8, 53, 56–8, 64–5, 96, 103–4, 106, 148n136, 107145n103
   Hutton's English interpretation of, 44
   neptunian, 31, 49, 60, 96, 102, 104–5, 114
   neptunic, 72
   Neptunist, 138n23, 145n96
   of orogenic phases, 101
   of orthogenetic evolution, 117
   oxygen (theory of), 138n23
   plutonian, 108–9, 112
   of rain (Hutton's), 93, 149n149
   of relativity, 4
   Stahl's (of phlogiston), 138n23
   volcanic, 73
   Vulcanian, 102
   water, 56
   Wegener's, 117
   Wernerian, 29, 31, 43–5, 76, 94, 103, 107
   Werner's, 5, 14, 31, 33, 65, 67, 72, 79–80, 84–6, 97, 101, 106–7
Time
   imaginary, of gods and prophets full of miracles, 22
   infinite extent of, 24
   of ordinary human beings and events, 22
Titans, 21
Toadstone(s), 50, 81
Topography
   dynamic, 138n17
   irregularities, 138n17
   present, 14, 18
   terrestrial, 14
      age of, 17
Torrents, 10, 54
Tourmaline, 51
Trachyte, 47
Transition
   Period, 146n113
   rocks, 14, 101
   trap, 102
Trap rock, *See* Rock(s), trap
Trinity College, Cambridge, 20
Truth
   fundamental, 34
   ultimate, 34

U

Unconformity(ies), 5, 42, 85, 101
Uniformitarianism, 2, 42–3, 63, 87, 148n121
   attack on (by religion), 22
Uniformitarian(s), 20, 22, 43, 61–2, 87, 101, 148n132, 117148n131
   thinking, 22
Uniformity of earth processes, 20
United States Army, 106
United States of America, 101–11
Universal
   conflagration, 56
   correlatability (in stratigraphy), 44, 116
   formations, 14, 44, 107, 121
Universe
   eternal, 138n17
   evolution of, 87
   formative force in, 22
   system of, 56
   unified, homogeneous, 22
University College London, 5, 119
University of Paris, 25
University of Strathclyde, 34
Uplift
   of entire continents, 85
   magma-propelled, 44
   magmatically driven vertical, 89

V

Valleys, fluvial origin of, 9
Vapor, 69, 96
   elasticity of, 96
Vegetable substances, *See* Substance(s), vegetables
Vein(s)
   of gold, 8, 137n4
   granite, 31
   metallic, 7, 40, 98, 108, 138n23
      origin of, 98
   mineral, 15, 32, 41, 43, 58, 108
   quartz, 30
   of silver, 38, 147n4
   of spar, 55
Veracity of the *Genesis* narratives, 15
Vertebrates
   Quaternary, 17
Volcano(es)
   active, 60, 85, 87, 94
   basaltic, 72
   as breathing holes, 81
   of central France, 34, 52
   current, 72
   dormant, 56
   magmatic uplift theory of, 94
   subterranean, 50
Volcanology, 7, 138n16
Vulcanism, 76, 82, 84–5, 114
Vulcanist-Neptunist debate (also Neptunist-Vulcanist Debate), 76
Vulcanist(s), 38, 59–60, 65, 67, 74–7, 80, 82, 85, 88, 98, 100, 102, 118, 148n146

W

Wakke, 58
Water, agency of, 31, 38, 55–6, 58, 104
Waterfalls, 23
Welteislehre, 117
Wernerian Era (in America), 101, 108
Wernerian Era in American geology, 101–8
Wernerian hypothesis, *See* Hypothesis, Wernerian
Wernerian Natural History Society, 45, 48–9
Wernerian nomenclature, 101
Wernerian(s), 15, 29–30, 32–7, 39–40, 43–5, 48, 58, 61, 67, 71, 75–6, 80, 84, 87–8, 90, 94, 101–9, 113, 116, 137n13, 138n23, 145n91
Wernerism, 43, 49, 97, 100–2, 137n13
Werner's nomenclature, *See also* Wernerian terms
Werner's terms, 49, 101, *See also* Wernerian nomenclature
Whin, *See* Whinstone
Whinstone, 5, 50, 68, 81
Widmanstätten pattern in iron meteorites, 148
Witch trials, 145n105
Witchcraft act, 145n101
Witches, 145n101
World
   divine, 22, 69
   human, 22
   natural, 22
   past behavior of, 141n54
   present (behavior of), 141n54
World Wide Standardized Seismograph Network, 2

X

Xenolith(s), 5, 23

Y

Yom (= יום =day), 34

Z

Zeolite, 29, 70
Zinc, 47
Ziusudra myth, 20
Zoology, 20, 53, 88, 108

Index of Titles of Publications

A

A geological account of the southern district of Stirlingshire, commonly called the Campsie Hills, with a few remarks relative to the two prevailing theories as to geology, and some examples given illustrative of these remarks, 45, 46
A Manual of Geology, 40, 131
A New System of Geology in Which the Great Revolutions of the Earth and Animated Nature are Reconciled at once to Modern Science and Sacred History, 34, 135
A Tour in Scotland, 132, 147n125
A Treatise on a Section of the Strata from Newcastle-upon-Tyne to the Mountain of Cross Fell in Cumberland, 32, 127
A Treatise on a Section of the Strata, commencing near Newcastle upon Tyne, and concluding on the West Side of the Mountain of Cross-Fell. With remarks on Mineral Veins in General, and Engraved Figures of Some of the different Species of those Productions to which are added Tables of the Strata in Yorkshire and Derbyshire the whole intended to Amuse the Mineralogist and to Assist the Miner in his Professional Researches, 32, 127
A Treatise on Primary Geology, 39, 123
Abhandlungen der königlichen Akademie der Wissenschaften zu Berlin, 87, 133
Aepitome omnis Philosophieae..., 143n65
Allgemeine Erdkunde—Astronomische und Physische Geographie, Geologie und Biologie, 89
Allgemeine Geschichte der Philosophie, 22, 126
American Journal of Science, 105, 125
American Journal of Science and Arts, 105, 111, 125, 129, 130
An Elementary Treatise on Geology, 15, 16, 27, 125, 124
An Elementary Treatise on Mineralogy and Geology, being an Introduction to the Study of these Sciences, and designed for the use of pupils,—for persons attending lectures on these subjects, —and as a companion for travelers in the United States of America, 122, 124
An Introduction to Geology, 105, 123
An Introduction to Geology, and its associated sciences Mineralogy, Fossil Botany, and Palæontology, 40, 133
Anleitung zum Studium der Geognosie und Geologie, besonders für Deutsche Forstwirthe, Landwirthe und Techniker, 83, 125
Anleitung zur Geognosie, insbesondere zur Gebirgskunde—nach Werner für die k. k. Berg-Akademie, 89, 133
Arte de los Metales, 116

B

Babylonian Talmud, 21
Bedeutung und Stand der Mineralogie, 74, 130
Berg-Büchlein, darinnen von der Metallen und Mineralien Generalia und Ursprung, 116
Bericht vom Bergwerck, 116
Beschreibung der Allerfürnemsten Mineralischen Erzt- und Bergwerksarten, 116
Bibliotheque Britannique, 129, 132, 149n163
Brevi Cenni sulla Storia della Geologia compilati per I suoi Allievi, 100, 132
Bursting the Limits of Time, ix, 120, 133

C

Chapters from the Physical History of the Earth—An Introduction to Geology and Palæontology, 42, 131
Characteribus Fossilium Externis, 6, 127
Chemical and Geological Essays, 110, 129
Common Sense, ix
Comparative View of the Huttonian and Neptunian Systems of Geology in Answer to the Illustrations of the Huttonian Theory of the Earth by Professor Playfair, 53, 131
Conjectures sur la Genèse, 142n63
Considerations on Volcanos, the probable Cause of their Phenomena, the Laws which determine their March, the Disposition of their Products, and their Connexion with the present State and past History of the Globe; leading to the Establishment of a New Theory of the Earth, 134, 145n98
Conversations on Geology; Comprising a Familiar Explanation of the Huttonian and Wernerian Systems; the Mosaic Geology as Explained by Mr. Granville Penn; and the Late Discoveries of Professor Buckland, Humboldt, Dr. Macculloch and others, 34, 132
Corso di Geologia, 99, 134
Cosmogonie, ou de la Formation de la Terre et de l'origine des Pétrifications, 60, 123
Cosmographia, 138n16
Coup-d'Œil Historique sur la Géologie et sur les Travaux d'Élie de Beaumont, 67, 133
Cours Élémentaire de Géognosie, 61, 133
Cours Élémentaire de Géologie, 61, 70, 129

D

De Aeternitate Mundi contra Aristotelem, 26
De Orthodoxa Fide, 21
De Ortu Venarum Metalliferarum, 134, 139n23
De Rerum Natura, 120
De Universo, 21
De Universo Creaturarum, 21
Der Bergmannslied, 128, 150
Der Tiefe Meissner Erbstollen—Der einzige, den Bergbau der Freyberger Reffer für die fernste Zukunft sichernde Betriebsplan, 78, 128
Descr. Of the Western Islands, 93
Die Alpen. Handbuch der Gesammten Alpenkunde, 91, 135
Die Geologie der Gegenwart Dargestellt und Beleuchtet, 83
Dinosaur Doctor, 125, 140n34
Discours Préliminaire, 21, 26, 53, 61
Dizionario Biografico degli Italiani, 100, 124
Dzieje Ziemi, 91

E

Edinburgh Review, ii, 56, 97, 122, 127, 140n27, 149n166
Einleitung in das Alte Testament, 135, 142n63
Einleitung zu der Mathematischen und Physikalischen Kenntnis der Erdkugel, 74
Elementi di Geologia ad Uso della Regia Universita' degli Studi in Catania, 97, 127
Elementi di Geologia Pratica e Teoretica Destinati Principalmente ad Agevolare lo Studio del Suolo dell' Italia, 98
Éléments de Géologie, 97
Elements of Geognosy, 145n94, n95, n96
Elements of Geology, 97, 109, 110, 128
Elements of Geology, for the Use of Schools, 106, 131
Ἑλένης Ἐγκώμιον (see *Encomium of Helen*), 143n68
Encomium of Helen, 143n68
Encyclopaedia Britannica, 14, 43, 127, 132
Encyclopædia Metropolitana, 43, 132
Encyclopédie, 50
Encyclopedie Méthodique, 17, 49, 61, 125
Entwicklungsgeschichte der Natur, 87, 123
Erdgeschichte, 90, 131
Essai Géologique sur l'Écosse, 48, 59, 123
Essai sur la Philosophie des Sciences, ou Exposition Analytique d'une Classification Naturelle de Toutes les Connaissances Humaines, 122, 148n140
Études et Expériences Synthétiques sur la Métamorphisme et sur la Formation des Roches Cristalliens, 63, 125
Études sur l'Histoire de la Terre et sur les Causes des Révolutions de sa Surface, 62, 123
Études sur les Glaciers, 92, 122
Explication de Playfair sur la Théorie de la Terre par Hutton, 77, 123

F

Faust, der Tragödie, 88
Fizicheskaya Geologiya, 73, 131
Founders of Geology, 2, 9, 111, 113, 127

G

Geognosie, 89, 131, 133
Geognosie und Geologie, 77

Note: A lowercase "n" indicates the note number, so 143n65 indicates page 143, note 65.

Geognostical Essay on the Superposition of Rocks in both Hemispheres, 34
Géographie Physique, 17, 49, 50, 125, 135
Γεωγραφιχή Υφήγησις (see *Geographike Uphegesis*), 26
Geographike Uphegesis (see *Manual of Geography*), 26
Geologia Sive Philosophemata De Genesi Ac Structura Globi Terreni. Oder: Natürliche Wissenschafft / Von Erschaffung und Bereitung der Erd-Kugel, 144n84
Geologia: or, a Discourse Concerning the Earth before the Deluge, 144n84
Geologica Norvegica, 144n84
Geological Sketches and Glimpses of the Ancient Earth, 38, 128
Geological Text-Book, 104, 126
Géologie Contemporaine—Histoire des Phénomènes Actuels du Globe, 66, 124
Géologie du Granite, 4, 136
Géologie et Paléontologie, 66, 122
Geologie von Brasilien, 117, 123
Geologiya—Obshchiy Kurs Lektsiy, 73, 129
Geology for Beginners, 40, 133, 146n106
Geology of England and Wales, 34, 124
Geology—Chemical, Physical, and Stratigraphical, 45, 132
Geschichte der Mineralogie von 1650–1860, 118
Geschichte der Urwelt mit Besonderer Berücksichtigung der Menschenrassen und des Mosaischen Schöpfungsberichtes, 84, 135
Geschichte und Litteratur der Geognosie, ein Versuch, 81, 130
Glückauf, 128, 131, 150n177
Gornyy Zhurnal, 71, 72, 130, 134
Grundriß der Geognosie und Geologie, 83, 125, 130
Grundzüge der Geognosie für Bergmänner zunächst für die des Österreichischen Kaiserstaates, 89, 128
Grundzüge der Geognosie und Gebirgskunde für Praktische Bergmänner, 89, 128

H

Hannöverisches Magazin, 74
Harry Beaufoy, or The Pupil of Nature, 38
Helga, or the Rival Minstrels, A Tragedy Founded On An Icelandic Saga, 30, 37
Herbarium Diluvianum, 141n47
Histoire de la Botanique de la Minéralogie et de la Géologie Depuis les Temps les Plus Reculés Jusqu'à Nos Jours, 67, 128
Histoire de la Progrès de la Géologie de 1834 à 1845, 66
Histoire de la Terre—Origines et Métamorphoses du Globe, 66, 134
Histoire Physique de la Mer, 18, 131
Historia Critica Philosophiae, 22, 126
Hutton's purpose, 8, 127
Hydrogéologie, 137n5
Hymns on Paradise, 21, 124, 143n64

I

Il Bel Paese—Conversazioni sulle Bellezze Naturali—La Geologia e la Geografia Fisica d'Italia, 99
Illustrations of the Huttonian Theory of the Earth, 53, 131, 132, 148n136
Index to the Geology of the Northern States, with a Transverse Section from Catskill Mountain to the Atlantic, 103, 126, 141n47
Introduction à l'Ancien Testament, 150n182
Introduction à la Géologie—première partie des Éléments d'Histoire Naturelle Inorganique Contenant des Notions d'Astronomie, de Météorologie et de Minéralogie, 150n170
Introduzione alla Geologia, 76, 94, 122, 123
Is the Present the Key to the Past or the Past the Key to the Present? James Hutton and Adam Smith versus Abraham Gottlob Werner and Karl Marx in Interpreting History, x, 134
Istoriya Zemli, 90
Iter Extaticum II, 148n139

J

Johann Wolfgang Goethe Gedenkausgbe der Werke, Briefe und Gespräche, v. 1, *Sämtliche Gedichte Erster Teil: Ausgabe letzter Hand*, 149n152
Jordens Historia, 90
Journal de Mines, 81
Journal de Physique (see *Journal de Physique, de Chimie, d'histoire Naturelle et des Arts*), 15, 81
Journal de Physique, de Chimie, d'histoire Naturelle et des Arts, 125, 141

K

Kosmos, 4, 83, 85, 125, 129
Kritik der Geologischen Theorie besonders der von Breislak und Jeder Ähnlichen, 75, 122
Kritik der Reinen Vernunft, 15, 129
Kurs geognozii, Sostavlennyy Korpusa Gornykh Inzhenerov Polkovnikom, Sanktpeterburgskogo Universiteta professorom D. Sokolovym, 72, 134
Kurs Geologii, Chitannyy v Gornom Institute S. Petersburg, 73, 131

L

La Géologie—Son Objet, Son Développement—Sa Méthode, Ses Applications, 94, 135
La Jeunesse de la Science Grecque, 133, 143n68
Leçons de Géologie données au Collège de France, 53, 125
Lectures on Geology; Being Outlines of the Science, 104, 133
Lehrbuch der Chemischen und Physikalischen Geologie, 82, 123
Lehrbuch der Geologie und Petrefactenkunde. Zum Gebrauche bei Vorlesungen und Selbstunterricht. Theilweise nach L. Elie de Beaumont's Vorlesungen an der Ecole des Mines, 83, 135
Lehrbuch der Mineralogie nach des Herrn O.B.R. Karsten Mineralogischen Tabellen, 89
Lehrbuch der Physikalischen Geographie, 88, 128
Lehrbuch der Physikalischen Geographie und Geologie, 93, 134, 135
Les Époques de la Nature, 17, 133
Lettres Physiques et Morales sur l'Histoire de la Terre et de l'Homme adressées à la Reine de la Grande Bretagne, 75, 118
Lettres sur les Révolutions du Globe, 70, 123
Logik der Forschung, 15, 132

M

Magazin für das Neueste aus der Physik und Naturgeschichte, 149n149
Manual of Geography, 26, 138n16
Manual of Geology, 109, 125, 150n172
Manual of Geology: Practical and Theoretical, 42, 132
Manual of Mineralogy, 125, 150n172
Manual of Mineralogy and Geology, 105, 126
Margarita Philosophica Totius Philosophiae Rationalis, Naturalis & Moralis Principia Dialogice Duodecim Libris Complectens, 21
Medals of Creation, 131, 145n106, 146n106
Meißnische Land- und Berg Chronica..., 116
Mém. Géol. Sur l'Allemagne, 93
Memoir on the Geology of Central France; including the Volcanic Formations of Auvergne, the Velay and the Vivarais, 34, 130, 134
Mémoires Géologiques et Paléontologiques, 60, 123
Memoirs of the Wernerian Natural History Society, 45, 48, 49, 123, 125, 126, 128, 129, 131, 132, 135, 136
Mémoirs Pour Servir à l'Histoire Naturelle de l'Italie, 17
Meteorologikon (see *Meteorology*), 19, 24, 25
Meteorology, 24, 25
Mineral Physiology and Physiography, 110, 129
Mineralogical observations in Galloway, 46, 128
Mineralogical queries, proposed by Professor Jameson, 45, 129
Mineralogical Travels Through the Hebrides, Orkney and Shetland Islands, and Mainland of Scotland, with Dissertations upon Peat and Kelp, 89, 129
Mineralogy, 58
Monthly Review, 34, 145n98
Mundus Subterraneous, 148n139

N

Naturgeschichte der drei Reiche. Zur allgemeinen Belehrung bearbeitet, 77
Naturgeschichte des Erdkörpers in Ihren ersten Grundzügen, 81, 130
Naturgeschichte des Mineralreichs für Schule und Haus, 87, 133

Neue Theorie von der Entstehung der Gänge mit Anwendung auf den Bergbau besonders den Freibergischen, 136, 139n23
Ninth Bridgewater Treatise, 99, 122, 123
Notes on the geognosy of the Crif-Fell, Kirkbean, and the Needle's Eye in Galloway, 48, 129
Notice in regard to the Trap Rocks in the Mountain Districts of the West and North-west of the Counties of York, Durham, Westmoreland, Cumberland, and Northumebrland, 49, 136

O

Observations Relative Chiefly to Picturesque Beauty Made in the Year 1776 on Several Parts of Great Britain; particularly the High-Lands of Scotland, 127, 147n125
Oevures de Lavoisier, 128, 130, 142n56
Old Testament, 4, 27, 103, 137n11, 143n64, 144n86, 145n86
On the Eternity of the Universe, 138n17
On the Geognosy of Germany, with observations on the igneous origin of trap, 46, 123
On the mineralogy and local scenery of certain districts in the Highlands of Scotland, 45, 131
On the Transition Greenstone of Fassney, 45, 132
On the veins that occur in the newest flætz-trap formation of East Lothian, 45, 132
Oratio de Telluris Habitabilis Incrementro, 20, 130
Outlines of Geology, 107, 123, 124

P

Part I Physical Geology and Palæontology, 44, 132
Pentateuch, 27, 142n64
Peshitta, 142n64
Philobiblon, 136, 144n84
Philosophie der Geologie und Mikroskopische, 85, 135
Principles of Geology, ix, x, 2, 9, 20, 61, 89, 111, 130, 134, 136
Prodrome de Géologie, 64, 135
Protogaea, 25, 130, 146n109, 148n139

Q

Quantum Geographia Novissimis Periegesibus et Peregrinationibus Profecerit, 82, 128
Questiones super tres primos libros meteororum et super majorem partem quarti a magistro, 25

R

Recherches sur les Ossemens Fossiles de Quadrupèdes, 26, 53
Reflections on Duelling, 146n115

Reise eines Gallo-Amerikaners (m. Simond's) durch Großbritannien in den Jahren 1810-1811, 54
Relazioni d'alcuni Viaggi Fatti in Diverse Parti della Toscana, 17, 135
Reliquiæ Diluvianæ, 34

S

Sarepta, 6, 131, 140n26
Science and Civilisation in China, 137n9
Sententiae, ix
Septuagint, 142n64
Skizzen aus dem Leben und der Natur. Vermischte Schriften, 79, 128
Speculum Metallurgiæ Politissimum oder: Hell-Polierter Berg-Bau-Spiegel, 118
Storia della Terra, 90
Sudelbücher, 130, 139n23
System of Chemistry, 141n50, 141n51
System of Mineralogy, 125, 150n172

T

Tergumim, 142n64
The Cyclopædia; or, Universal Dictionary of Arts, Sciences, and Literature, 32, 133
The Decline and Fall of the Roman Empire, 124, 127
The Earth in Decay, 7, 125
The Edinburgh New Philosophical Journal, 145n104
The Granite Controversy, 4, 133
The Historical Journal, 146n115
The Histories, 18, 24, 25, 144n75
The History of Philosophy, 22, 126
The Natural History of the Mineral Kingdom, 14, 74, 87, 136, 145n97
The Nature and Origin of Granite, 4, 132
The Principall Navigations, Voiages, Traffiques and Discoveries of the English Nation, 26
Theoretische Geologie, 91, 133
Théorie Analytique des Probabilités, 119, 130
Théorie de la Terre, 51, 53, 125
Theories of the Earth 1644-1830—The History of A Genre, 122, 141
Theory of the Earth, 1, 10, 42, 52, 64, 66, 70, 73, 74, 77, 81, 84, 85, 93, 100, 108, 109, 112, 120, 122, 125, 126, 127, 129, 131, 132, 135, 141n39, 141n47, 141n50, 146n114, 147n119
Three Dispensations, 34
Time's Arrow Time's Cycle, 8, 127
Times Literary Supplement, 21
Torah, 142n63, 142n64
Traité de Géologie, 69, 130
Traité Élémentaire de Chimie, 26

Traité Élémentaire de Géologie, 61, 125, 133
Transactions of the Geological Society, 59, 123
Transactions of the Royal Society of Edinburgh, 30, 127, 128, 129, 132, 147n119, 149n149
Trattato Storico-Scientifico sull'Origine, su'Progressi, sullo stato della Geologia in Sicilia, 99, 130
Travels in the Island of Iceland During the Summer of the Year MDCCCX, 28, 131
Treatise on Human Nature, 15
Tusculanae Disputationes, 143n68

U

Über den Ursprung der von Pallas gefundenen und anderer ihr ähnlicher Eisenmassen und über einige damit in Verbindung stehende Naturerscheinungen, 124, 139n23
Über die Lehre vom Metamorphismus und die Entstehung der kristallinischen Schiefer, 87, 133
Ueber die Theorien der Erde, den Amorphismus Fester Körper und den Gegenseitigen Einfluß der Chemie und Mineralogie, 127, 149n151
Untersuchungen vom Meere, 18
Uspekhi geognozii, 71, 134

V

Versuch einer Mineralogie, 130, 136, 139n23
Von den Äußerlichen Kennzeichen der Fossilien, 136, 139n23
Voyage au Centre de la Terre, 135, 138n23
Voyage d'un Français en Angleterre Pendant les Années 1810 et 1811: Avec des Observations sur l'État Politique et Moral, les Arts et la Littérature de ce Pays, et sur les Mœurs et les Usages de ses Habitans, 54
Voyage en Ecosse, 93
Voyage Physique et Lithologique dans la Campanie, 149n166
Voyages dans les Alpes, 12, 22
Vulgate, 142n64

W

Wells's First Principles of Geology. A Text-Book for Schools, Academies and Colleges, 109, 136
Wikipedia, x, 115
Wilhelm Meisters Wanderjahre, 88
Wonders of Geology, 39, 131
Worlds Before Adam, 4, 136, 137n9

Z

Zahme Xenien VI, 88, 149n152